AF812570

Grinding of Single-Crystal Superalloys

Grinding of Single-Crystal Superalloys

Fundamentals and Technologies

Wenfeng Ding, Qing Miao, Yao Sun, Ning Qian, Biao Zhao,
Yadong Gong, and Jiuhua Xu

WILEY-VCH

Authors

Prof. Wenfeng Ding
Nanjing University of Aeronautics and Astronautics
Nanjing
China

Prof. Qing Miao
Associate Professor
Suzhou University of Science and Technology
Suzhou
China

Prof. Yao Sun
Associate Professor
Northeastern University
Shenyang
China

Prof. Ning Qian
Associate Professor
Nanjing University of Aeronautics and Astronautics
Nanjing
China

Prof. Biao Zhao
Associate Professor
Nanjing University of Aeronautics and Astronautics
Nanjing
China

Prof. Yadong Gong
Professor
Northeastern University
Shenyang
China

Prof. Jiuhua Xu
Professor
Nanjing University of Aeronautics and Astronautics
Nanjing
China

Cover Design: Wiley
Cover Image: Courtesy of the Authors

Typesetting: Lumina Datamatics, Madurai, India
Printing and Binding: CPI Group (UK) Ltd, Croydon, CR0 4YY

Library of Congress Card No.: applied for
British Library Cataloguing-in-Publication Data
A catalogue record for this book is available from the British Library.

Bibliographic information published by the Deutsche Nationalbibliothek
The Deutsche Nationalbibliothek lists this publication in the Deutsche Nationalbibliografie; detailed bibliographic data are available on the Internet at http://dnb.d-nb.de.

Print ISBN 9783527355228
ePDF ISBN 9783527852475
ePub ISBN 9783527852468
oBook ISBN 9783527852482

© 2026 Wiley-VCH GmbH, Boschstraße 12, 69469 Weinheim, Germany

C9783527355228_221125

All rights reserved (including those of translation into other languages, text and data mining and training of artificial technologies or similar technologies). No part of this book may be reproduced in any form – by photoprinting, microfilm, or any other means – nor transmitted or translated into a machine language without written permission from the publishers. Registered names, trademarks, etc. used in this book, even when not specifically marked as such, are not to be considered unprotected by law.

The manufacturer's authorized representative according to the EU General Product Safety Regulation is Wiley-VCH GmbH, Boschstr. 12, 69469 Weinheim, Germany, e-mail: Product_Safety@wiley.com.

In view of ongoing research, equipment modifications, changes in governmental regulations, and the constant flow of information relating to the use of experimental reagents, equipment, and devices, the reader is urged to review and evaluate the information provided in the package insert or instructions for each chemical, piece of equipment, reagent, or device for, among other things, any changes in the instructions or indication of usage and for added warnings and precautions. While the publisher and the authors have used their best efforts in preparing this work, including a review of the content of the work, neither the publisher nor the authors make any representations or warranties with respect to the accuracy or completeness of the contents of this work and specifically disclaim all warranties, including without limitation any implied warranties of merchantability or fitness for a particular purpose. No warranty may be created or extended by sales representatives, written sales materials, or promotional statements for this work. The fact that an organization, website, or product is referred to in this work as a citation and/or potential source of further information does not mean that the publisher and authors endorse the information or services the organization, website, or product may provide or recommendations it may make. This work is sold with the understanding that the publisher is not engaged in rendering professional services. The advice and strategies contained herein may not be suitable for your situation. You should consult with a specialist where appropriate. Further, readers should be aware that websites listed in this work may have changed or disappeared between when this work was written and when it is read. Neither the publisher nor the authors shall be liable for any loss of profit or any other commercial damages, including but not limited to special, incidental, consequential, or other damages.

Contents

Foreword

Single-crystal nickel-based superalloys have extensive application prospects in diverse industry fields such as aerospace, energy, and power, renowned for their outstanding mechanical properties at elevated temperatures. Nevertheless, they are classified as typical high-strength, high-toughness, and difficult-to-machine materials. The pursuit of high-quality and high-efficiency machining of single-crystal nickel-based superalloys has long been a bottleneck issue, severely hindering the broader industrial utilization of this material.

Among various machining techniques, grinding stands out as one of the primary methods for processing single-crystal nickel-based superalloys. Currently, there remains a lack of in-depth understanding regarding the material removal mechanism. Simultaneously, insufficient knowledge about the parameters involved in the grinding process exists, making it difficult to control the integrity of the ground surface and enhance machining efficiency.

In recent years, the author team has dedicated extensive efforts to researching the highly efficient and precision grinding technology of single-crystal nickel-based superalloys. This book provides a systematic exposition of the material removal mechanisms of single-crystal nickel-based superalloys in conventional grinding, high-efficiency grinding, and micro-grinding, respectively. It also delves into the analysis of forces, temperature, and tool wear during the grinding process. Additionally, it thoroughly investigates the integrity of the machined surface achieved through different grinding processes, encompassing aspects such as the subsurface micro–nano structure, micro-hardness, residual stresses, and wear resistance. Overall, this book exhibits remarkable scientific rigor, logical coherence, and practical applicability.

Finally, I sincerely hope that this book can make significant contributions to the innovative development of grinding processing technology and facilitate the extensive utilization of single-crystal nickel-based superalloys.

Sincerely,
Di ZHU
PhD, Professor
Member of Chinese Academy of Sciences
Nanjing University of Aeronautics and Astronautics

Preface

This book, *Grinding of Single-crystal Superalloys: Fundamentals and Technologies*, is a comprehensive exploration of the grinding processes and techniques applied to single-crystal nickel-based superalloys. It aims to bridge the gap between fundamental theory and practical applications in the field of precision machining.

The book is divided into five key parts with 15 chapters, each addressing critical aspects of grinding technology for these advanced materials. In Part I, it provides the grinding mechanism of single-crystal nickel alloys. Part II discusses the grindability of single-crystal nickel alloys, while Part III delves into the surface integrity during high-efficiency grinding. In addition, Part IV explores the surface integrity of micro-grinding. Finally, Part V includes simulation, optimization, and control in the grinding of single-crystal turbine blade roots as case studies.

This book summarizes the latest progress in the grinding theory and technology of single-crystal nickel-based superalloys. With the development of aerospace, energy, and other fields, the demand for high-temperature alloy materials is constantly increasing, making it crucial to gain an understanding of the grinding theory and technology of single-crystal nickel-based superalloys. This book aims to provide the latest theory and technologies in this field, further providing valuable references for future research and development. In addition, this book also provides a comprehensive reference for researchers and students in the fields of materials science, mechanical engineering, and aerospace engineering, helping them gain a deeper understanding of the latest developments in the grinding theory and technology of single-crystal nickel-based superalloys. By reading this book, readers can gain a deeper knowledge of the grinding theory and technology of single-crystal nickel-based superalloys and understand how to apply these theories and technologies to solve practical problems.

I hope that this book will contribute to the advancement of grinding technology and the development of high-performance materials for various applications in aerospace, power generation, and other industries.

With gratitude to all contributors and reviewers who have helped shape this work, I present this monograph for the benefit of the broader scientific community.

Sincerely,
Wenfeng Ding
PhD, Professor
Nanjing University of Aeronautics and Astronautics

Part I

Grinding Mechanism of Single-crystal Nickel Alloy

The aeroengine is hailed as the crown of modern industry and has become an important indicator of a nation's development and comprehensive strength. Single-crystal turbine blades are the most critical components, usually operated in extreme environments of high temperature, complex stresses, and worst running conditions, which determine the service performance of an advanced aeroengine. With the increasing demand for large thrust-to-weight ratio engines, the high-temperature capability of turbine blades has improved significantly, from 750 °C in the 1940s to the present 2000 °C, posing great challenges for the metal material preparation and manufacturing technologies of turbine blades. As a critical process to guarantee the machined quality of the workpiece, grinding is widely used to manufacture single-crystal components and attracts extensive attention worldwide, especially for high-efficiency grinding and micro-grinding methods. This part is dedicated to reviewing the development of grinding technology of single-crystal nickel alloy, describing the removal mechanism of single-crystal nickel alloy in high-efficiency grinding, and clarifying the plastic deformation mechanism of single-crystal nickel alloy in micro-grinding.

1

Introduction

1.1 Development and Practical Application of Single-crystal Nickel Alloy

Nickel alloys possess excellent comprehensive properties, such as high-temperature strength, high oxidation resistance, high fatigue resistance, and high thermal stability, and therefore have been employed widely in manufacturing the hot-end components of aeroengines [1, 2]. The development of nickel alloys goes through three generations, e.g., wrought nickel alloy, solidified directionally nickel alloy, and single-crystal nickel alloy (Figure 1.1). The temperature capability is enhanced from generation to generation. For example, compared to directionally solidified nickel alloy, a single-crystal turbine blade allows an operating temperature higher by around 25–50 °C, equivalent to an increase in turbine blade service life by up to 3 times from the aspect of working efficiency.

The turbine inlet temperature has been improved significantly for advanced aeroengines. Using the single-crystal hollow turbine blades can make the turbine inlet temperature reach 2100 K for the F119 aeroengine. At such a high temperature, the polycrystalline turbine blade formed by the conventional casting method can melt and cease to work. However, single-crystal turbine blade works well and is becoming the preferred choice of first-stage turbine blades of aeroengines with a higher thrust-weight ratio of above 10. This is mainly because the whole turbine blade grows from a single grain without any grain boundaries, which often induce defects of micropores and microcracks in polycrystalline nickel alloy (Figure 1.2). Meantime, the elements enhancing grain boundaries as well as decreasing the metal melting point can be reduced effectively (Figure 1.3). Thus, these promote the high temperature resistance of single-crystal nickel alloy. By using the advanced material preparation technique for single-crystal nickel alloy and the double-wall air cooling/casting-in-chill manufacturing technique for single-crystal turbine blade, the turbine inlet temperature can be increased to 2200 K, which makes the extreme machining of single-crystal nickel alloy one of the cutting-edge research topics in fabricating the military and commercial aeroengines.

Grinding of Single-Crystal Superalloys: Fundamentals and Technologies, First Edition.
Wenfeng Ding et al.
© 2026 Wiley-VCH GmbH. Published 2026 by Wiley-VCH GmbH.

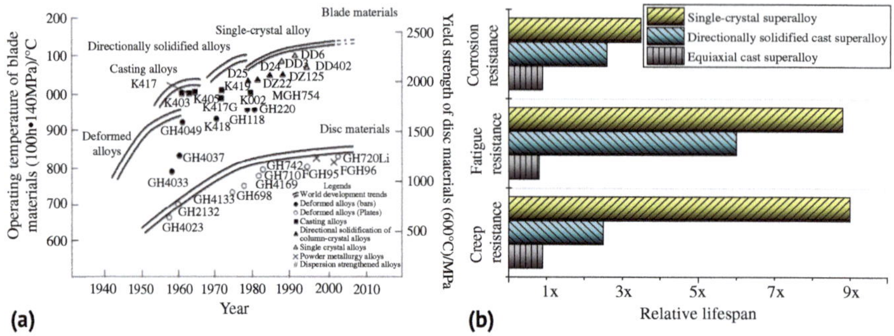

Figure 1.1 Development of (a) temperature capability and (b) mechanical properties of nickel alloys.

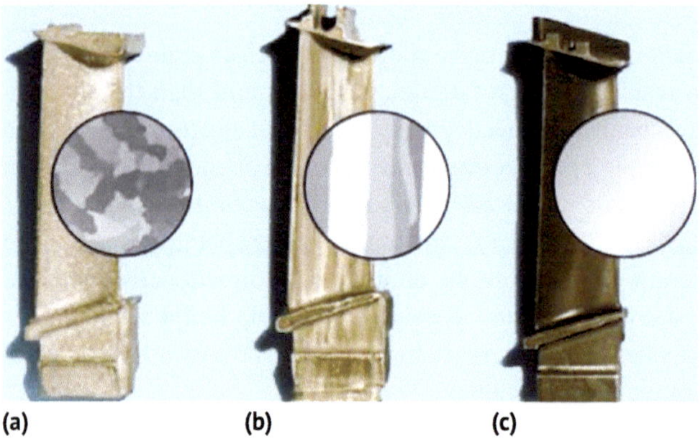

Figure 1.2 Microstructures in different casting nickel alloys: (a) equiaxial cast; (b) directionally solidified cast; and (c) single crystal. [2]/*Journal of Mechanical Engineering*.

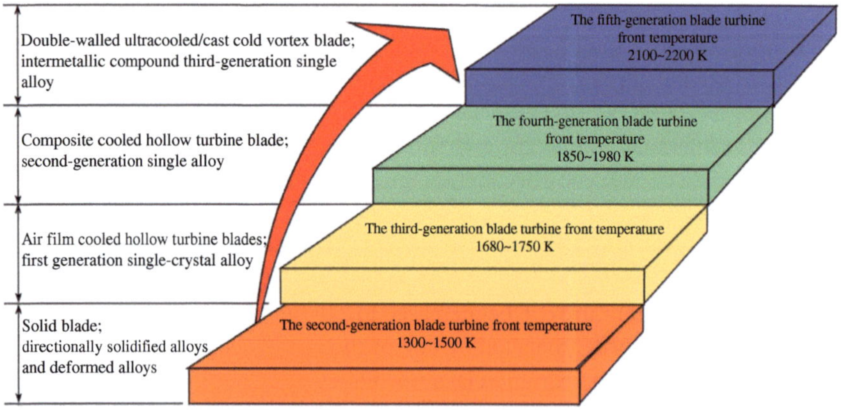

Figure 1.3 Application of single-crystal turbine blades in an advanced aeroengine. Adapted from [3].

1.2 Advantages of Grinding Technology of Single-crystal Nickel Alloy

Relying on the micro-cutting behaviors of thousands of abrasive grains on the wheel surface, the grinding process can fabricate a high-quality surface with low surface roughness, compressive stresses, and high accuracy (Tables 1.1 and 1.2), regardless of workpiece materials. This nature is different from the conventional turning, broaching, and milling methods, which often face major challenges in difficult-to-cut materials (i.e., nickel alloys) because of the rapid tool wear and the huge difficulty in controlling surface integrity and accuracy (Table 1.3). Currently, various grinding techniques have been developed to fulfill the requirements of different practical applications for almost any workpiece material (e.g., nickel alloys and ceramics). For instance, surface grinding is usually employed in manufacturing the automobile components and machine tools with flat surface requirements; profile grinding is usually combined with the creep-feed deep machining method that permits high efficiency in grinding the aeroengine parts with large stock removals and complex shapes; and micro-grinding works as one of the effective methods to fabricate the micro features in metals. The common characteristic of the abovementioned grinding techniques is that the grinding process always serves as the final process to guarantee the workpiece surface quality.

In the case of aeroengine manufacturing, the typical features of nickel alloy turbine blades include high machining accuracy, good surface quality, and high fatigue resistance. Though the nontraditional machining methods (e.g., electrochemical

Table 1.1 Material removal rates achieved in the common grinding process.

Contents	Shallow grinding	Creep-feed grinding	High-speed grinding	High-efficiency deep grinding
Grinding depth a_p (mm)	Low 0.001–0.05	High 0.1–30	Low 0.003–0.05	High 0.1–30
Workpiece speed v_w (m/min)	High 1–15	Low 0.05–0.5	High 1–15	High 0.5–10
Grinding speed v_s (m/s)	Low 15–40	Low 15–40	High 60–200	High 60–200
Material removal rate Q_w' ($mm^3/(mm \cdot s)$)	Low 0.05–2	Low 1–10	Medium ≤ 60	High 50–2000

Table 1.2 Surface roughness achieved in the common grinding process.

Grinding methods	Conventional grinding	Precision grinding	Ultraprecision grinding	Mirror grinding
Surface roughness R_a (μm)	0.16–1.25	0.04–0.16	0.01–0.04	≤ 0.01

Table 1.3 Main methods and related characteristics for machining single-crystal nickel alloy components.

Machining method	Machining efficiency	Machining accuracy	Machined surface quality	Other characteristics
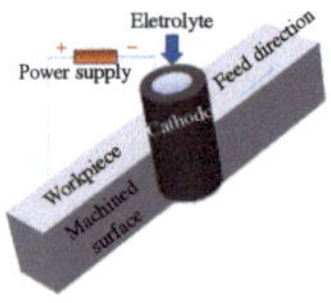 Electrochemical machining (ECM)	High	±50 μm	$R_a \leq 1.6$ μm, with slight/without affected layer	With slight/without electrode wear, suitable for mass production, large initial investment, and not environmentally friendly
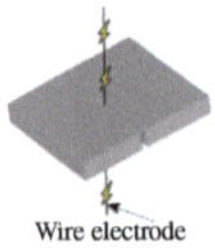 Electrical discharge machining (EDM)	Medium	±5 μm	$0.2 \leq R_a \leq 3.2$ μm, with affected layer and even microcrack	Suitable for any workpiece materials, having electrode wear, suitable for mass production, large initial investment, and not environmentally friendly
 Additive manufacturing	Low	±350 μm	$3.2 \leq R_a \leq 12.5$ μm, easy to deform	Limited workpiece size, difficulty in fixturing and supporting workpiece, and large initial investment
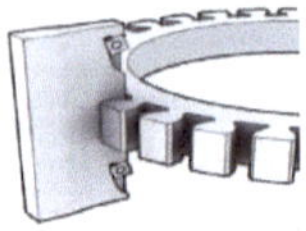 Broaching	Medium	≤ 10 μm	$0.8 \leq R_a \leq 1.6$ μm	Rapid tool wear and high cost
 Milling	Low	≤ 10 μm	$0.4 \leq R_a \leq 3.2$ μm	Rapid tool wear, limited workpiece size, and low machining efficiency
 Grinding	Medium	≤ 5 μm	$R_a \leq 0.6$ μm, with compressive stresses	Rapid tool wear if using conventional abrasive wheels, good surface quality

machining [ECM] and electrical discharge machining [EDM]) have achieved rapid developments in the past decades, the extremely strict requirements of tight tolerance (< 10 μm), low surface roughness (R_a < 0.6 μm), and high surface integrity still make the grinding process, particularly the profile grinding, dominate the main positions in machining aeroengine components. Therefore, both conventional grinding (including surface grinding and profile grinding) and micro-grinding are still attracting great attention in machining difficult-to-cut materials, especially single-crystal nickel alloys, and deserve further exploration.

1.3 High-efficiency Grinding Technology Development of Single-crystal Nickel Alloy

1.3.1 High-efficiency Grinding Mechanism and Development

During grinding, the workpiece material is removed in the form of tiny wedge-shaped chips by the effective cutting behavior of an individual abrasive grain. In the grinding zone, the interaction between abrasive grain and workpiece is complicated and always changing due to the varied penetration depth of abrasive grain into the workpiece. Thus, the grinding chip formation can be divided into three stages (Figure 1.4): (i) sliding, (ii) plowing, and (iii) cutting. The cutting stage is expected to be induced because of its effective material removal behavior. The other two states are expected to be avoided owing to their low efficiency in removing workpiece material, together with high grinding energy consumption. However, for each kind of workpiece material, the duration time of such three stages might be different even under identical grinding conditions. Understanding the underlying material removal mechanism of each workpiece material is essential to optimize the grinding process.

Furthermore, in surface grinding, the grinding loads are distributed uniformly along the grinding surface, and thus, the formation of surface integrity is relatively easy to clarify. Nevertheless, in profile grinding, there are various factors that limit the improvement of machining accuracy and surface quality, such as the long,

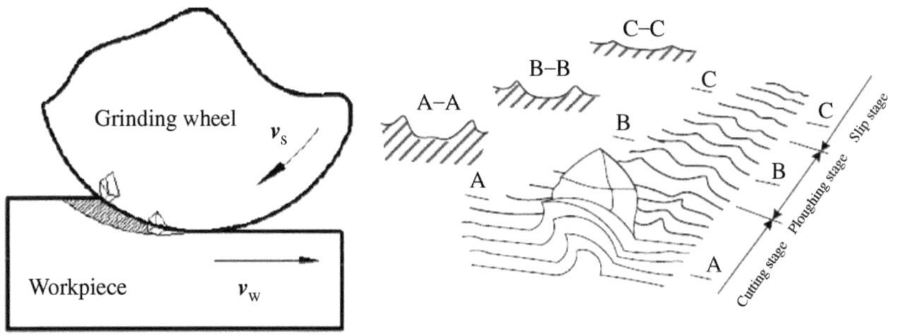

Figure 1.4 Schematic illustration of the material removal process in the grinding process.

narrow, and large contact zone, the complex workpiece shape, the varied grinding loads along workpiece shape, the insufficient and uneven cooling, the varied grinding wheel wear, etc. Jin et al. [4, 5] pointed out that by combining the theoretical model, experimental testing, and simulation method, the material removal process in profile grinding can be further understood. Because of the low thermal coefficient and high temperature strength, the grinding of nickel alloys often produces a large amount of grinding heat, which is distributed ununiformly (Figure 1.5) in the grinding zone. In other words, the strongly coupled thermo-mechanical-structural effects induce easily local grinding burn and microcracks on the workpiece surface.

To further improve the grinding efficiency, many scholars are dedicated to developing new ideas according to the classical formula of the maximum undeformed chip thickness of a single grain (Eq. 1-1).

$$a_{\text{gmax}} = \left[\frac{4}{C \cdot N_{\text{d}}} \cdot \frac{v_{\text{w}}}{v_{\text{s}}} \cdot \left(\frac{a_{\text{p}}}{d_{\text{s}}} \right)^{\frac{1}{2}} \right]^{\frac{1}{2}} \tag{1.1}$$

where a_{gmax} is the maximum undeformed chip thickness of a single grain, N_{d} is the active grain number, C is a constant related to grain edge, and d_{s}, a_{p}, v_{w}, and v_{s} are grinding wheel diameter, grinding depth, workpiece speed, and grinding speed.

Based on the above idea, creep-feed grinding process (CFDG) was developed in the 1950s, whose typical features are using large-porosity conventional abrasive grinding wheels, large grinding depths, and small workpiece speeds. This method can keep a small value of a_{gmax}, while improving the grinding efficiency remarkably and simultaneously maintaining the high surface quality. However, the sudden burning during grinding hinders the application of the creep-feed grinding method until Andrew et al. proposed the famous theory of film boiling and nucleate boiling of the grinding zone in the 1960s [6]. The problem of sudden burning is being overcome to some extent, promoting CFDG, which is widely employed in machining nickel alloys.

Recently, because of the improvement in the material properties of silicon carbide, alumina, cubic boron nitride (CBN), and diamond abrasive grains, various new grinding techniques (i.e., high-speed grinding, ultrahigh-speed grinding,

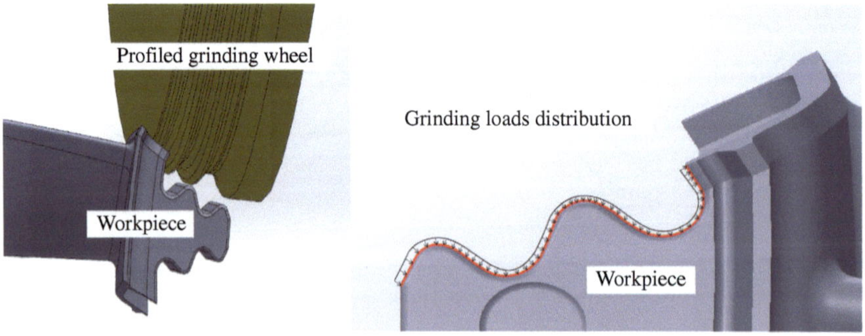

Figure 1.5 Schematic illustration of grinding loads along the turbine blade root shape.

high-efficiency deep grinding, composite energy-assisted grinding, and continuous dressing grinding) are developing vigorously [7, 8] (Figures 1.6 and 1.7). These provide effective methods to improve the grinding performance of abrasive wheels and the grindability of difficult-to-cut materials. However, the studies on such aspects are insufficient, such as the material removal mechanism, the process stability and adaptability, the formation and control mechanism of grinding surface integrity, the relationship between subsurface structure and mechanical properties at multi-scale size, etc. There still exists a big gap between practical application and laboratory study.

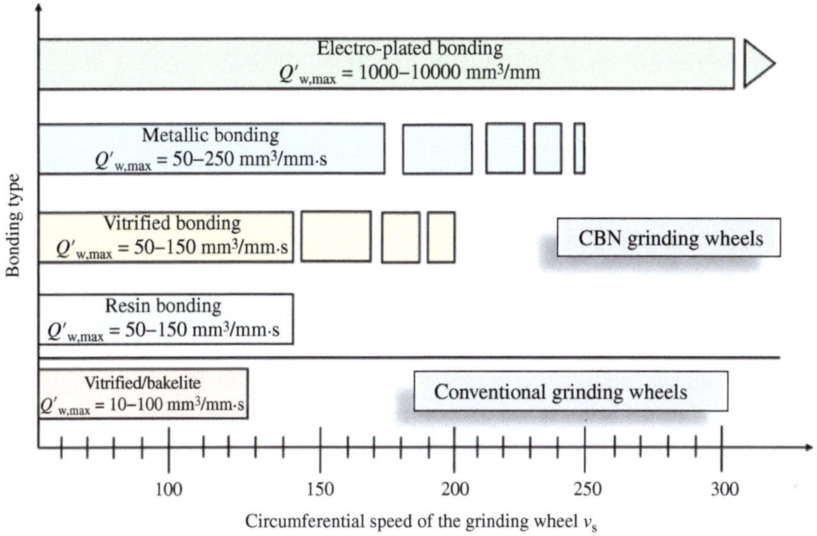

Figure 1.6 Development of abrasive grinding wheels.

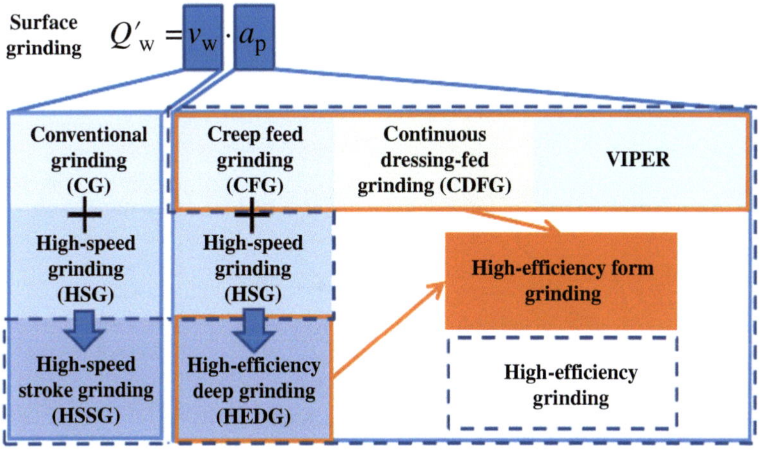

Figure 1.7 Development of grinding techniques.

1.3.2 High-efficiency Grinding Characteristics and Challenges

Numerous investigations have been conducted on the grindability of nickel alloys in the past decades. The related research contents can be as follows: (i) A series of advanced grinding machine tools and process parameters are developed, as well as various methods to improve the grinding performance of abrasive wheels, material removal rates (MRRs), and surface integrity of nickel alloys. (ii) The influencing factors on the grindability of nickel alloys are clarified (Figure 1.8), including the workpiece material properties, cooling behavior, machine performance, tool specification, dressing parameters, and grinding parameters. In particular, the challenges concerning low grinding efficiency, difficulty in controlling grinding burn, and insufficient understanding of micro-/nano-scaled surface integrity in nickel alloys are prominent. These are the main bottlenecks that restrict the realization of high-quality and high-efficiency grinding of nickel alloys and therefore become one of the cutting-edge research points in extreme manufacturing fields.

1. Low grinding efficiency

   Low grinding efficiency is one of the main problems in the grinding of nickel alloys. In general, the grinding efficiency is highly dependent on MRR, and MRR is decided by workpiece speed and grinding depth. Obviously, the tool wear and grinding surface quality play an important role in extending the grinding parameters.

   For wrought superalloys, a complete evaluation of the effects of grinding parameters, i.e., grinding speed v_s, workpiece speed v_w, and grinding depth a_p, on the machined surface integrity of Inconel 718 alloy was performed by Zeng et al. [9]. It was found that the machined surface integrity was highly sensitive to the grinding depth. If a small grinding depth was employed, the residual stress, microhardness, and microstructure alternations in ground surface layer could be much more localized. The optimized parameters were finally determined as $v_s = 20$ m/s, $v_w = 22$ m/min, $a_p = 0.015$ mm. Yao et al. [10] conducted a further

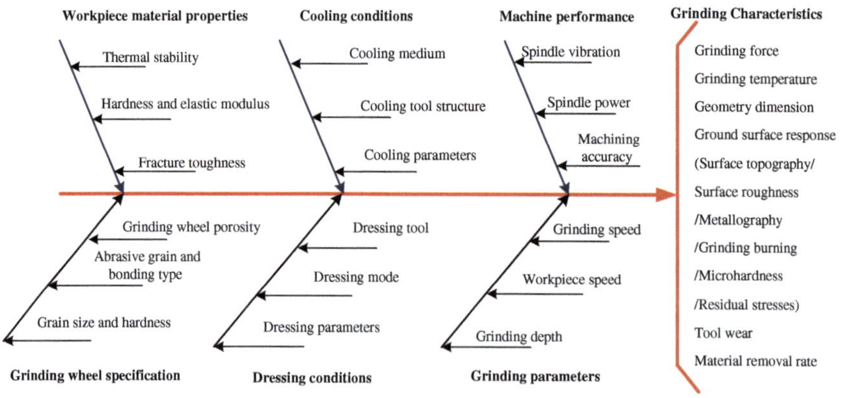

Figure 1.8 Fishbone diagram of influencing factors on grindability of nickel alloys.

investigation on the surface grinding of Inconel 718 alloy by using alumina wheels. The surface roughness of $R_a = 0.11$ μm, the softened ground surface layer of 30 – 40 μm, and the plastic deformation layer of 10 – 15 μm were identified when the MRR reached 1.3 $mm^3/(mm \cdot s)$ ($v_s = 25$ m/s, $v_w = 16$ m/min, $a_p = 0.005$ mm). At an MRR of 3.7 $mm^3/(mm \cdot s)$ ($v_s = 20$ m/s, $v_w = 15$ m/min, $a_p = 0.015$ mm), it was found that the chemical reactions between the abrasive grains and workpiece material led to strong workpiece material adhesion on the machined surface and the sharply increasing grinding force.

In the case of cast superalloys, Xu et al. [11–14] presented detailed studies on the grinding temperature and the tool wear mechanism, and their effects on surface integrity parameters in grinding of K417 alloy. Their observations indicated a clear effect of grinding temperature on the color evolution of the machined surface and on the microstructure evolution of the machined surface layer. The increasing ductile deformation of the workpiece was found to contribute to the chemical bonding between the abrasive grains and the workpiece at the elevated temperatures. Furthermore, Österle and Li [15] studied the subsurface microstructural changes of workpiece in creep-feed grinding of IN738LC alloy by using alumina wheels. Analysis of the microstructure evolution of workpiece cross section denoted the presence of local melting at contact spots that led to the formation of white layer when the MRR was 8 $mm^3/(mm \cdot s)$ ($v_s = 20$ m/s, $v_w = 0.48$ m/min, $a_p = 1$ mm). Chen et al. [16, 17] explored the grinding characteristics of DZ4 alloy using the MRR ranging from 0.24 to 3.75 $mm^3/(mm \cdot s)$. It was observed that the machined surface hardness was weakened remarkably by 25% and the depth of the affected layer exceeded 0.5 mm when the grinding temperature reached 1100 °C.

From the review of the aforementioned literature, the grindability upon different nickel-based superalloys using alumina wheels can be detailed in Table 1.4. It can be seen that the machined surface integrity and tool wear are influenced significantly by the grinding conditions and workpiece materials. Specifically, as more advanced superalloys are developed, i.e., from wrought to single-crystal nickel alloys, their comprehensive properties are further enhanced and improved. However, machining of such alloys becomes more difficult. High mechanical and thermal loads are usually associated with the grinding of nickel-based superalloys, leading to these undesirable features of machined surface (summarized in Table 1.4).

Furthermore, a practical result found in previous tests is that the rapid rate of wheel wear and low ability to maintain the dimensional accuracy are always the bottleneck for conventional abrasive wheels to reach a desired grinding efficiency. Increasing the dressing frequency tends to solve these practical problems, while the tool service life has to be greatly shortened. For this reason, the superhard abrasive wheels, e.g., CBN wheels, are proposed to be used for machining the fir-tree blade root, as reported by Zhao et al. [12], where a very high MRR of up to 50 $mm^3/(mm \cdot s)$ was achieved by using CBN wheels in the grinding of Inconel 718 alloy. Nevertheless, the high cost

Table 1.4 Grinding behavior of nickel alloys using conventional wheels.

Workpieces	MRR (mm^3/(mm · s))	Grinding parameters	Grinding wheel and its wear behavior	Ground surface integrity	References
Inconel718	6.67	$v_s = 25$ m/s, $v_w = 10$ m/min, $a_p = 0.04$ mm	Single Al_2O_3(80#)	$R_a = 0.89$ μm, surface crack	[9]
Inconel718	3.75	$v_s = 20$ m/s, $v_w = 15$ m/min, $a_p = 0.015$ mm	White Al_2O_3 (60#), strong material adhesion	$R_a = 0.6$ μm, grinding burn and feed marks	—
Inconel718	6.67	$v_s = 25$ m/s, $v_w = 16$ m/min, $a_p = 0.025$ mm	Single Al_2O_3 (80#)	$R_a = 0.18$ μm, many feed marks	[10]
K417	1.40	$v_s = 25$ m/s, $v_w = 0.07$ m/min, $a_p = 1.2$ mm	White Al_2O_3 (100#), strong chemical bonding	$R_a = 1.2$ μm, grinding burn with puce color on ground surface and remarkable smeared materials	[11–14]
DZ4	3.75	$v_s = 25$ m/s, $v_w = 1.5$ m/min, $a_p = 0.15$ mm	White Al_2O_3 (60#)	Recrystallization near ground surface	[16, 17]
IN738LC	2.00	$v_s = 20$ m/s, $v_w = 0.12$ m/min, $a_p = 1$ mm	White Al_2O_3 (60#)	White layer and internal microcracks	[15]

and the low potential concerning the dressing capability are generally major problems for superhard abrasive wheels to profile grinding, even though it is demonstrated that the active surfaces of superhard abrasive wheels can be specially designed and textured, and are capable of creep-feed grinding deep fir-tree slots and fir-tree root forms in the difficult-to-cut nickel alloys with free workpiece burning. On the contrary, due to the low hardness and cost-effective price compared to the superhard abrasive wheels, conventional abrasive wheels are evidenced with an excellent profiling ability and have become widely applied in the turbine blade root manufacture. High surface quality, e.g., the surface roughness of R_a 0.65 μm and the compressive stress of up to 490 MPa, was achieved. Guo et al. [18, 19] developed the process to realize the continuous dressing creep-feed profile grinding of the fir-tree blade root forms using conventional abrasive wheels. Control of the blade root grinding process was further optimized by monitoring the power signature and adjusting the dress infeed rate and workpiece speed to improve the grinding efficiency. The results of the above tests indicate that the relatively short service life of conventional abrasive wheels can be offset by high profiling ability and acceptable surface quality.

In addition, the newly developed microcrystalline alumina abrasive wheels have a higher potential for profile grinding as evidenced by the large grinding ratios, high MRRs, low grinding forces, low geometrical wear, and good ground surface integrity compared to the conventional abrasive wheels. This type of grinding wheel is therefore very suitable for creep-feed profile grinding blade root forms though further investigations have to be done to optimize the process results. Note that the best option for grinding of nickel alloys can be determined usually by comprehensively considering the machining cost, machined surface quality, and machining complexity. Therefore, it is still of utmost importance to study grinding techniques to further improve the grinding efficiency of nickel alloys.

2. Difficult controlling of grinding burn
Grinding burn is induced by massive heat accumulation in the grinding zone and usually leads to a very high temperature together with the alternation of appearance and microstructure of workpiece surface layer. As mentioned earlier, the high grinding force and temperature can be generated even though a small MRR is used for grinding with alumina wheels. Under such conditions, grinding burn is difficult to control. Moreover, the poor cooling effectiveness or/and severe wheel wear could also result in the occurrence of grinding burn with an instantaneous temperature of more than 1000 °C (Figure 1.9). Grinding burn limits the improvement of grinding efficiency significantly and is deleterious to the grinding surface integrity. Because of the sensitivity of single-crystal nickel alloy in aeroengine field, the components must be scrapped even if a slight grinding burn occurs. Thus, it is necessary to avoid the grinding burn through the detection and prediction methods.

Figure 1.9 Typical grinding burn on fir-tree turbine blade root surface.

Almost all indicators of surface integrity are affected by grinding burn. At present, the identification of grinding burn is performed mainly by the response features of the ground surface/subsurface, such as surface topography and appearance, microstructure, microhardness, and noise detection. In addition, the monitoring of sudden changes in grinding force, temperature, and grinding power can also be used to identify the grinding burn. An effective and early warning of grinding burn is of significant importance to ensure the grinding quality and save the economic loss due to scrapped components.

It is notable that the grinding temperature is critical to monitor grinding burn. To prevent the workpiece from burning, the theoretical and simulated models should be developed to figure out the evolution and distribution of temperature under typical grinding conditions (Figure 1.10). Then the thermal-induced surface/subsurface damage can be predicted, which can provide useful guidance for the process design in the following grinding arrangements. Although the possibility of achieving the dimensional tolerance of less than 10 μm and surface roughness R_a of less than 0.6 μm has been reported in creep-feed grinding (CFG), diverse thermal damage has still been reported as the biggest issue, including large tensile residual stress, thermal softening, burns, and microcracks. Although substantial efforts were made to understand grinding temperature even with the consideration of stochastic nature, the high temperature in CFG of blade roots seemed to be different due to the limited coolant transportation in the grinding zone, the very high specific energy in CFG, and the low thermal conductivity of nickel-based superalloys [20]. To gain a more in-depth understanding, multitudes of efforts and attentions from both academic and engineering aspects have already been made, and most of them can be categorized into two groups: (i) theoretical and (ii) numerical studies.

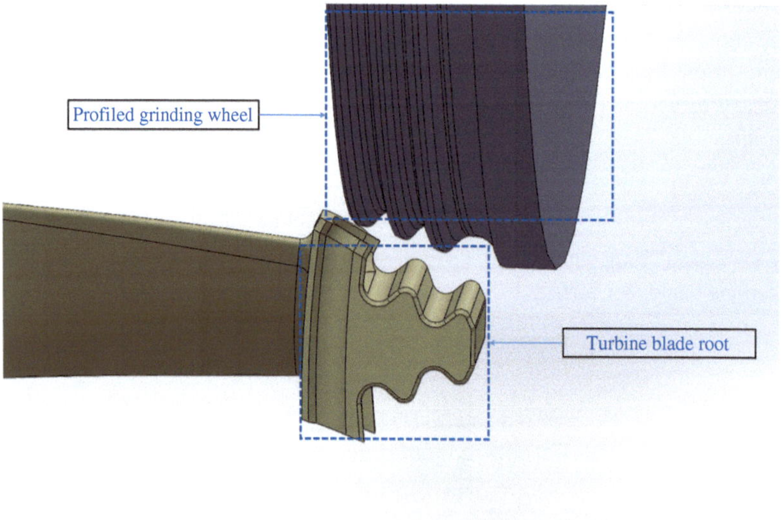

Figure 1.10 Grinding of the fir tree-shaped/profile of turbine blade root.

For the theoretical aspect, the pioneer work started with the one-dimensional (1D) analytical modelling of the heat transfer within the grinding zone where the maximum temperature in CFG considering the grinding fluids was modeled and the critical temperature for film boiling to occur in the grinding zone was determined on the basis of the analytical results. Guo and Malkin [19] further developed the model so as to predict the critical heat flux for burns in CFG, and the energy partition in CFG was estimated as 4.5% for up grinding and 3% for down grinding during certain conditions. The best match between theoretical and experimental results was found when using the heat source with the scalene-triangular-shaped heat flux, and the energy partition to the workpiece was found to be 4.3% for up CFG and 3.5% for down CFG. Except for all the above studies focusing on the CFG process of flat surfaces, Jin et al. [21, 22] presented the three-dimensional (3D) analytical model in creep-feed grinding of curved surfaces of gears. The effect of the contact geometry along the gear tooth profile on the grinding temperature distribution was analyzed while taking into account gear teeth number, module, and rolling angle. The heat source with the triangular-shaped heat flux was used to calculate the temperature profiles, and the simulated results indicated a good agreement with the measured temperature, with the maximum relative error being only 10.3%. The more comprehensive 3D grinding temperature model can be the stochastically grain-discretized temperature model, where the grain-workpiece micro-interactions (including rubbing, plowing, and cutting) and the corresponding varied thermal effects were considered. The relative difference between the calculated and experimental results of less than 20% was reported, proving the model's accuracy to a large extent. More importantly, both the two-dimensional (2D) and 3D grinding temperature domains with highly localized details at the grain scale were obtained for the first time. More details about theoretical grinding temperature work can also be found in Refs. [23] and [24].

Parallel to the theoretical studies, extensive numerical studies on the grinding temperature were also carried out based on finite element method (FEM) due to their good visualization and powerful ability to solve multiphysics problems. The 2D simulation results were found to be generally consistent with the experimental ones, with the relative error of less than 20%. The detailed effects of various grinding parameters (e.g., wheel speed, workpiece feed rate, grinding depth, and fluid properties) on the temperature distribution were investigated comparatively based on the 2D finite element (FE) model with a rectangular heat source. The simulated results matched the experimental data where an error of around 10% was found. The effect of dynamic viscosity and surface tension of the grinding fluid on the grinding temperature domain was found to be much more significant than that of the specific heat capacity of grinding fluids. The obtained results were consistent with the reports from Jin [5], indicating the parabolic heat source might be more suitable to estimate grinding temperature in the high-efficiency deep grinding due to the high geometrical similarity of the workpiece–wheel contact zone to the parabolic surface.

With the availability of the aforementioned studies, it might be noted that very few studies focus on the temperature modelling in the CFG of single-crystal turbine blade roots where the workpiece–wheel contact region has a relatively complex profile. The non-flat workpiece–wheel contact zone might make the temperature field even more complicated due to the temperature concentration, thanks to the lack of heat convection. Therefore, in practice, the factories for manufacturing the aeroengine components take mild process parameters to grind the nickel alloys in order to ensure a burn-free grinding surface. In addition, the grinding wheels are dressed frequently to maintain the sharpness and shape accuracy. This operation inevitably decreases the grinding efficiency. As a result, there exist great challenges to obtain the desirable surface integrity of nickel alloys as well as the high geometry accuracy of the workpiece.

3. Insufficient understanding of micro-/nano-scaled surface integrity
   Due to the complex thermo-mechanical interaction between grinding tools and workpiece, severe plastic deformation (SPD) always takes place in the superficial layer of workpieces, usually leading to local alterations of microstructure (e.g., grain refinement) and material properties (e.g., work hardening) (Figure 1.11). Understanding these alterations and the underlying formation mechanism has been a prominent topic in previous studies, because of their significant effects on surface integrity, which determines the fatigue life, creep, fracture, and tribological properties of nickel-based superalloy components [14]. However, because of the difficulty in detection methods and the lack of analytical knowledge, the studies on micro-/nano-scaled surface integrity are insufficient.

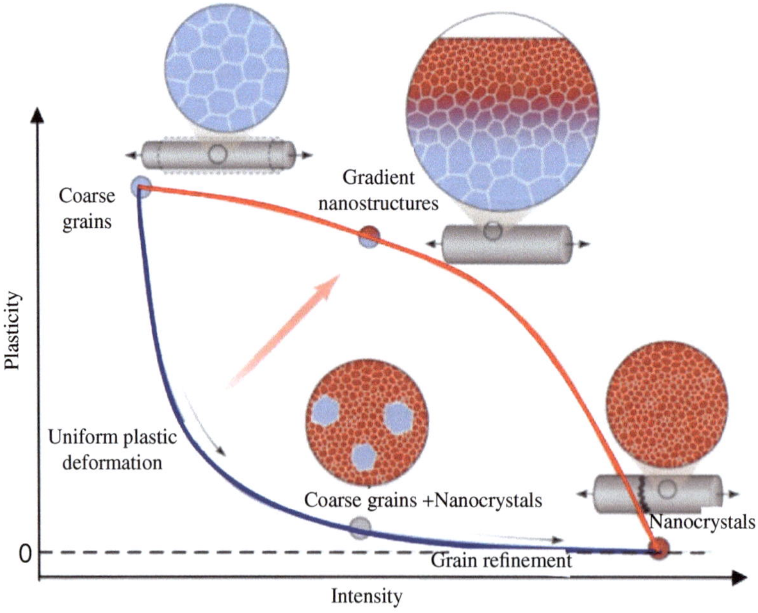

Figure 1.11 Relationship between mechanical properties and microstructures.

For this reason, numerical investigations are dedicated to clarifying the effects of mechanical grinding on the workpiece microstructure and material properties at the micro- and nano-scale. For example, Ding et al. [25] examined the subsurface metallurgical alteration of the ground groove of K424 cast nickel-based superalloy produced by creep-feed grinding. Crystal lattice composed of the γ, γ', and γ'' phases was observed with moderate plastic deformation and distortion (up to 30 μm in thickness). High work hardening (up to 46% above the bulk hardness) was found in the workpiece superficial layer. This change in microstructure of nickel-based superalloys is owing to the high thermal energy and high mechanical deformation during grinding, whose competing influence determines the transition between hardening and softening in material properties. Furthermore, under a given machining condition, the white layer, which exhibits featureless and white appearance using an optical microscope, can be formed in the machined surface. It is usually detrimental to the fatigue life of the machined workpiece because of its hard and brittle nature. A full exploration on such a white layer was conducted by Liao et al. [26] who studied in detail the formation mechanism, microstructure, and mechanical properties of the white layer of S135H nickel-based superalloy. It was indicated that the deterioration of the white layer to material properties was attributed to the dissolution of γ' strengthening phase, although the grain refinement, which was beneficial for improving material performance to some extent, can be obtained within the white layer.

However, it is worth underlining that the mechanical grinding has been proved as an effective way to modify the material properties by introducing refined grains to the workpiece surface. As presented by Ding et al. [27–29], a gradient microstructure (i.e., surface nano-laminated structure, deformation-twinned structure, and severely deformed structure) of Hastelloy C-22HS nickel-based superalloy could be produced by mechanical grinding. By virtue of micro-pillar compression tests, the yield strength of about 1650 MPa for both nano-laminated structure and deformation-twinned structure was achieved, which was greater by around 3.7 times than that of the bulk material. In addition, Liu et al. [22] reported an ultrahard and ultrastable nano-laminated structure (about 6.4 GPa at 504 °C) induced by high-rate shear deformation during grinding of polycrystalline nickel alloy. It was further found that the dislocation activities determined the formation mechanism of such a special structure.

Interestingly, it can be found in the above analysis that mechanical grinding, on one hand, can enhance material properties through grain refinement, but on the other hand, it can degrade the service performance of the workpiece under high temperature and stress conditions by inducing local detrimental modifications. For such a paradox, a deep understanding is of utmost importance for controlling material properties through grain refinement due to mechanical grinding.

It is noted that the single-crystal nickel-based superalloy grows from one single grain and eliminates the grain boundaries thoroughly. Therefore, it is capable of withstanding higher thermal and mechanical loads than the polycrystalline one. This makes them one of the key materials used for the manufacture of turbine blade root, which works under extremely high temperature and pressure. However, such a single-crystal alloy is very sensitive to the stress concentration during mechanical grinding. Up to now, great efforts have been devoted to purposely achieve some special structures via mechanical grinding, which could allow for the tailoring of high-temperature performance of polycrystalline nickel-based superalloys [30, 31]. Nevertheless, the studies for the grinding-induced surface alterations have rarely been reported for the single-crystal nickel-based superalloy. Hence, questions are raised about how the grinding-induced surface alteration behaves in the microstructure, and whether the mechanical grinding could introduce a beneficial impact on the material performance via grain refinement; the answers are still not clear for the single-crystal nickel-based superalloys. As a consequence, it is urgent to conduct an investigation into the micro-/nano-scaled surface integrity of single-crystal nickel alloys in the grinding process.

1.4 Micro-grinding Technology Development of Single-crystal Nickel Alloy

1.4.1 Micro-grinding Mechanism of Single-crystal Nickel Alloy

With the rapid development of industries such as aerospace, defense, bioengineering, modern medicine, and microelectronics, the demand for 3D micro-components ranging from micrometers to millimeters, with high precision in size and surface quality, is becoming increasingly urgent, including micro-optical elements and microsensors (Figure 1.12).

Currently, machining methods like micro-milling, photolithography, EDM, and laser processing can create microstructures such as micro-holes and microgrooves ranging from several micrometers to hundreds of micrometers. However, issues like edge burrs, thermal damage, and electrochemical corrosion layers on the surface/subsurface exist. These problems exacerbate the contradiction between

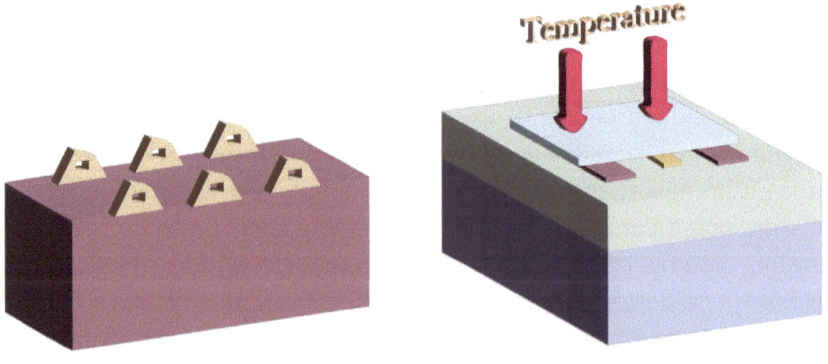

(a) (b)

Figure 1.12 Microproducts and microstructures: (a) aerospace microproducts and (b) microsensors.

machining quality, efficiency, and demands. Achieving high-quality, efficient machining of complex-shaped 3D micro-parts has become a pressing challenge in mechanical manufacturing.

At present, micro-grinding technology has drawn widespread attention from domestic and international researchers. However, research on micro-grinding tool design and preparation methods, micro-grinding mechanisms, and micro-grinding process equipment is still in the exploratory stage. A complete theoretical system and technical application capabilities have not been formed yet. Thus, research on improving the design, preparation, and processing technology of micro-grinding tools is of great theoretical and engineering application value. It can enrich the preparation methods of micro-grinding tools, extend their service life, and enhance the quality of micro-grinding processes.

Micro-grinding technology is a machining method that achieves micro-/nano-scale mechanical removal of workpiece surfaces using micro-sized grinding tools. It has a competitive edge over other fine mechanical machining technologies in obtaining 3D micro-parts and microstructures with high dimensional accuracy and surface quality.

The micro-grinding tools used in the micro-grinding process are mainly divided into ultrathin grinding blades (with a grinding part thickness of less than 1 mm) and micro-grinding rods (with a grinding part diameter less than 1 mm). Among them, cylindrical micro-grinding rods can perform micro-grinding on various complex microstructures and are the main micro-grinding tools [32].

The abrasive particles on the surface of the micro-grinding tool are very fine and closely distributed, resulting in limited chip accommodation space. Moreover, the chip size is close to the abrasive particle size, which easily causes chip clogging between the abrasive particles, leading to premature failure of the micro-grinding tool. Additionally, due to the small size of the micro-grinding tool, secondary dressing is difficult, severely affecting the surface quality of micro-grinding and the tool's service life. Therefore, the service life of existing micro-grinding tools has become one of the key factors restricting the widespread application of micro-grinding technology. Overall, research on micro-grinding mechanisms, processes, and tools is still in the exploratory stage, lacking a complete and mature technical system and manufacturing application capability.

Micro-grinding can achieve high machining precision and excellent surface quality for micro-components or microstructures through micro–nano mechanical material removal on the workpiece surface. This is accomplished using a micro-grinding tool embedded with ultrafine abrasive particle. Unlike conventional macroscale grinding, micro-grinding operates under different mechanisms.

The micro-abrasive tools used in micro-grinding are mainly classified into ultrathin grinding segments with a grinding section thickness of less than 1 mm and micro-abrasive rods with diameters under 1 mm. Given the significant reduction in the diameter of these micro-abrasive tools, factors typically ignored in conventional grinding, such as grinding wheel deformation and ploughing forces, become crucial and cannot be overlooked in micro-grinding. Furthermore, in micro-grinding, the depth of cut by abrasive grains is generally less than the particle size of the workpiece material. This situation leads to significant changes in the material removal mechanism due to size effects and microstructural recrystallization phenomena [33].

Material micro-grinding processes macroscopically appear as the removal of workpiece material under the external force of the cutting tool, resulting in a machined surface and grinding debris. Microscopically, this process is characterized by irreversible permanent plastic deformation or even fracture of the workpiece material.

The fracture modes of polycrystalline metal materials include intragranular (within the grain) and intergranular (between grains) fractures. Under the external stress of abrasive particles, slip systems within the grains start and accumulate to form slip bands that extend toward the grain boundaries. Due to the orientation differences between adjacent grains, the primary slip initiation in one grain makes it difficult for adjacent grains to initiate their main slip, resulting in strengthening. This strengthening effect shows the interaction between grain boundaries and dislocations within the grain.

The short-range stress field at the grain boundaries of polycrystalline materials makes it difficult for lattice dislocations to enter or cross the grain boundary. The hindrance of grain boundaries to the movement of dislocations within the grains causes uneven and discontinuous deformation at the grain boundaries, leading to stress concentration. As the grinding process progresses, a large number of dislocations generated within the grains can be absorbed by the grain boundaries, releasing the corresponding stress concentrations. The accumulation of dislocations at the grain boundaries leads to grain boundary migration and fracture, thus forming grinding shear slip bands and grinding debris.

Unlike polycrystalline materials, single-crystal materials consist of a single grain without any grain boundaries. The atoms of nickel-based single-crystal superalloy are arranged periodically and repetitively in 3D space. The unit cell of the nickel-based single-crystal superalloy is shown in Figure 1.13a, with a face-centered cubic (FCC) structure. Figure 1.13b and c compare the 2D atomic arrangements of single-crystal and polycrystalline materials: the atomic arrangement in single-crystal materials forms a long-range ordered structure, while polycrystalline materials are composed of numerous grains with different atomic orientations. In Figure 1.13a, the black dots represent metal atoms, and the straight lines indicate the bonds between atoms.

(a) (b) (c)

Figure 1.13 (a) Face-centered cubic lattice, and atomic arrangement of (b) single-crystal and (c) polycrystalline material.

Under the action of grinding force, dislocations are generated within the lattice of a single-crystal material, initiating slip. When internal slip and fracture occur within the lattice, they encounter various resistances. The most fundamental inherent resistance is the lattice resistance, commonly known as the Peierls stress, as shown in -Eq. 1.2. This resistance comes from the crystal lattice's hindrance to the movement of dislocations within the lattice, mainly due to the attractive forces between the atoms on the two layers corresponding to the slip plane.

$$\tau_{\mathrm{p}} = \frac{2G}{1-\nu} \mathrm{e}^{-\frac{2\pi m}{n(1-\nu)}} \tag{1.2}$$

where G represents the material's shear modulus, ν is Poisson's ratio, m is the interplanar spacing, and n is the atomic spacing along the slip direction. Evidently, its magnitude is directly determined by interplanar spacing and atomic spacing. In a crystal, different crystallographic planes have different interplanar spacings. According to Eq. 1-2, the largest interplanar spacing corresponds to weaker atomic bonding forces between the planes, resulting in the smallest lattice resistance to slip, which often makes slip more likely to occur. The slip direction in a crystal is usually along the direction of close-packed atoms, and the interplanar spacing increases with the atomic density within the plane. Lattice dislocation slip first occurs in the direction of closely packed planes because the atomic bonding forces are stronger within high-density planes, making slip less likely, whereas the bonding forces between closely packed planes are weaker.

During the microscale grinding process, the nickel-based single-crystal superalloy DD98 undergoes slip and fracture on the (111) plane due to the shear forces of abrasive particles. The direction of slip is often along the most closely packed crystallographic orientation in the crystal. Figure 1.14 shows the various slip planes within the {111} family of lattice planes and their corresponding slip directions, including: the [101], [011], and [110] crystallographic directions within the (111) plane, as shown in Figure 1.14a; the [110], [011], and [101] crystallographic directions within the (111) plane, as shown in Figure 1.14b; the [101], [011], and [110] crystallographic directions within the (111) plane, as shown in Figure 1.14c; and the [110], [101], and [011] crystallographic directions within the (111) plane, as shown in Figure 1.14d.

The essence of nanomachining single-crystal nickel is to break the bonds between nickel atoms within the original single crystal, achieving atomic-level removal.

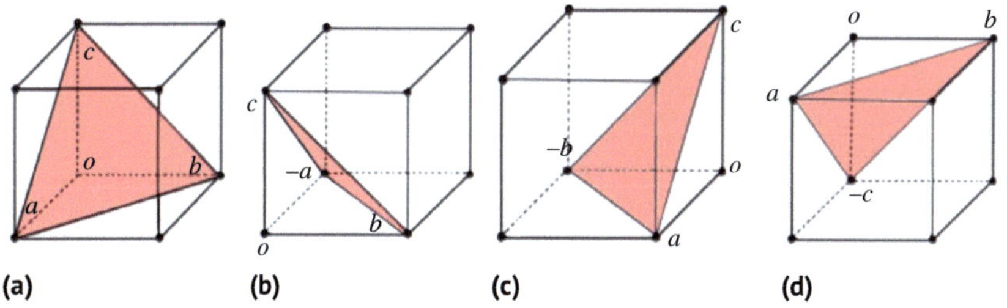

Figure 1.14 Gliding plane and direction of face-centered cubic lattice: (a) (111) plane; (b) $(\bar{1}11)$ plane; (c) $(1\bar{1}1)$ plane; and (d) $(11\bar{1})$ plane. Adapted from [4].

Therefore, during the machining process, the breaking of atomic bonds and the generation of defects should be considered as the starting point for studying material failure.

By using the atomic displacement vector method, one can track the magnitude and direction of each atom's displacement when the workpiece material deforms, allowing for a more direct observation of the flow direction and deformation extent of the material atoms. The displacement vectors of the defect atoms at machining distances of 0.4 nm and 0.7 nm are shown in Figure 1.15a and b [16], respectively, which show the displacement vectors of surface atoms corresponding to the above-mentioned machining distances.

As shown in Figure 1.15a, when the machining distance is 0.4 nm, the workpiece atoms in contact with the tool are compressed toward the interior of the workpiece by the tool. As the force at the contact point gradually increases, defect atoms gradually accumulate and form clusters of defect atoms. These atoms exhibit a displacement trend toward the [110] direction due to the breaking of atomic bonds. At this time, the displacement vector direction of the defect atoms is not on the slip plane of the single crystal. This is because, at the initial stage of machining, the slip system of the crystal has not yet been activated, and the displacement of defect atoms within the material has not yet reflected the characteristics of single-crystal materials.

Figure 1.15b shows the atomic displacement vector situation on the upper surface of the workpiece. It can be seen that under the pushing effect of the tool, the displacement change of surface atoms is relatively uniform, and the overall displacement decreases with the increase of the distance between surface atoms and the tool. Moreover, the displaced surface atoms only remain on the workpiece surface near the tool, indicating that the force acting on the workpiece is relatively concentrated. Furthermore, the displacement direction of surface atoms is consistent with the machining direction, and the displacement difference between adjacent atoms

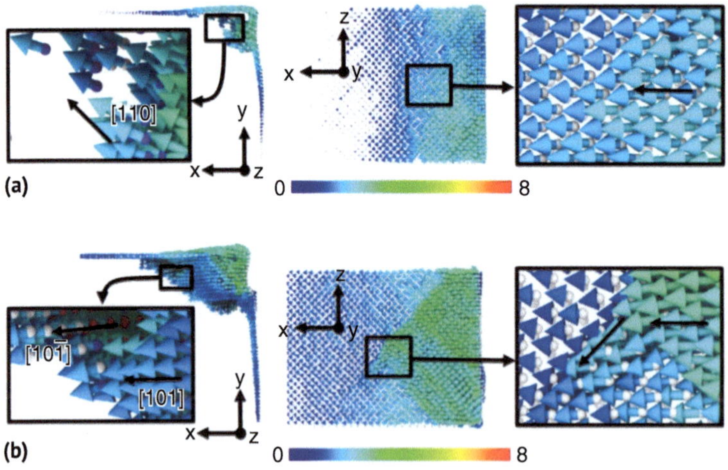

Figure 1.15 Defect atomic displacement vectors at cutting distances of (a) 0.4 nm and (b) 0.7 nm. *Source*: Zhu et al., (2016) / Springer Nature.

is not significant, indicating that the workpiece surface has not yet experienced cracks or other forms of damage caused by machining.

In the micro–nano machining of single-crystal nickel, the process by which atoms in the machined part detach from the substrate is not a rigid slip on one side relative to the other as a whole, but rather it is achieved through the movement of dislocations. When a dislocation moves from within the material to the surface, it is equivalent to the entire region swept by its dislocation line having slipped one unit distance (the distance between adjacent lattice planes) along the direction of its Burgers vector. Thus, as dislocations continuously originate from within the material and move to the surface, they can provide the interplanar slip required for continuous plastic deformation.

In the micro–nano machining of single-crystal nickel, the process by which atoms in the machined part detach from the substrate is not a rigid slip on one side relative to the other as a whole, but rather it is achieved through the movement of dislocations. When a dislocation moves from within the material to the surface, it is equivalent to the entire region swept by its dislocation line having slipped one unit distance (the distance between adjacent lattice planes) along the direction of its Burgers vector. Thus, as dislocations continuously originate from within the material and move to the surface, they can provide the interplanar slip required for continuous plastic deformation.

The machining process of the crystal, as well as the conditions under which the crystal was formed and the thermal motion of atoms, has a direct impact on the generation of defects within the crystal. Moreover, crystal defects evolve continuously with changes in various conditions. The nucleation and movement of crystal defects are an important form of plastic deformation in metallic materials, and the machining process of the crystal is accompanied by the nucleation, movement, interaction, and annihilation of dislocations.

During the nanomachining process of single-crystal nickel, special defect structures may form due to the interaction between dislocations as they move within the crystal. The internal defect evolution within the workpiece is shown in Figure 1.16.

Figure 1.16a shows the defect situation when the machining distance is 4 nm. It can be seen that since the abrasive particles have not fully entered the workpiece, the entire machining process has not reached a stable state. At this time, many larger volume stacking faults extend into the interior of the workpiece.

As machining progresses, the primary shear zone forms a high-temperature and high-pressure region dominated by amorphous atoms. The high temperature reduces the nucleation energy for dislocations, leading to dislocation proliferation. Once dislocations nucleate, they do not immediately move away but instead swirl around in the shear zone. Two Shockley partial dislocations evolve on different slip planes within the crystal, meet, and hinder each other, forming an L-C dislocation lock. Dislocation climb occurs perpendicular to the original extension direction, and ladder dislocations form at the dislocation junction lines. This ladder dislocation, along with two partial dislocations, collectively forms a "V"-shaped dislocation loop, as shown in Figure 1.16b. Since the "V"-shaped dislocation loop is supported by the L-C dislocation lock, it has a certain stability.

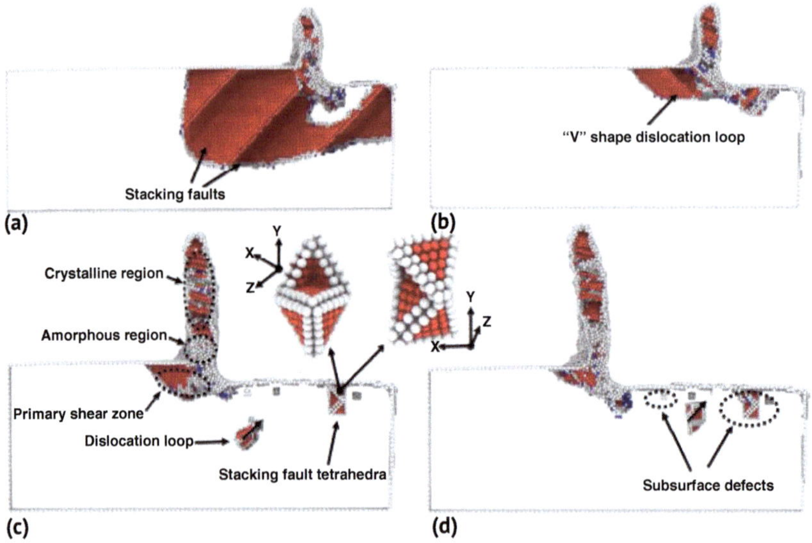

Figure 1.16 Snapshots of workpiece defects during nanometric machining at a machining speed of 200 m/s at different cutting distances of (a) 4.0, (b) 4.8, (c) 12, and (d) 15 nm.

From Figure 1.16c, it is evident that there is a clear layered phenomenon in the atom types in the upper part of the chip, the lower part, and the primary shear zone, mainly due to different forces and temperatures experienced by various regions during machining. Atoms entering different regions change into different structural types due to changes in external forces and temperatures.

Due to the extrusion effect of the rear tool face, dislocations nucleated below the tool inside the workpiece do not have enough energy to expand to the free surface and instead form closed dislocation loops within the workpiece, as shown in Figure 1.16c. When the tool leaves and the external force disappears, the stored energy inside the workpiece begins to release, causing the dislocation loops to move toward the machined surface and eventually disappear at the workpiece surface.

Due to the extrusion effect of the rear tool face, dislocations nucleated below the tool inside the workpiece do not have enough energy to expand to the free surface and instead form closed dislocation loops within the workpiece, as shown in Figure 1.16c. When the tool leaves and the external force disappears, the stored energy inside the workpiece begins to release, causing the dislocation loops to move toward the machined surface and eventually disappear at the workpiece surface.

In the micro-grinding process, material removal can be classified into three stages: rubbing, ploughing, and chip formation. Each stage corresponds to different surface generation mechanisms and exhibits different mechanical behaviors. The shape of the abrasive grit significantly influences the micro-grinding process.

During the rubbing stage (l_e), elastic and local plastic deformations are mainly governed by elastic forces acting on the workpiece. Some larger abrasive particles exceed the strength limit and form chips [34].

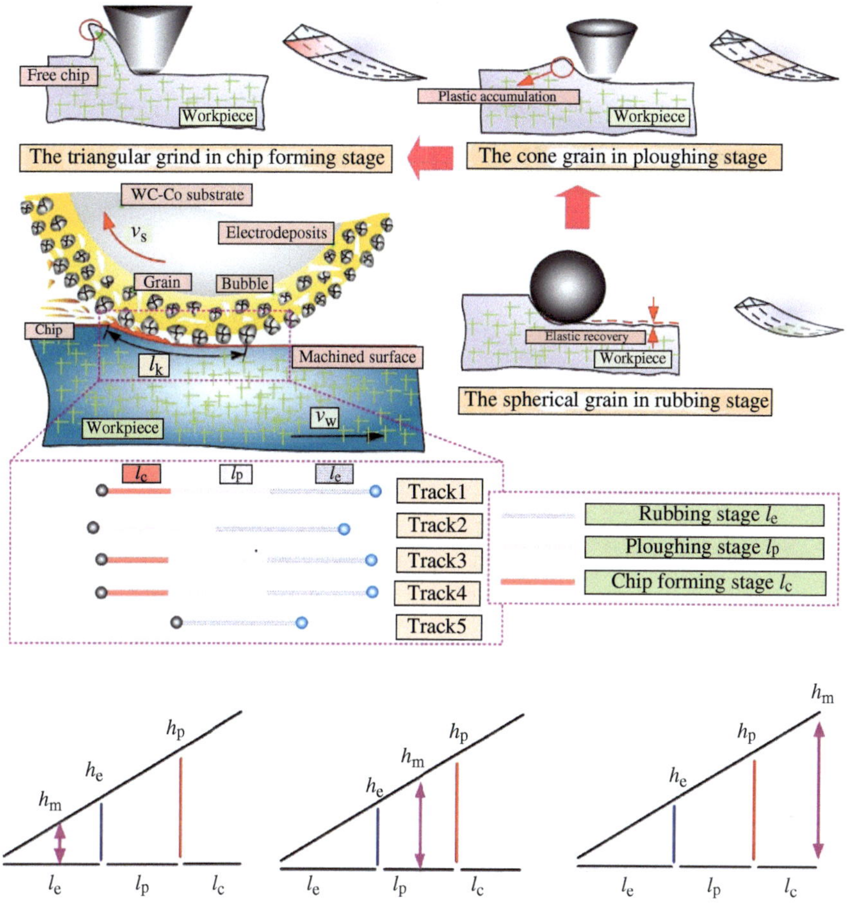

Figure 1.17 Schematic diagram of micro-grinding stage division for single-crystal superalloy.

The micro-grinding process of single-crystal superalloy material using the micro-grinding tool is illustrated in Figure 1.17. As the cutting depth increases, the contact effect between abrasive particles and workpiece continuously intensifies, eventually surpassing the yield strength and transitioning into the ploughing stage (l_p). In this stage, the number of abrasive particles forming chips rises compared to the rubbing stage, with material removal mainly manifested as plastic flow under the influence of cultivated pears.

With further increases in cutting depth, the micro-grinding force exerted by most abrasive grains surpasses the ultimate strength, ultimately leading to free chip formation and reaching the chip-forming stage (l_c). However, a small number of abrasive particles remain in the elastic contact stage.

The actual cutting depth of the abrasive grains differs from the grinding depth and is instead the undeformed chip thickness h_m, as shown in Figure 1.18. To calculate the undeformed chip thickness, it is necessary to know the interval between

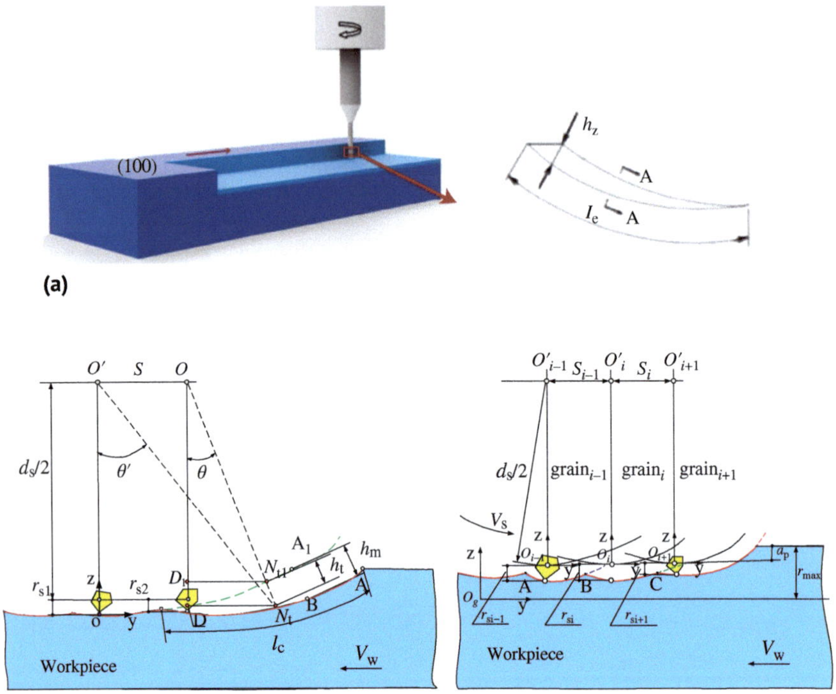

Figure 1.18 (a) Undeformed chip thickness and chip in surface grinding, and motion trajectories of (b) single abrasive particle and (c) multi-abrasive particles.

two consecutive cutting edges L, the number of cutting edges per unit wheel area C, the average effective cutting width per edge b_c, the contact arc length l_c, the average undeformed chip thickness h_a, the grinding depth a_p, the spindle speed v_s, the feed rate v_w, and the grinding rod diameter d_s. Contact arc length between grinding wheel and workpiece [112]:

$$l_c = \frac{d_s\theta}{2} = \frac{d_s \arccos\left(1 - \frac{2a_p}{d_s}\right)}{2} \tag{1.3}$$

In traditional macroscale grinding, $2a_p << d_s$. For microscale grinding, it is essential to account for the influence of size effects. The quantity of chips produced per unit of time during the grinding process of the rod is determined by the product of the volume of each chip and the MRR of the rod and the h_m can be obtained as:

$$h_m = \left[\frac{4}{Cr}\left(\frac{v_w}{v_s}\right)\left(\frac{a_p}{d_s}\right)^{\frac{1}{2}}\right]^{\frac{1}{2}} \tag{1.4}$$

In traditional conventional-size grinding, the diameter of micro-grinding rods is typically less than 1 mm, with some micrometer-level rods developed internationally.

In microscale grinding, the grinding mechanism significantly differs from that of conventional grinding. During the process, the contact arc length between the grinding rod and the workpiece is reduced, often approaching the same order of magnitude as the maximum undeformed cutting thickness of the abrasive grains. Therefore, when modelling the maximum undeformed cutting thickness for micro-grinding, it is crucial to consider the influence of the size effect. In the micro-grinding process, where h_m is nearly equal to l_m, as illustrated in Figure 1.18, the maximum undeformed cutting thickness for microscale grinding is:

$$h_m = \frac{2a_p v_w}{KC \cdot v_s b_c d_s \arccos\left(1 - \frac{2a_p}{d_s}\right)} \tag{1.5}$$

1.4.2 Characteristics and Challenges of Micro-grinding Process

Micro-grinding achieves high machining accuracy and excellent surface quality for micro-parts or structures through micro–nano mechanical removal on the workpiece surface using micro-abrasive tools with ultrafine abrasive particles. These micro-abrasive tools are generally classified into ultrathin grinding discs with a grinding section thickness of less than 1 mm and micro-abrasive rods with diameters under 1 mm.

Due to the significant reduction in the diameter of these micro-abrasive tools, factors typically negligible in conventional grinding, such as grinding wheel deformation and ploughing forces, have a substantial impact on the micro-grinding process and cannot be ignored. Additionally, in micro-grinding, the grinding depth of the abrasive particles often falls below the grain size of the workpiece material, resulting in size effects and microstructure recrystallization that significantly alter the material removal mechanism. Consequently, the micro-grinding mechanism differs markedly from that of traditional grinding, as summarized in the comparison of conventional and micro-grinding in Table 1.5.

Therefore, the mechanisms underlying micro-grinding processes are markedly distinct from those of traditional grinding. The comparison of highlighting key differences and characteristics between conventional and micro-grinding processes is emphasized in Table 1.5.

Table 1.5 Comparisons of micro-grinding and conventional grinding processes.

Content	Conventional grinding	Micro-grinding
Ratio of grinding radius to grinding depth	50~100	0.1~1
Plough effect	Not significant	Significant (20–30%)
Front rake angle	Constant negative	Variable negative
Feed rate (v_w)	1~30 m/min	0.000045~0.18 m/min
Cutting speed (v_s)	20~60 m/s	0.018~6 m/s
Material removal rate	10^2~10^{-1} mm^3/mm · s	10^{-1}~10^{-3} mm^3/mm · s

Micro-grinding can produce micro-parts and components with high dimensional accuracy and surface quality. It is a common choice for final finishing and can handle a wide range of materials, including hard and brittle substances like glass and ceramics, which are difficult to machine by traditional cutting methods. Microscale grinding can yield excellent results and is an effective way to improve micromachining quality. It typically uses small-sized grinding tools and processes the workpiece at a specific linear speed via a high-speed spindle.

Compared to micro-milling, research on micro-grinding started relatively late. Despite growing interest and progress, it remains an area with limited research. Brittle materials, which are prone to fracture under mechanical stress, are often machined by grinding. Achieving a transition to a ductile mode of removal can significantly improve machining quality. The current research progress of various scholars on micro-grinding is as follows:

Cheng et al. [35] conducted micro-grinding experiments on soda-lime glass to explore the material removal mechanism in brittle materials. They proposed a predictive model for brittle material removal in micro-grinding, considering size effects and abrasive tool surface topography. Based on the undeformed chip thickness, they classified the removal mechanisms into ductile, ductile-brittle transition, and brittle removal. Chip-type experiments showed that the edge chipping length varies significantly with the undeformed chip thickness. Coolant was found to be essential for ductile removal in soda-lime glass micro-grinding. The critical chip thicknesses for ductile-brittle transitions were determined to be 2 and 5 nm, respectively, and this was validated through grinding force variations.

Park [36] studied the influence of material grain structure, integrating thermal-mechanical coupling and size effects, to develop an interaction model for single-abrasive-grain removal at the microscale. This model incorporated a heat-transfer analysis based on a moving heat source to predict micro-grinding forces. By combining plowing and friction effects, the model evaluated the thermal influences on micro-grinding forces. Experimental calibration of heat distribution in the workpiece, using thermocouples and analytical calculations, confirmed the model's accuracy. Sensitivity analysis highlighted micro-grinding wheel and grain size as key influencing factors.

Wen et al. [32] systematically investigated the thermal characteristics of micro-grinding forces, surface roughness, and tool wear using electroplated micro-grinding tools. They established a mathematical model relating grinding forces to surface roughness and validated it against experimental data. FE simulations and thermocouple methods were used to analyze the workpiece temperature distribution during micro-grinding. Incorporating factors such as the minimum chip thickness and size effect, they created a model for the microscopic wear of grinding grains and studied the relationship between tool wear and workpiece surface roughness.

To improve micro-grinding performance, researchers have incorporated ultrasonic vibration-assisted technology into micro-grinding processes. Zhang et al. [18] conducted theoretical and experimental studies on ultrasonic vibration-assisted face micro-grinding. They developed a theoretical model based on how ultrasonic vibration geometry affects the instantaneous cutting thickness of abrasive grains.

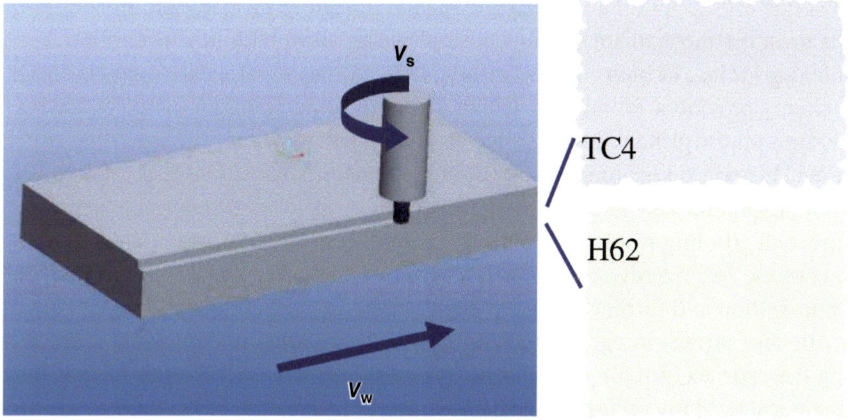

Figure 1.19 Micro numerical control machine, micro-grinding tool, and micro-manufacturing feature of Northeastern University.

Experiments on quartz glass showed that ultrasonic assistance significantly reduced grinding forces and enabled ductile-mode machining, improving surface quality. Moreover, the decrease in grinding forces expanded the range of grinding parameters, allowing the use of larger parameters to enhance processing efficiency.

Lubrication is crucial in grinding as it reduces friction, lowers processing temperatures, and improves both grinding performance and abrasive lifespan.

Dull abrasives significantly reduce grinding performance. While dressing dull grinding wheels is relatively easy in conventional grinding, the small size and fine abrasives of micro-grinding tools make dressing more challenging. Feng et al. [17] investigated the micro-grinding of ceramic end-faces and examined how tool wear and stiffness affect grinding forces, system vibration, acoustic emission signals, and spindle load. They introduced a tool-wear monitoring approach that combines grinding force and system vibration signals and validated its feasibility under various processing conditions. Although the study focused on ceramics, challenges remain due to the low stiffness of micro-tools and the high grinding forces required for hard, brittle materials.

Gong et al. [14] developed a micro-grinding machine and produced micro-feature components, as shown in Figure 1.19, while analyzing how different parameters affect the surface quality of TC4 and H62 in micro-grinding.

1.5 Contents of this book

The main contents of this book are as follows:

Part I: A brief description of grinding technology (including conventional and micro-grinding) for single-crystal nickel alloys is provided. Particularly, the development of understanding in grinding mechanisms, grinding characteristics, grinding process simulation, and control for single-crystal nickel alloys is analyzed, and some crucial issues in high-efficiency and high-quality grinding of single-crystal nickel alloys are provided and need more attention. In addition, the grinding mechanism of single-crystal nickel alloy, both in conventional and micro-grinding, is presented by means of FEM and theoretical analysis. These contents provide a foundation for an in-depth understanding of the following analysis on the grindability of single-crystal nickel alloy.

Part II: This section evaluates the grinding characteristics of single-crystal nickel alloys comparatively and comprehensively, in terms of grinding force, temperature, and grinding wheel wear behavior in both conventional and micro-grinding processes. The massive data on such aspects is valuable and could be used for optimization and further control of single-crystal nickel alloy grinding.

Part III: The surface integrity of single-crystal nickel alloy in conventional grinding (i.e., surface grinding and profile grinding) is examined both at micro- and nano-scales. Many indicators of surface integrity, such as grinding surface topography, surface roughness, microhardness, subsurface microstructure, residual stresses, and even fretting behavior, are assessed in detail.

Part IV: The micro-grinding-induced surface integrity of single-crystal nickel alloy is studied. In particular, the effects of micro-grinding parameters on surface integrity indicators (i.e., surface roughness, surface topography, subsurface damage, recrystallization, and so on) are analyzed comprehensively and deeply.

Part V: This section develops the simulation and theoretical models for predicting the grinding temperature and residual stresses in profile grinding of single-crystal nickel alloy components. The obtained results agree well with the experimental ones, indicating the feasibility of the present models to be used for optimization and controlling the grinding process of the single-crystal nickel alloy component with complex shape/structure.

References

1 Grimmert, A., Florian, P., and Petra, W. (2023). Temperature modeling of creep-feed grinding processes for nickel-based superalloys with variable heat flux distribution. *CIRP Journal of Manufacturing Science and Technology* 41: 477–489.

2 Ding, W.F., Miao, Q., Li, B.K. et al. (2019). Review on grinding technology of nickel-based superalloys used for aero-engine. *Journal of Mechanical Engineering* 55: 189–215.

3 Das, N. (2010). Advances in nickel-based cast superalloys. *Transactions of the Indian Institute of Metals* 63: 265–274.

4 Liang, C., Gong, Y., Li, P. et al. (2023). Subsurface deformation and burr formation in nickel-based single-crystal superalloy under grinding. *Archives of Civil and Mechanical Engineering* 23: 126.

5 Sun, Y., Jin, L., Gong, Y. et al. (2022). Experimental evaluation of surface generation and force time-varying characteristics of curvilinear grooved micro end mills fabricated by EDM. *Journal of Manufacturing Processes* 73: 799–814.

6 Parrott, E. (1983). Development and application of continuous dress creep feed grinding. *Proceedings of the Institution of Mechanical Engineers, Part B: Management and engineering manufacture* 197 (4): 231–235.

7 Gu, Y., Li, H., Du, B. et al. (2019). Towards the understanding of creep-feed deep grinding of DD6 nickel-based single-crystal superalloy. *The International Journal of Advanced Manufacturing Technology* 100: 445–455.
8 Meira, F., Renato, P., and Jose, C.P. (2020). Optimization of creep feed grinding using CBN wheels on nickel-based superalloys: a case study. *Journal of Manufacturing Technology Research* 12: 1–17.
9 Cai, M., Zhu, T., Gao, X. et al. (2022). Study on machining performance in grinding of Ni-base single crystal superalloy DD5. *The International Journal of Advanced Manufacturing Technology* 120 (11): 7657–7671.
10 Yao, C., Luo, J., Tan, L. et al. (2023). Experimental investigation on surface integrity and fatigue of nickel-based single-crystal superalloy DD6 during grinding-shot peening composite manufacturing. *The International Journal of Advanced Manufacturing Technology* 125 (7): 3449–3463.
11 Zhang, W., Gong, Y., Zhao, X. et al. (2023). Modeling and analysis of tangential force in robot abrasive belt grinding of nickel-based superalloy. *Archives of Civil and Mechanical Engineering* 23: 124.
12 Zhao, B., You, H., Miao. Q. et al. (2025). Surface integrity characterization of third-generation nickel-based single crystal blade tenons after ultrasonic vibration-assisted grinding. *Chinese Journal of Aeronautics* 38: 103138.
13 Miao, Q., Ding, W., Gu, Y. et al. (2019). Comparative investigation on wear behavior of brown alumina and microcrystalline alumina abrasive wheels during creep feed grinding of different nickel-based superalloys. *Wear* 426: 1624–1634.
14 Xu, Y., Gong, Y., Zhang, W. et al. (2022). Microstructure evolution and dynamic recrystallization mechanism induced by grinding of Ni-based single crystal superalloy. *Journal of Materials Processing Technology* 310: 117784.
15 Li, J., Xiong, D., Zhang, Y. et al. (2011). Friction and wear properties of MoS2-overcoated laser surface-textured silver-containing nickel-based alloy at elevated temperatures. *Tribology Letters* 43: 221–228.
16 Chen, W., Sun, Y., Feng, X. et al. (2023). Atomic-scale study of linear reciprocating friction of TCP/γ phase in nickel-based single crystal alloy. *Physica Scripta* 98: 125021.
17 Feng, X., Zhu, Z., Wu, Z. et al. (2023). Atomic-scale study of the repeated friction processes of γ/γ' phase nickel-based single crystal alloys. *Tribology International* 185: 108538.
18 Zhang, M., Cai, M., Gong, Y. et al. (2024). Experimental study on the effect of minimum quantity lubrication on the grinding surface/subsurface of nickel-based single-crystal superalloys. *The International Journal of Advanced Manufacturing Technology* 134: 5375–5390.
19 Guo, C., Campomanes, M., Mcintosh, D. et al. (2003). Optimization of continuous dress creep-feed form grinding process. *CIRP Annals* 52 (1): 259–262.
20 You, H., Zhao, B., Ding, W. et al. (2024). Wear behavior of white alumina and seeded gel abrasive wheels in ultrasonic vibration-assisted grinding of nickel-based single-crystal alloy. *Wear* 556: 205505.
21 Jin, Q., Cao, S., Wang, W. et al. (2022). Study on surface integrity of DD5 nickel-based single crystal super-alloy in creep-feed grinding. *Xibei Gongye Daxue Xuebao/Journal of Northwestern Polytechnical University* 40 (1): 189–198.

22 Jin, Q., Liu, H., Zhang, S. et al. (2022). Experimental study on grinding surface quality of DD9 nickel-based single crystal super-alloy. *Advances in Aeronautical Science and Engineering* 14 (1): 73–80.
23 Gong, Y.D., Zhou, J., Zhou, Y.G. et al. (2018). Micro-grinding temperature simulation for nickel-based single crystal superalloy. *Journal of Northeastern University (Natural Science)* 39: 82.
24 Zhou, Y., Ma, L., Gong, Y. et al. (2019). Study on force and temperature characteristics of micro-grinding nickel-based single-crystal superalloy. *Journal of the Brazilian Society of Mechanical Sciences and Engineering* 41: 1–18.
25 Ding, Q., Lao, Z., Wei, H. et al. (2020). Site occupancy of alloying elements in γ' phase of nickel-base single crystal superalloys. *Intermetallics* 121: 106772.
26 Wang, C., Yan, X., Liao, H. et al. (2021). Process parameter optimization and anisotropy sensitivity study for abrasive belt grinding of nickel-based single-crystal superalloy. *Archives of Civil and Mechanical Engineering* 21: 1–18.
27 Miao, Q., Ding, W., Kuang, W. et al. (2022). Microstructure evolution and formation of gradient structures in single crystal nickel-based superalloy by surface mechanical creep-feed grinding treatment. *Materials* 16: 321.
28 Miao, Q., Ding, W., Xu, J. et al. (2021). Creep feed grinding induced gradient microstructures in the superficial layer of turbine blade root of single crystal nickel-based superalloy. *International Journal of Extreme Manufacturing* 3: 045102.
29 Miao, Q., Ding, W., Kuang, W. et al. (2022). Indentation behavior of creep-feed grinding induced gradient microstructures in single crystal nickel-based superalloy. *Materials Letters* 306: 130956.
30 Qian, N., Jamil, M., Ding, W. et al. (2024). Thermal management in grinding of superalloys–A critical review. *Journal of Intelligent Manufacturing and Special Equipment* 5: 3–33.
31 Ge, Y., Zhu, Z., and Zhu, Y. (2019). Electrochemical deep grinding of cast nickel-base superalloys. *Journal of Manufacturing Processes* 47: 291–296.
32 Zhou, Y., Wen, X., Yin, G. et al. (2021). Study on theoretical model of roughness and wear of the microgrinding tool in microgrinding nickel-based single crystal superalloy. *Journal of the Brazilian Society of Mechanical Sciences and Engineering* 43: 317.
33 Gong, Y., Li, Q., Li, J. et al. (2018). Tool wear performance and surface integrity studies for milling DD5 Ni-based single crystal superalloy. *Science China Technological Sciences* 61: 522–534.
34 Huang, M. and Jing, Z. (2016). An overview of rhenium effect in single-crystal superalloys. *Rare Metals* 35 (2): 127–139.
35 Gong, Y., Zhou, Y., Wen, X. et al. (2017). Experimental study on micro-grinding force and subsurface microstructure of nickel-based single crystal superalloy in micro grinding. *Journal of Mechanical Science and Technology* 31: 3397–3410.
36 Park, K.H. and Paul, W. (2018). Observation of the dissolution of topologically close packed phases by an additional heat treatment in third generation nickel-based single crystal superalloy turbine blades. *Advanced Engineering Materials* 20: 1700987.

2

Removal Mechanism of Single-crystal Nickel Alloy in High-efficiency Grinding

2.1 Yield Criterion and Failure Criterion of Single-crystal Nickel Alloy

Yield criterion indicates the moment that the material undergoing continuous stresses starts to yield. This process governs the initial state during plastic deformation. Because the single-crystal nickel alloy is orthotropic, Hill's yield criterion was used as the yield function for plastic strain for the mechanical analysis [1].

$$
\begin{aligned}
R_{11} &= \frac{\overline{\delta}_{11}}{\delta^0}; R_{22} = \frac{\overline{\delta}_{22}}{\delta^0}; R_{33} = \frac{\overline{\delta}_{33}}{\delta^0} \\
R_{12} &= \frac{\overline{\delta}_{12}}{\tau^0}; R_{13} = \frac{\overline{\delta}_{13}}{\tau^0}; R_{23} = \frac{\overline{\delta}_{23}}{\tau^0}
\end{aligned} \tag{2.1}
$$

where δ^0 and τ^0 are the reference values for yield stress, while $\overline{\delta}_{11}$, $\overline{\delta}_{12}$, $\overline{\delta}_{13}$, $\overline{\delta}_{22}$, $\overline{\delta}_{23}$ and $\overline{\delta}_{33}$, are the yield stress values for main crystal orientations, respectively.

The failure criterion was decided based on orthotropic strain method, where the equivalent plastic strain was the result induced by the coupled effects of shear stress and strain rate, as calculated by [2]:

$$
D = \mathrm{Max}\left(d_{ijl}\right) = \mathrm{Max}\left(\frac{\epsilon_{ijf\_l}}{\epsilon_{ijl}} \cdot \frac{\epsilon_{ijl} - \alpha \cdot \epsilon_{ijd\_l}}{\epsilon_{ijf\_l} - \epsilon_{ijd\_l}}\right) \tag{2.2}
$$

When the monotony increment of plastic strain was equal to 1 in the grinding process, the workpiece material in front of an abrasive grain was considered to achieve the failure criterion and started to damage. The damage evolution of workpiece material was determined by the damage displacement. In other words, if the displacement reached the critical value of damage failure, the unite nodes used in the simulation were separated and led to the formation of grinding chips [3, 4].

Grinding of Single-Crystal Superalloys: Fundamentals and Technologies, First Edition.
Wenfeng Ding et al.
© 2026 Wiley-VCH GmbH. Published 2026 by Wiley-VCH GmbH.

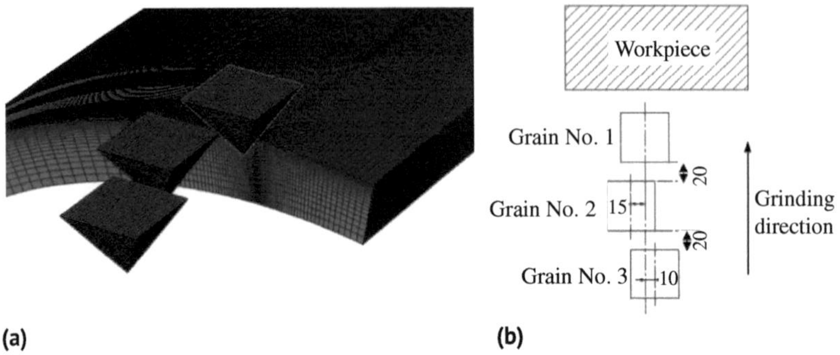

Figure 2.1 Distribution of multi-abrasive grains: (a) simulation model and (b) top view.

2.2 Simulation Model and Experiment Conditions

In order to save computing time and improve the calculation efficiency, the material removal process in grinding of the workpiece with typical convex arc was studied. Furthermore, the abrasive grain distribution maintained unchanged when multi-abrasive grains were used in simulation. The undeformed chip thickness was consistent with that calculated according to the actual grinding conditions. The grinding direction was along [001] crystal orientation. The undeformed chip thickness of one single grain can be calculated in the range of 0.4–1.2 μm, if the grinding speed, workpiece speed, and grinding depth varied from 20 to 35 m/s, 100 to 300 mm/min, and 0.2 to 1 mm, respectively. Therefore, three typical undeformed chip thicknesses in simulation were used: 0.2, 0.5, and 1 μm. The three abrasive grains were arranged on the top and side surfaces, as shown in Figure 2.1. Detail parameters in experiment and simulation were shown in Table 2.1.

2.3 Simulation Results on Material Removal by Multi-abrasive Grains

2.3.1 Evolution of Material Removal Process

Figure 2.2 shows the comparison in material removal behavior induced by both one single grain and multiple abrasive grains at different time steps. Note that the grinding speed and workpiece speed were fixed at 25 m/s and 200 mm/min respectively, and the undeformed chip thickness was maintained at 0.5 μm. It can be found that for grinding with one single grain (Figure 2.2b, d, f, and h), the workpiece material was removed in the form of chips with the maximum Mises stresses of 886.2 MPa. Obvious plastic pileups flowed upwards along the grain edge tip, forming a deformed area.

For grinding with multi-abrasive grains (Figure 2.2a, c, e, and g), the abundant material pileups appeared on the side of the grain tips at time step 35 (Figure 2.2c). The maximum Mises stresses of grains No. 1 to No. 3 were 888.6, 852.1, and

Table 2.1 Experimental and simulation parameters.

Contents	Values
Workpiece profile	Convex arc
Grinding speed v_s (m/s)	20–35
Workpiece speed v_w (mm/min)	100–300
Grinding depth a_p (used in experiment) (mm)	0.2–1
Grinding depth a_p (used in simulation) (μm)	0.2, 0.5, 1
Constant related to grain edges C	4
Active grain number per wheel area N_d (mm^{-2})	5.1
Wheel diameter d_s (mm)	400
Undeformed chip thickness a_{gmax} (μm)	0.4–1.2
Distribution patterns	Equal spacing

814.2 MPa, respectively. Under such conditions, the grain-moving trajectory began to overlap, and some material pileups in front of grain tips formed the chips. When the time step was 80 (Figure 2.2g), the grain trajectories presented continuous overlaps. In other words, the grinding marks were covered by the later grain trajectories, and the material pileups induced by the former grains became the long grinding chips under the cutting effect of the later grains. New surfaces were generated. This process is more complicated than that produced by one single grain. In addition, the topography of grinding marks showed great difference. For example, the grooves by single grain were narrow and deep, while they presented flat and wide surfaces when grinding with multi-abrasive grains.

Because the material removal process was affected significantly by the distribution and moving trajectory of multi-abrasive grains, the grinding force of each abrasive grain was different [5]. For example, as shown in Figure 2.3, the grinding force for grain No. 1 changed from 1.5 – 1.75 N, while it was around 0.5 N for grain No. 2. This was because the material removal volume by grain No. 2 was lower than that by grain No. 1, originating from the interaction of moving trajectories of grains. Thus, compared to grain No. 1, the lower grinding loads of grain No. 2 produced smaller grinding force. However, the biggest grinding force was found when grain No. 3 was involved in grinding. The possible reason was the work hardening of workpiece material induced by the former grains. The material pileups and plastic deformation also influenced the material removal process [6]. Therefore, the grinding forces were always changing with the motion of abrasive grain tips.

The cross section of the grinding trajectory of multi-abrasive grains is shown in Figure 2.4a. When the grain No. 1 first made contact with the workpiece material, there were obvious material pileups on the sides of the grain tip, and one groove was created (Figure 2.4b). With the motion of grinding, the grain No. 2 penetrated into the workpiece and produced another deep groove. The overlaps were caused in the middle

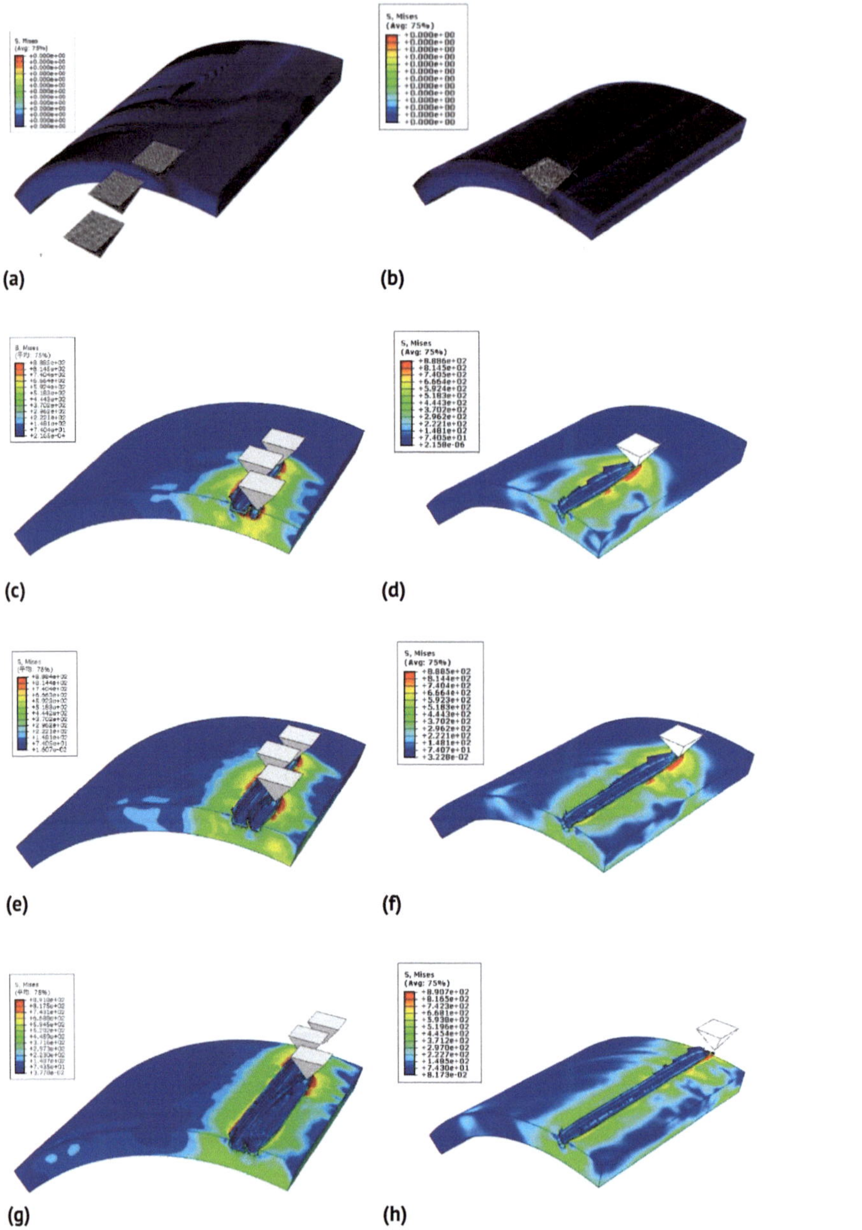

Figure 2.2 Evolution of material removal process by single grain and multi-grains: (a) Step 0 (multi-grains); (b) Step 0 (single grain); (c) Step 35 (multi-grains); (d) Step 35 (single grain); (e) Step 50 (multi-grains); (f) Step 50 (single grain); (g) Step 80 (multi-grains); and (h) Step 80 (single grain).

position between the two grooves (Figure 2.4c). The slight interferences in grinding trajectory were because of the relatively large distance between two abrasive grains. However, with the grain No. 3 cutting-in (Figure 2.4d), its trajectory caused interference with previous ones. In addition, the residual pileups were removed quickly, and a new grinding trajectory was generated. With the progression of grinding, the surface

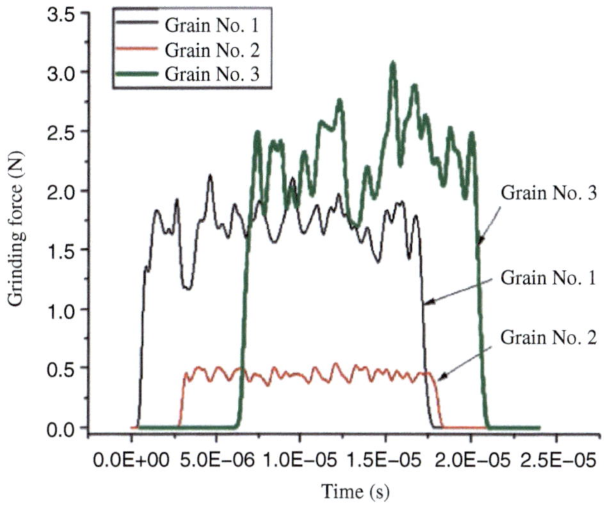

Figure 2.3 Grinding forces of multi-abrasive grains.

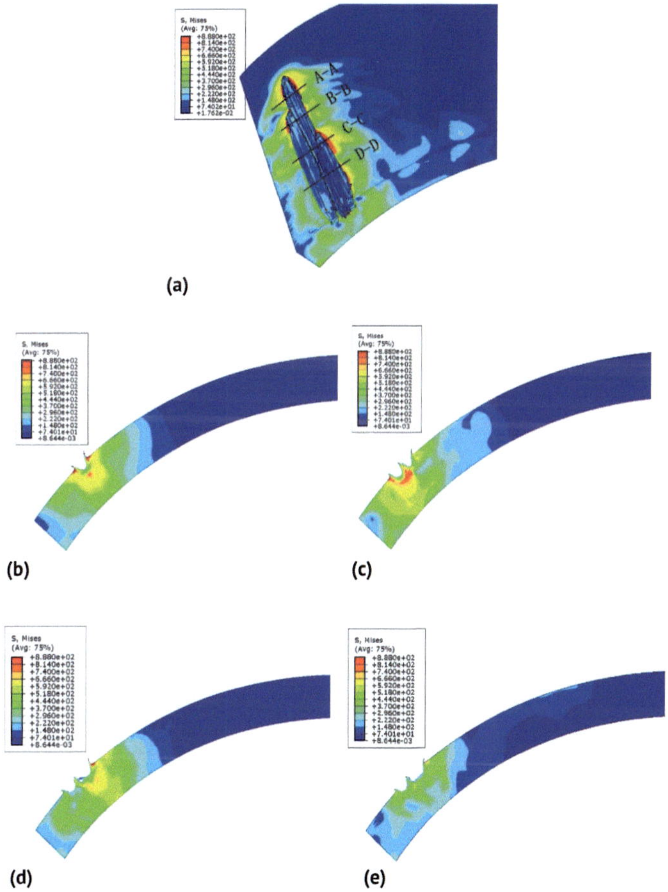

Figure 2.4 The cross-sectional features of grinding trajectory of multi-abrasive grains: (a) the whole image; (b) A–A view; (c) B–B view; (d) C–C view; and (e) D–D view.

was created under the interactions of three abrasive grains, as shown in Figure 2.4e. The stress concentration of around 518 – 592 MPa can be found in the surface layer. Therefore, the final surface formation was induced by the complex plastic deformation and overlaps by numerous abrasive grains in the grinding.

2.3.2 Effect of Grinding Speed on Material Removal Behavior

Figure 2.5 shows the effect of grinding speed on material removal behavior of single-crystal nickel alloy. The grinding depth was fixed at 0.5 μm. It can be found that when $v_s = 20$ m/s (Figure 2.5a and b), the workpiece material was deformed plastically under the pressure of abrasive grains. When v_s was further increased to 25 m/s

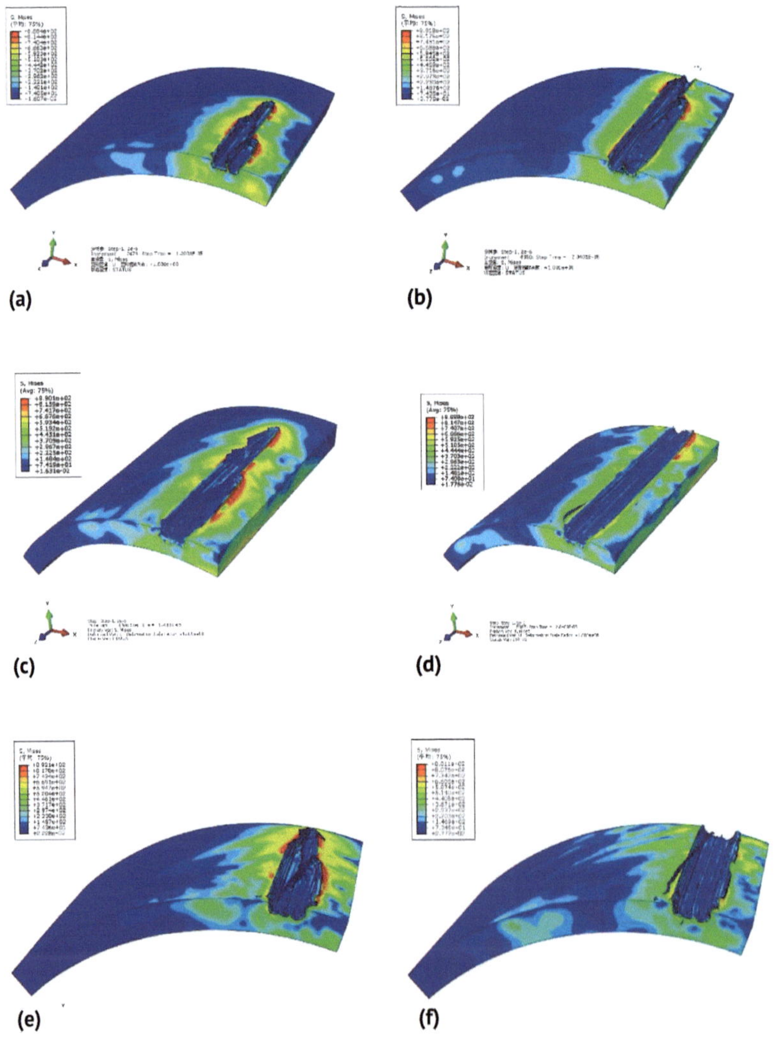

Figure 2.5 Stress map of grinding trajectory by multi-abrasive grains with grinding speed: (a) $v_s = 20$ m/s, Step 50; (b) $v_s = 20$ m/s, Step 85; (c) $v_s = 25$ m/s, Step 50; (d) $v_s = 25$ m/s, Step 85; (e) $v_s = 35$ m/s, Step 50; and (f) $v_s = 35$ m/s, Step 85.

(Figure 2.5c and d), the grinding marks became clear with few burrs generated on the edges. The long grinding chips were formed, and the grinding surface appeared smoothly when the grinding speed was 35 m/s (Figures 2.5e and f). Furthermore, it can be found that increasing grinding speed can reduce the material pileups and overlaps, promoting the chip formation and thus leading to the fine surface texture.

2.3.3 Effect of Grinding Depth on Material Removal Behavior

Because the workpiece has convex arc surface, similar to the shape of turbine blade root partially, the actual grinding depth (i.e., the equivalent grinding depth) varies along the workpiece profile (Figure 2.6). If grain No. 1 was in the tip region of convex arc surface, the grinding depth was equal to the equivalent grinding depth. If grain No. 2 was in the side region, the relationship between the grinding depth a_p and the equivalent grinding depth a_e can be as follows:

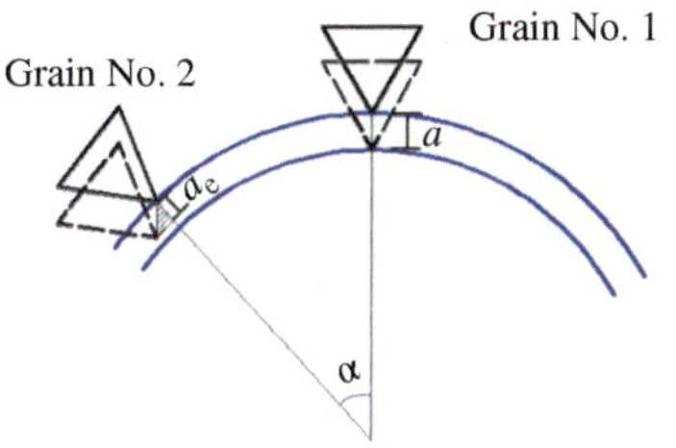

Figure 2.6 Relationship between the grinding depth and the equivalent grinding depth.

$$a_e = a_p \cos\alpha \quad (2.3)$$

where α is the central angle between grain No. 1 and grain No. 2 (i.e., $\pi/3$).

Therefore, a_e was calculated according to Eq. (2-3). The material removal behavior of the side region with grinding depth is shown in Figure 2.7. When $a_p = 0.2$ μm (Figure 2.7a), the

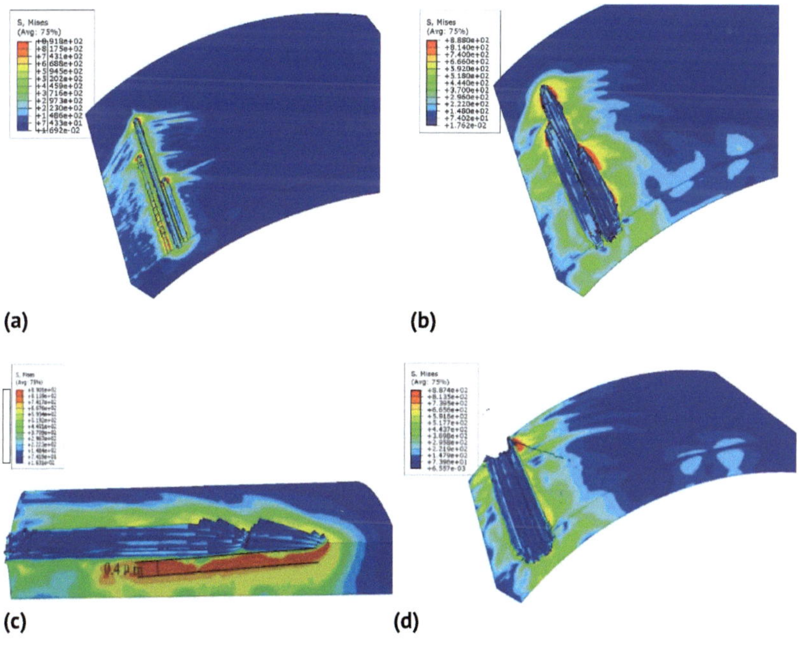

Figure 2.7 Stress map of grinding trajectory by multi-abrasive grains with grinding depth: (a) $v_s = 25$ m/s, $a_p = 0.2$ μm; (b) $v_s = 25$ m/s, $a_p = 0.5$ μm; (c) cross section along grinding direction for (b); and (d) $v_s = 25$ m/s, $a_p = 1$ μm.

grinding groove was so shallow that obvious sliding and plowing were produced because of the rather low a_e (0.05 μm). The workpiece material cannot be removed effectively. When a_p was increased to 0.5 μm (Figure 2.7b), the majority of workpiece material was removed in the form of grinding chips, and the clear surface texture was generated. The stress distribution map on the cross section of the workpiece indicated that the maximum Mises stress reached 890.1 MPa and extended to a depth of 0.4 μm. The stress concentration layer was created below and in front of the abrasive grain during grinding (Figure 2.7c). More plastic deformation and overlaps can be observed if a_p was 1 μm (Figure 2.7d).

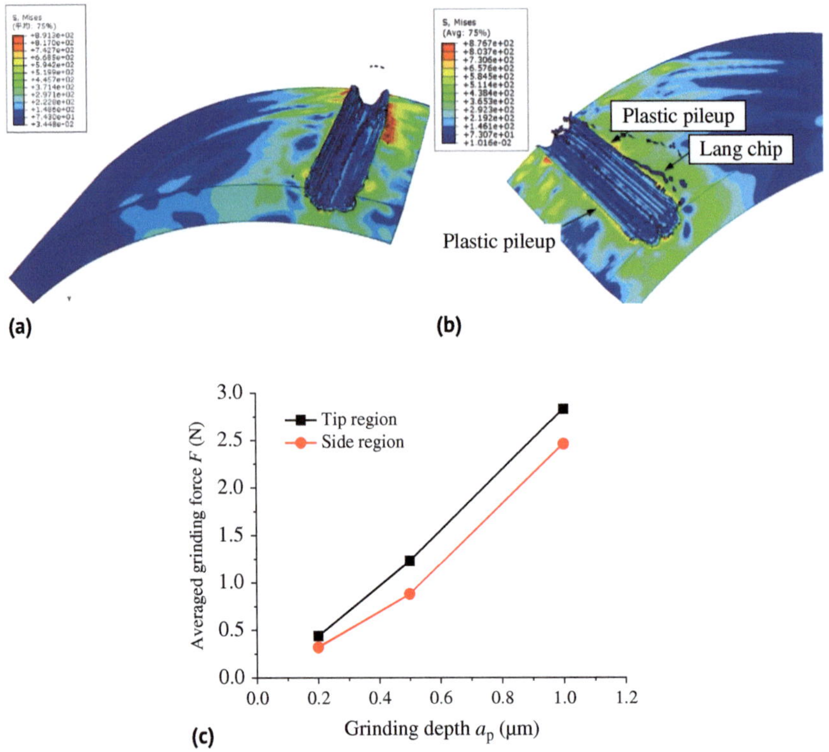

Figure 2.8 Effect of workpiece profile on material removal behavior (v_s = 35 m/s, v_w = 200 mm/min, and a_p = 0.5 μm): (a) tip region; (b) side region; and (c) variation of grinding force with grinding depth.

The effect of workpiece profile on material removal behavior is displayed in Figure 2.8. For the workpiece tip region (Figure 2.8a), the material can be removed uniformly by multi-grains. The formed surface seems to be much smoother than that in the workpiece side region, where plastic pileups and residual grinding chips were observable (Figure 2.8b). This difference can be attributed to the bigger equivalent grinding depth a_e in the tip region than that in the side region, resulting in the higher efficiency in material removal of the former. The results were also demonstrated by the grinding forces in the workpiece tip and side region (Figure 2.8c).

2.4 Experimental Verification of Simulation Results

For comparison, the experimental results were provided as shown in Figure 2.9. The main features of material removal can be as follows: (i) In Figure 2.9a and b, the workpiece material was removed mainly by strong friction between abrasive grain and workpiece due to the small a_{gmax} (0.2 μm). Thus, the cutting texture in ground surface appeared unclearly. (ii) In Figure 2.9c and d, the majority of workpiece material was removed in the form of grinding chips when $a_{gmax} = 0.5$ μm. The cutting texture was fine without material coating. Good grinding quality was achieved. (iii) In Figure 2.9e and f, the a_{gmax} was 1 μm. The grinding loads increased as the large material removal rate increased. Thus, rich plastic deformations occurred frequently, such as material pileups, coatings, and tearing. The grinding quality was degraded possibly. (iv) In Figure 2.9g and h, the typical slender and long grinding chips were produced in creep feed grinding. The chip topography in simulation agreed well with that in experiment.

Using finite element method to simulate the material removal behavior provides an effective way to study the grinding mechanism of single-crystal nickel alloy theoretically [7, 8]. The optimization of grinding process can be improved further based on the deep understanding of material removal of single-crystal nickel alloy.

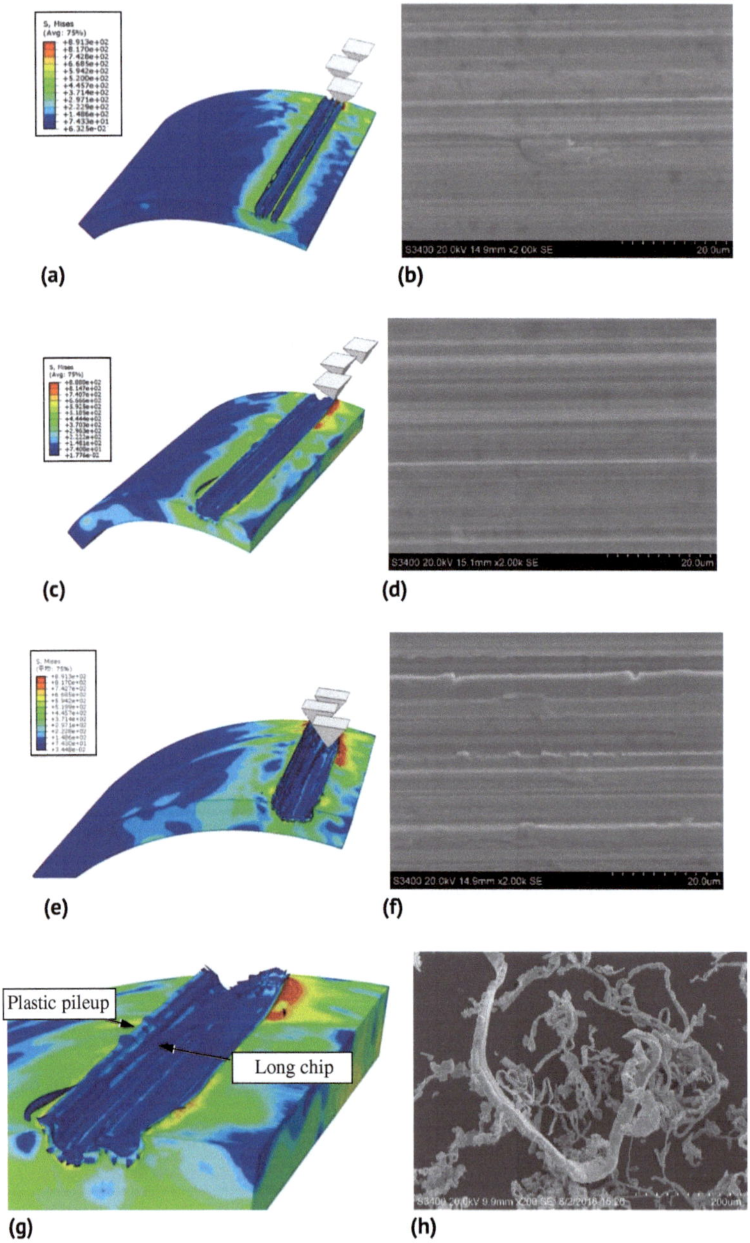

Figure 2.9 Comparison between simulation and experiment on material removal of single-crystal nickel alloy in grinding: (a) $v_s = 25\,\text{m/s}$, $a_p = 0.2\,\mu\text{m}$; (b) $v_s = 25\,\text{m/s}$, $a_{gmax} = 0.2\,\mu\text{m}$; (c) $v_s = 25\,\text{m/s}$, $a_p = 0.5\,\mu\text{m}$; (d) $v_s = 25\,\text{m/s}$, $a_{gmax} = 0.5\,\mu\text{m}$; (e) $v_s = 25\,\text{m/s}$, $a_p = 1\,\mu\text{m}$; (f) $v_s = 25\,\text{m/s}$, $a_{gmax} = 1\,\mu\text{m}$; (g) simulated grinding chips for (e); and (h) experimental grinding chips for (f).

References

1 Hill, R. (1948). A theory of the yielding and plastic flow of anisotropic metals. *Proceedings of The Royal Society A* 193: 281–297.

2 Zhu, T., Cai, M., Gong, Y. et al. (2023). Study on the chip formation in grinding of nickel-based single-crystal superalloy DD5. *The International Journal of Advanced Manufacturing Technology* 125: 5781–5798.

3 Brinksmeler, E. and Glwerzew, A. (2003). Chip formation mechanisms in grinding at low speeds. *CIRP Annals-Manufacturing Technology* 52 (1): 253–258.

4 Yang, Z., Zou, P., Zhou, L. et al. (2023). Research on the influence of stress wave on crack and chip formation mechanism in radial ultrasonic vibration assisted grinding. *Ultrasonics* 133: 107054.

5 Chu, Y., Yan, S., Yang, Z. et al. (2024). Grain shape-protrusion-based modeling and analysis of material removal in robotic belt grinding. *Journal of Manufacturing Processes* 110: 211–223.

6 Dai, C., Ding, W., Xu, J. et al. (2017). Influence of grain wear on material removal behavior during grinding nickel-based superalloy with a single diamond grain. *International Journal of Machine Tools and Manufacture* 113: 49–58.

7 Hao, Z., Liu, Z., and Fan, Y. (2024). Work hardening of Ni-based single crystal alloy in vibration grinding based on molecular dynamics method. *Archives of Civil and Mechanical Engineering*. 24: 39.

8 Liu, M., Gong, Y., Zhang, W. et al. (2024). Theoretical investigation of anisotropic material removal mechanism in robotic belt grinding single crystal superalloy process. *The International Journal of Advanced Manufacturing Technology* 135: 3487–3509.

3

Plastic Deformation Mechanism of Single-crystal Nickel Alloy in Micro-grinding

3.1 Verification of Plastic Deformation Mechanism in Micro-grinding Materials

In polycrystalline materials, dislocations face difficulties crossing grain boundaries and moving between grains during grinding. Grain boundaries act as barriers, causing dislocations to accumulate at these interfaces. This accumulation creates stress concentrations, which can trigger new dislocation sources and induce dislocation slip in neighboring grains.

When a metal crystal undergoes plastic deformation under the action of abrasive particles, slip is the primary mode of deformation. This involves adjacent crystal sections sliding relative to each other along a specific crystal plane and direction due to the cutting stress from the abrasive particle. Figure 3.1a and b shows the atomic arrangements in a metal crystal before and after deformation. Similarly, during grinding, slip in the subsurface material forms distinct slip marks or slip lines. However, slip is not a rigid sliding of the metal on either side of the slip plane. Dislocations in the metal crystal can be activated at low stress levels, and as grinding progresses, slip lines become visible throughout the grinding-affected zone.

Figure 3.2 presents high-magnification scanning electron microscopy (SEM) images of the subsurface and base microstructure of the nickel-based single-crystal high-temperature alloy DD98, and Figure 3.2a and b shows plastic deformation layer's microstructure near the grinding subsurface, and Figure 3.2c displays the unaffected matrix microstructure of DD98, without any shear slip lines. The zone of intense plastic deformation on the micro-grinding surface is clearly visible in Figure 3.2b, where the square-shaped γ' phase in the deformation zone exhibits pronounced slip lines, along with a slip band on the grinding subsurface. This is because the face-centered cubic lattice structure of nickel-based single-crystal alloys has low stacking-fault energy. This makes cross-slip and dislocation climb difficult, facilitating shear-band formation. Under micro-grinding forces, lattice slip occurs, and the accumulation of these slips results in micro-shear slip bands, as shown in Figure 3.2b. The slip directions of these bands generated during the micro-grinding process are generally aligned along the diagonal of the square-shaped γ' phase,

Grinding of Single-Crystal Superalloys: Fundamentals and Technologies, First Edition.
Wenfeng Ding et al.
© 2026 Wiley-VCH GmbH. Published 2026 by Wiley-VCH GmbH.

composed of lattices arranged in the same direction, where the accumulated micro-lattice slip effect appears as plastic shear slip bands within the γ' phase of the grinding subsurface. Thus, it can be concluded that during micro-grinding of nickel-based single-crystal high-temperature alloys with a face-centered cubic

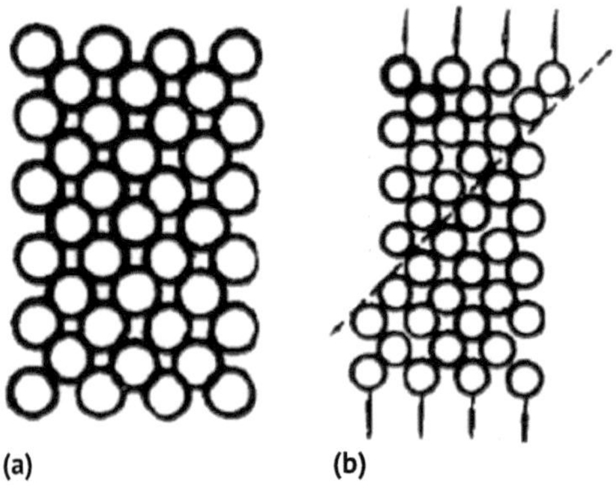

Figure 3.1 Movement of atoms in the deformed metal: (a) after deformation and (b) before deformation. Adapted from [1].

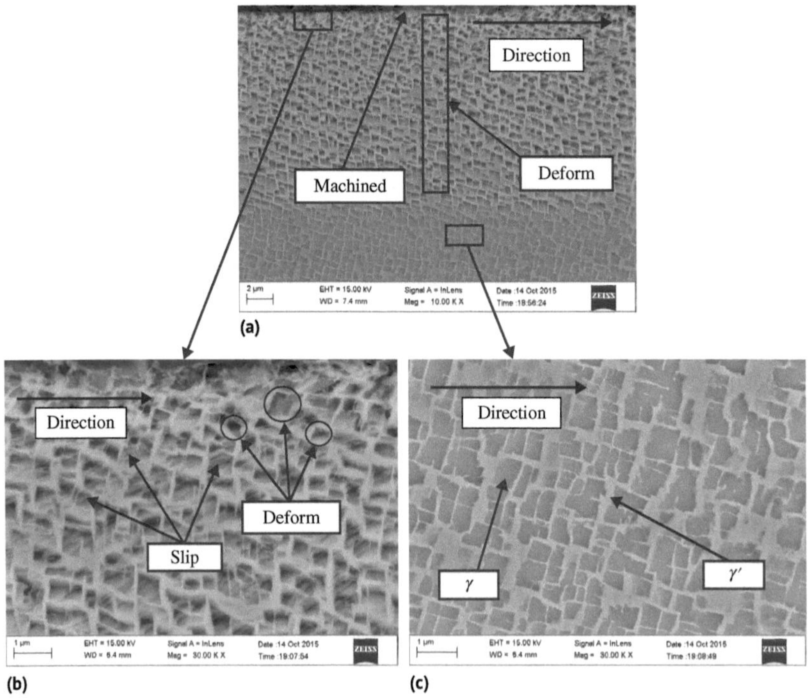

Figure 3.2 Microstructure of single-crystal superalloy DD98 matrix and the grinding subsurface: (a) surface after micro-grinding; (b) enlarged view; and (c) enlarged view. [2]/Elsevier.

structure, the shear slip plane of lattice deformation is the (111) crystal plane. Additionally, as shown in Figure 3.2, it is evident that γ and γ' phases near the grinding surface in the subsurface region undergo twist deformation. This deformation occurs primarily because, as the workpiece progresses along the feed direction, the lattice in γ and γ' phases shifts along the slip plane and direction under the shearing force of abrasive particles. The unequal forces on either side of the slip plane within γ and γ' phases generate a moment, resulting in the rotation of these phases.

The main difference between polycrystalline and single-crystal materials is the presence of numerous grain boundaries in polycrystals. These grain boundaries significantly affect the plastic deformation of metal in the grinding and subsurface zones. As a result, the plastic deformation process in polycrystalline materials is more complex than in single crystals, as depicted in Figure 3.3, which shows the activation of slip systems on the grinding surface of a polycrystal [3–5].

In the grinding process, plastic deformation in polycrystalline metals is also mainly governed by slip. However, each grain in polycrystalline materials has a unique orientation and is surrounded by neighboring grains. During deformation, each grain is constrained by its neighbors, and the orientation and direction of the slip system vary among grains. Slip cannot directly transfer from one grain to another, as illustrated in Figure 3.3a. Molecular dynamics simulations of the micro-grinding process in polycrystalline materials, shown in Figure 3.3b, indicate that the slip direction within individual grains is relatively uniform. But slip propagation between grains is challenging due to the disordered atomic arrangement and lattice distortion at grain boundaries, which increase deformation resistance. Thus, the plastic deformation mechanism in polycrystalline materials differs from that in single crystals.

Dislocations formed within a crystal will move until they glide out of the crystal if there are no obstacles. The movement of many dislocations can form slip lines on the crystal surface, leading to macroscopic deformation. However, during processing, dislocations are often hindered by obstacles within the crystal, causing them to accumulate in front of these obstacles, forming pileups. These dislocation pileups play a crucial role in the initiation and evolution of plastic deformation [4]. The more dislocations in a pileup, the stronger the force the leading dislocation exerts on the obstacle. When this force reaches a certain level, the obstacle may break, allowing dislocations to continue sliding and causing plastic deformation of the workpiece.

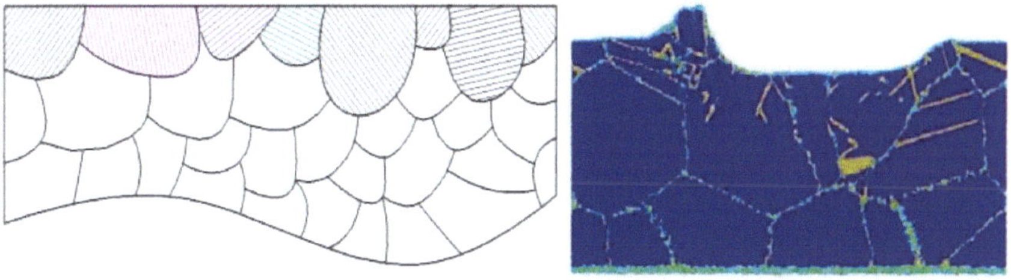

(a) (b)

Figure 3.3 Grain slipping in grinding polycrystalline material: (a) schematic diagram and (b) molecular dynamics simulation.

From Figure 3.4, it can be observed that as the cutting tool progressively penetrates the workpiece, several dislocations form near the tip of the tool (Figure 3.4a). These dislocations extend along the $(\bar{1}11)$and $(1\bar{1}1)$planes, respectively. The stacking fault structures enclosed behind these dislocations also extend along their respective slip planes (Figure 3.4b). Due to a temporal difference in the nucleation of dislocations on these two different slip planes, dislocations on the $(\bar{1}11)$plane develop more rapidly. When the movement of dislocations on the $(1\bar{1}1)$plane reaches the slip plane where dislocations on the $(\bar{1}11)$plane are located, they are obstructed by the stacking fault structure on the $(\bar{1}11)$plane, hindering further movement along their original slip plane.

As the cutting energy from the tool increases, several dislocations with the same slip system as the blocked ones are generated near the tool tip. These new dislocations move toward the blocking location, causing a pileup of dislocations in front of the obstacle (Figure 3.4c). The accumulation hardens the obstructed part of the workpiece, and the tool's force on the workpiece starts to transfer to the atoms behind the pileup, seeking energy release. This leads to the breaking of atomic bonds at the far end of the workpiece, where new dislocations begin to nucleate, as shown in position 1 of Figure 3.4.

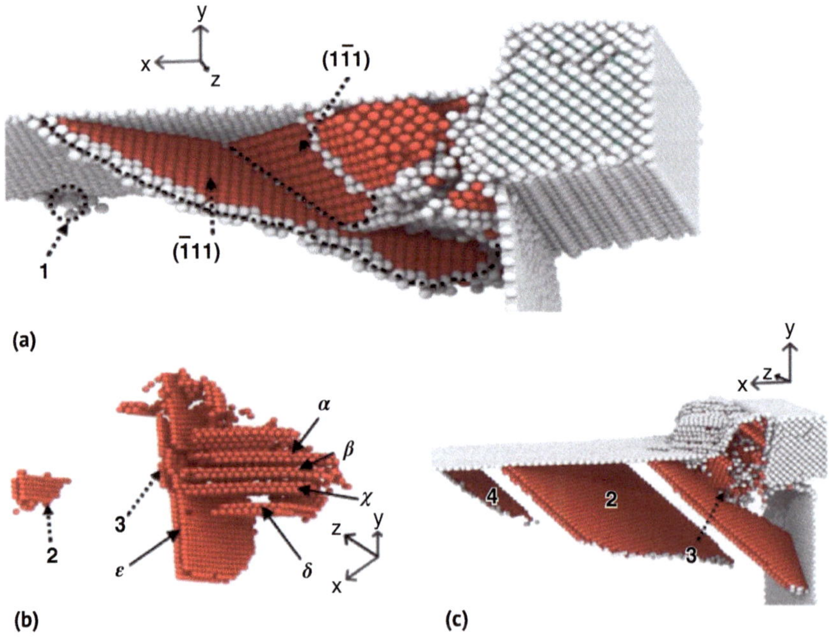

Figure 3.4 The formation of a dislocation block: (a) several dislocations near the tool tip; (b) buildup forms at the stagnation zone; and (c) obstructing stacking faults.

3.2 Microscale Debris in Micro-grinding of Single-crystal Nickel Alloy

Studying the microstructure and morphology of grinding debris is one of the most straightforward ways to investigate the material removal mechanism in microscale grinding. However, in high-speed and ultrahigh-speed grinding, the debris generated during grinding is tiny and is quickly ejected from the grinding zone by high-speed abrasive particles. This small debris scatters in various directions, making it difficult to collect. In contrast, microscale grinding has a lower linear velocity of abrasive particles, facilitating debris collection.

Domestic and international research on microscale grinding has mainly focused on surface quality, grinding forces, grinding temperatures, and subsurface damage. However, there are no reported studies on the microscopic morphology of grinding debris in the microscale grinding of nickel-based single-crystal superalloys. Analyzing the microstructure and morphology of grinding debris is crucial for understanding the material removal mechanism in microscale grinding, reducing costs, ensuring processing quality, and improving production efficiency. The grinding force, heat generation, wheel wear, and surface quality all fundamentally rely on the formation of grinding debris during the process.

Regarding the definition of the grinding deformation zone, the micro-grinding rod's surface has extremely sharp abrasive particles, making the grinding process akin to a negative-angle cutting process on the material. The formation of grinding debris can be regarded as three interrelated deformation processes:

1. First deformation zone: When abrasive particles contact the workpiece, the pre-cutting face of each particle compresses the metal in the grinding zone, causing it to experience strain and elastic deformation. As grinding progresses, the compressive force exerted by the abrasive particles on the workpiece material intensifies, leading to increased stress and deformation within the material. Once the stress surpasses the material's yield strength, internal slip occurs, resulting in plastic deformation.
2. Second deformation zone: As grinding progresses, the debris is expelled along the pre-cutting face of the CBN abrasive particles, where it undergoes additional squeezing and frictional forces. This causes the grinding debris to experience further deformation due to compression and friction. There is a strong interaction between the first and second deformation zones: when friction between the abrasive particle's pre-cutting edge and the grinding debris is high, debris expulsion becomes more challenging, intensifying material compression and deformation in the first deformation zone and increasing shear slip deformation within this zone.
3. Third deformation zone: The machined surface experiences friction and compression from tops of abrasive particles, resulting in deformation and rebound in the subsurface area of the grinding zone. This area is known as the third deformation zone, where friction occurs between the subsurface and the abrasive particle tops.

During the abrasive chip formation stage in the grinding process, each abrasive grain has a large negative rake angle. The formation of abrasive chips can be divided into the following three stages:

1. In the slip friction stage, due to the minimal depth of cut of the abrasive grain, the grain slides and rubs along the workpiece surface, resulting in only slight elastic deformation of the grinding surface without generating abrasive chips.
2. In the elastic–plastic deformation stage, as the abrasive grain penetrates deeper into the workpiece, it comes into greater contact with the workpiece material, causing localized plastic deformation on the workpiece surface. This results in a plowing effect on the grinding surface, with material accumulating and forming raised areas on both sides and at the front of the abrasive grain.
3. In the micro-cutting stage, as the depth of cut of the abrasive grains continues to increase, the grinding force exerted by a single abrasive grain also rises. When this grinding force surpasses the atomic bonding strength of the material, internal slip occurs, multiple slip systems are activated, and the abrasive grain advances, expanding the slip bands and ultimately forming abrasive debris.

The formation of abrasive debris in nickel-based single-crystal high-temperature alloys mainly results from plastic deformation through slip. During grinding, the grinding grain presses onto the metal in the grinding layer, causing the chip to experience bending torque and curling. The creation of abrasive chips is essentially due to the combined effects of extrusion and friction from abrasive particles, which induce shear slip in the metal and expose it to the bending moments of plastic deformation.

According to dislocation theory, the compression and deformation by abrasive grains in the grinding zone generate numerous dislocations that move along slip surfaces to produce slip. In micro-grinding, this slip starts at the contact point between the grit cutting edge and the workpiece surface, where dislocations move along slip planes under external forces. As grinding continues, material deformation intensifies, increasing the dislocation density. Dislocations propagate from the cutting edge through the grinding layer to the material's surface, forming a shear array. This causes plastic deformation in the grinding zone metal, resulting in wear chips that flow along the front blade surface of the abrasive grain.

In microscale grinding of nickel-based single-crystal high-temperature alloys, the brief contact time between the abrasive grain and the workpiece and the alloy's low thermal conductivity limit heat dissipation, leading to the formation of adiabatic shear in a narrow zone.

The critical conditions for adiabatic shear slip are mainly explained by two theories: periodic fracture theory and adiabatic shear theory. For difficult-to-machine materials like nickel-based high-temperature alloys, adiabatic shear is widely considered the cause of shear slip within the alloy. During microscale grinding, these alloys produce serrated abrasive chips due to concentrated shear slip. The abrasive grains separate the chips without allowing sufficient plastic deformation, leading to instability at the deformed end and resulting in thermoplastic shear instability or adiabatic shear. The constitutive relationship effectively describes the connection between flow stress and thermodynamic parameters, including strain, strain rate, and temperature. The thermoplastic eigenstructure equation for one-dimensional shear in materials is presented in Eq. (3-1).

$$\tau = f(\varepsilon, \dot{\varepsilon}, T) \tag{3.1}$$

where τ is the shear stress, $\dot{\varepsilon}$ is the shear strain rate, ε is the shear strain, and T is the plastic deformation temperature.

A comprehensive analysis of abrasive chips is essential for understanding the grinding mechanism. Studying the formation patterns of abrasive chips provides valuable insights. To further explore how micro-grinding process parameters affect the grinding mechanism of nickel-based single-crystal high-temperature alloys, this chapter focuses on analyzing the abrasive chip morphology in the microscale grinding of these alloys.

The nonuniform distribution of abrasive particles on the micro-abrasive rod leads to different protrusion heights of these particles on the rod's surface, resulting in various abrasive chip shapes under the same conditions. In this study, a representative set of abrasive chips was analyzed under consistent machining parameters to understand the overall characteristics of abrasive chips.

Figure 3.5 shows the abrasive chips generated during the microscale grinding of a nickel-based single-crystal high-temperature alloy. The free surface of the chips is discontinuous with typical serrated features (Figure 3.5a). This discontinuity indicates intense shear slip in the alloy, induced by the cutting action of the negative-rake-angle grits, which occurs in a highly localized region between serrations. The degree of serration on the abrasive chip's free surface, indicated by the number of serrations per unit length, increases with higher serration values. In contrast, the contact surface of the abrasive chip with the abrasive grain is smooth and flat, showing distinct scratches from the abrasive grain's uneven cutting edge (Figure 3.5b). In the micro-grinding process of nickel-based single-crystal high-temperature alloys, most grinding chips are strip-like, with sawtooth-shaped chips being predominantly observed. Although a significant portion of the microscale abrasive chips have serrated features, careful observation reveals notable variations in chip morphology as grinding parameters change [6].

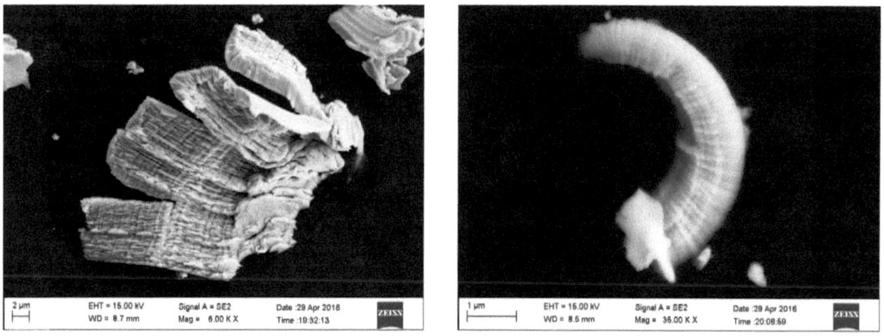

(a) (b)

Figure 3.5 Free surface of chip generated by micro-grinding tool electroplating CBN grinding material: (a) lamellar chips and (b) strip-like chips.

During the machining simulation of single-crystal nickel, similar to conventional machining, when the cutting depth does not reach the minimum machining thickness, the tool only plows the workpiece. When the cutting depth exceeds the minimum machining thickness, chips are formed. However, research has shown that the chips formed have two main morphologies as the cutting depth varies. Moreover, in the formation process of these two different chip morphologies, there are significant differences in how the tool interacts with the workpiece atoms and in the physical states of the chip atoms.

Research was conducted on the atomic displacement vectors of workpiece atoms at different stages: before chip formation, during the formation of spherical chips, and during the formation of strip-shaped chips. The study examined the deformation of internal atoms in the workpiece at various stages and conducted an in-depth investigation into the crystal structure and physical properties of atoms within chips of different morphologies.

As shown in Figure 3.6a, when the minimum cutting depth is not reached, there is a significant difference in the displacement of workpiece atoms on either side of the tool tip. The tool's plowing action on the workpiece surface compresses the atoms in front of the tool tip, causing a deformation trend toward the interior of the workpiece. The displacement vectors of these atoms extend downward into the workpiece interior, as indicated in Figure 3.6a. On the other hand, the atoms behind the tool tip start to recover from the compression due to the tool's departure and are also stretched forward due to the tool's scraping action on the workpiece surface. These combined deformation trends result in a diagonally upward displacement trend for the atoms behind the tool, with their displacement vectors pointing toward the machined surface.

As illustrated in Figure 3.6b, during the formation of spherical chips, the displacement vectors of atoms within the chips radiate outward in the direction of machining, which is related to the rolling motion of the spherical chips during the machining process. Additionally, the degree of deformation of internal atoms in the workpiece ahead of the tool is reduced compared to the stage before chip formation.

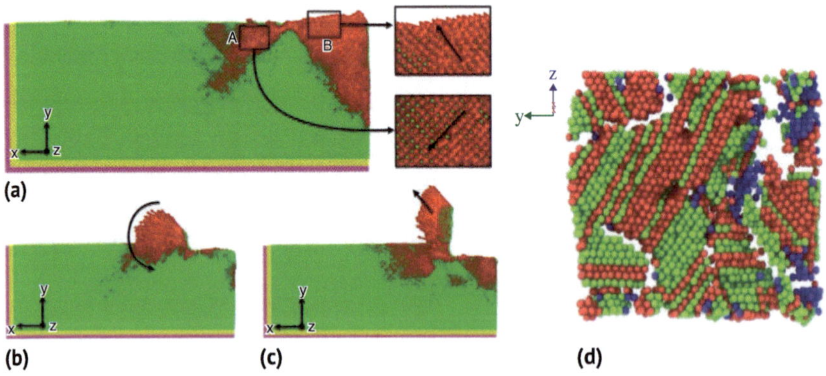

Figure 3.6 Atomic displacement vector of workpiece in different processing states: (a) the atomic displacement characteristics of the workpiece on both sides of the tool tip prior to reaching the minimum cutting depth; (b) the ball-shaped chips; (c) the formation of strip-like chips; and (d) microstructural characteristics inside the strip-like chips. [7]/IOP Publishing.

This indicates that when chips are formed, the energy applied by the tool to the workpiece is mainly absorbed by the atoms in the chips, reducing the energy impact on the workpiece substrate.

When the cutting depth increases to produce strip-shaped chips, as shown in Figure 3.6c and d, the displacement vectors of atoms within the chips become almost identical, and the tool's removal of chips mainly involves a shearing action. By examining the displacement vectors of atoms within the workpiece, it can be seen that the degree of deformation of internal atoms increases. This is due to two main factors: first, the increased cutting depth means more machining energy is applied by the tool to the workpiece; second, during the machining process, the repeated shearing action causes the chip to separate from the base material at its root. When the chip has not been completely sheared off at its root, the atoms at the chip root and those in the base are still connected by atomic bonds, and the chip root is relatively thick. Therefore, the tool's action on the chip atoms is transmitted to the base through the atoms at the chip root, causing a certain degree of displacement and deformation in the internal atoms.

References

1 Wang, C., Yan X., Liao, H., et al. (2021). Process parameter optimization and anisotropy sensitivity study for abrasive belt grinding of nickel-based single-crystal superalloy. *Archives of Civil and Mechanical Engineering* 21: 1–18.

2 Ding, Q., Lao, Z., Wei, H. et al. (2020). Site occupancy of alloying elements in γ' phase of nickel-base single crystal superalloys. *Intermetallics* 121: 106772.

3 Ding, W.F., Miao, Q., Li, B.K. et al. (2019). Review on grinding technology of nickel-based superalloys used for aero-engine. *Journal of Mechanical Engineering* 55: 189–215.

4 Yang, W.P., Li, J.R., Liu, S.Z. et al. (2019). Orientation dependence of transverse tensile properties of nickel-based third generation single crystal superalloy DD9 from 760 to 1100 °C. *Transactions of Nonferrous Metals Society of China* 29: 3558–568.

5 Feng, W.H., Chang, J.X., and Zhu, S.D. (2021). Nickel-based single crystal superalloys with different rhenium contents. *Chinese Journal of Rare Metals* 45: 3353–362.

6 Zhou, Y., Ma, L., Gong, Y. et al. (2019). Study on the mechanism of chip forming and the microhardness of micro-grinding nickel-based single-crystal superalloy. *The International Journal of Advanced Manufacturing Technology* 103: 281–295.

7 Chen, W., Sun, Y., Feng, X. et al. (2023). Atomic-scale study of linear reciprocating friction of TCP/γ phase in nickel-based single crystal alloy. *Physica Scripta* 98: 125021.

Part II

Grindability of Single-crystal Nickel Alloys

Grindability provides critical information for studying the material removal process and, therefore, fundamental guidance for grinding parameter optimization in the engineering application of single-crystal nickel alloys. Normally, the grindability of single-crystal nickel alloy concerns grinding force, grinding temperature, wheel wear behavior, and surface integrity. The grinding force, grinding temperature, and wheel wear behavior are highly dependent on the workpiece material properties and the used grinding conditions, while the grinding surface integrity is significantly influenced by the grinding force, grinding temperature, and wheel wear behavior. As the surface integrity will be described in Part III, this part mainly evaluates the grindability of single-crystal nickel alloy in three other aspects in the processes of creep-feed surface grinding, creep-feed profile grinding, and micro-grinding, respectively.

4

Grinding Force Evaluation

4.1 Grinding Force in Surface Grinding

4.1.1 Grinding Force Results

The experimental investigation employed DD6 second-generation single-crystal superalloy as the primary workpiece material, with GH4169 forged superalloy and DZ408 directionally solidified superalloy serving as comparative materials. All test specimens were machined to standardized dimensions of 40 mm (length) × 8 mm (width) × 50 mm (height). The study utilized two distinct alumina-based grinding wheels: a conventional brown alumina (BA) wheel (specification 80FF22V35, manufactured by Fenghang Co., Ltd., China) and a microcrystalline alumina (MA) wheel (specification 5SG80F45V, produced by Sisha Co., Ltd., China). Both grinding tools shared identical physical dimensions (outer diameter 400 mm × thickness 10 mm) and abrasive grit size (equivalent to US standard mesh #80). The straight surface grinding experiments were conducted on a Profimat MT-408 three-axis CNC high-speed grinding machine (manufactured by BLOHM, Germany), with grinding forces recorded using an HR-FP3407 triaxial force measurement system (H&T HORIZON Co., Ltd., China) (Figure 4.1a and b). Prior to experimental trials, each grinding wheel underwent precision dressing using a single-point diamond tool to optimize cutting performance. Process parameters included a constant grinding velocity of 25 m/s, with workpiece feed rates ranging from 100 to 400 mm/min and depth of cut varying between 0.2 and 2 mm.

The measurement of grinding forces provides essential insights into both workpiece machinability and wheel efficiency [1, 2]. As depicted in Figure 4.2a and b, the evolution of grinding forces follows a characteristic pattern with rising material removal rate (Q_w'): an initial sharp escalation $(0.83-1.67\,\text{mm}^3/(\text{mm}\cdot\text{s}))$ occurs and then transforms to a more moderated yet steady growth with intermittent variations at higher Q_w' values (up to $8\ \text{mm}^3/(\text{mm}\cdot\text{s})$), a phenomenon observed for both BA and MA wheel types. Material-specific analysis reveals an interesting transition in force requirements, while DD6 consistently demands the lowest forces and DZ408 the highest in the $0.83-3.3\,\text{mm}^3/(\text{mm}\cdot\text{s})$ range. GH4169 exhibits improved grindability relative to DD6 at elevated removal rates $(4-8\,\text{mm}^3/(\text{mm}\cdot\text{s}))$. The force

Grinding of Single-Crystal Superalloys: Fundamentals and Technologies, First Edition.
Wenfeng Ding et al.
© 2026 Wiley-VCH GmbH. Published 2026 by Wiley-VCH GmbH.

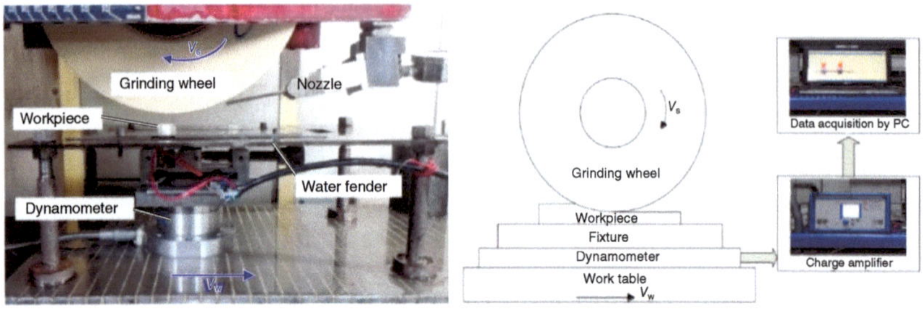

Figure 4.1 Grinding force measurement in the surface grinding process: (a) experimental setup and (b) schematic illustration.

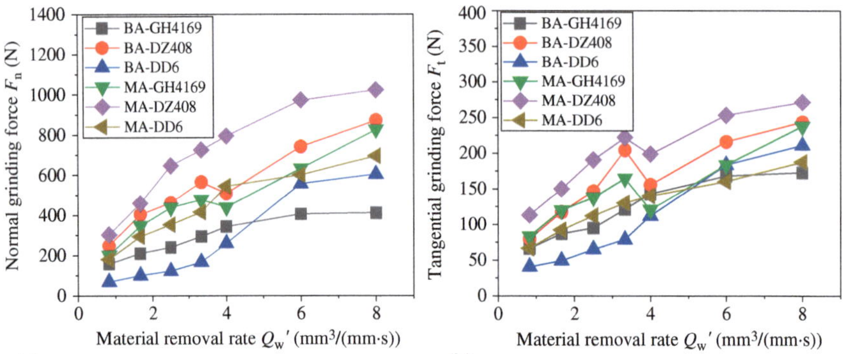

Figure 4.2 Grinding forces of different nickel alloys: (a) normal grinding force and (b) tangential grinding force.

amplification factors demonstrate significant wheel dependency, with BA wheels showing 2.1–5.9 times force multiplication versus MA wheels' more constrained 1.8–2.4 times increase, highlighting the superior force stability of microcrystalline alumina wheels attributable to the different dominant abrasive wear mechanisms.

4.1.2 Specific Grinding Energy

Using the tangential forces in Figure 4.2, the specific grinding energy e_s, which indicates the grinding wheel performance and the material removal difficulty of workpiece, can be calculated from the equation as follows:

$$e_s = \frac{F_t' v_s}{a_p v_w} = \frac{v_s}{Q_w'} F_t' \tag{4.1}$$

The relationship between e_s and F_t' is positively correlated when maintaining consistent grinding parameters, as demonstrated in Eq. (4.1). Figure 4.3 presents how the material removal rate (Q_w') influences specific grinding energy consumption. Observations reveal a notable reduction in specific grinding energy as Q_w' rises. In

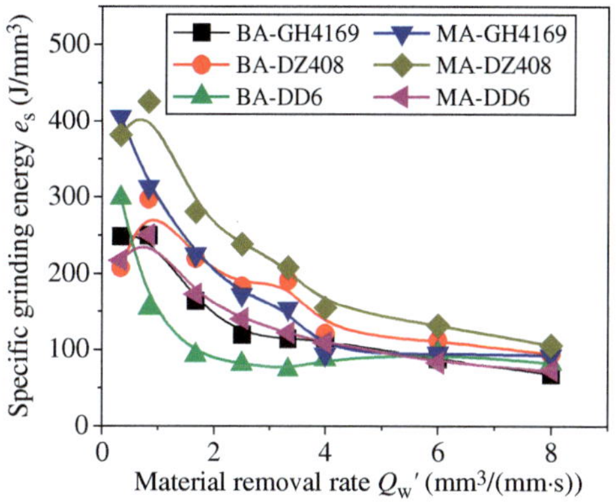

Figure 4.3 Effect of material removal rate on specific grinding energy.

particular, during DZ408 alloy processing, peak specific grinding energy reaches approximately 425.7 J/mm^3 at $Q_w' \approx 1$ mm^3/(mm · s). However, when Q_w' exceeds 4 mm^3/(mm · s), the variation in specific grinding energy among the three tested alloys diminishes significantly, stabilizing near 100 J/mm^3.

This phenomenon occurs because lower Q_w' values result in minimal undeformed chip thickness, causing friction and plowing effects between abrasive grains and workpiece materials to prevail in the grinding mechanism. According to Ghosh and Chattopadhyay [3], grinding energy primarily consists of friction energy and chip formation energy. Since most grinding energy transforms into heat, friction energy typically accounts for over 50% of total energy consumption, regardless of process parameters [4]. Based on Figure 4.3, it can be deduced that DZ408 alloy exhibits the highest friction energy during grinding, whereas DD6 alloy demonstrates the lowest.

Regarding the workpiece material, generally, regardless of the grinding wheel used, when $Q_w' < 4$ mm^3/(mm · s), the specific grinding energy follows the order of DD6 < GH4169 < DZ408. When $Q_w' > 4$ mm^3/(mm · s), the order becomes GH4169 < DD6 < DZ408. These orders correspond to those of the grinding forces.

In the case of workpiece material, it can be roughly stated that no matter which grinding wheel is used, when $Q_w' < 4$ mm^3/(mm · s), the specific grinding energy follows the order of DD6 < GH4169 < DZ408. In addition, when $Q_w' > 4$ mm^3/(mm · s), the order becomes GH4169 < DD6 < DZ408. These orders correspond to those of grinding forces.

Additionally, the maximum undeformed chip thickness a_{gmax} can be determined using the following equation:

$$a_{gmax} = \left[\frac{4}{CN_d} \cdot \frac{v_w}{v_s} \cdot \left(\frac{a_p}{d_s}\right)^{0.5}\right]^{0.5} \tag{4.2}$$

where C is the coefficient related to the cutting-edge angle (4.77 in this study), and N_d is the active cutting-edge density on the wheel surface. The changes of the specific grinding energy e_s against the maximum undeformed chip thickness a_{gmax} are shown in Figure 4.4.

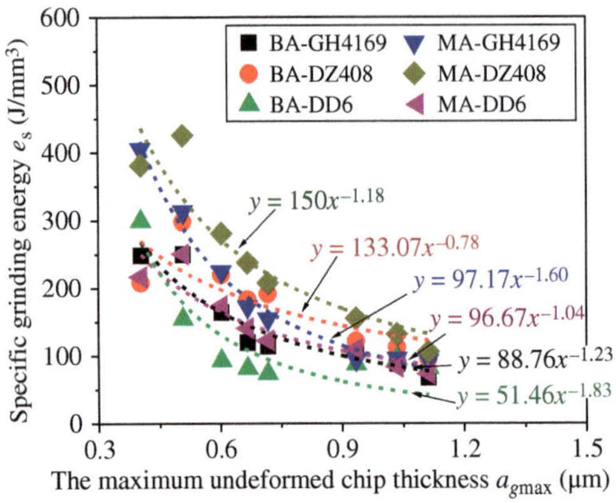

Figure 4.4 Changes of the specific grinding energy with the maximum undeformed chip thickness.

In general, the maximum undeformed chip thickness determines the material removal efficiency of an individual abrasive grain in grinding. That is, if a_{gmax} is smaller than the critical undeformed chip thickness, the abrasive grain mainly causes friction and plowing on the workpiece, leading to more plastic deformation of the workpiece material and high consumption of grinding energy. When a_{gmax} increases to a value greater than the critical undeformed chip thickness, grinding chips are generated with less energy required. This process is called the "size effect" when controlling the grinding energy using a_{gmax}. As shown in Figure 4.4, a final specific grinding energy of 100 J/mm^3 is obtained. For the grinding wheel, when a_{gmax} is smaller than 0.72 μm, grinding with the MA wheel requires more energy than with the BA wheel. However, when a_{gmax} is greater than 0.72 μm, the MA wheel requires less energy due to its high hardness and toughness. Based on the above data, the relationship between a_{gmax} and e_s can be obtained using the following empirical equations:

$$e_{s\,(BA-GH4169)} = 88.76{a_{gmax}}^{-1.23} \tag{4.3}$$

$$e_{s\,(BA-DZ408)} = 133.07{a_{gmax}}^{-0.78} \tag{4.4}$$

$$e_{s\,(BA-DD6)} = 51.46{a_{gmax}}^{-1.83} \tag{4.5}$$

$$e_{s\,(MA-GH4169)} = 97.17{a_{gmax}}^{-1.60} \tag{4.6}$$

$$e_{s\,(MA-DZ408)} = 150\,{a_{gmax}}^{-1.18} \tag{4.7}$$

$$e_{s\,(MA-DD6)} = 96.67{a_{gmax}}^{-1.04} \tag{4.8}$$

Therefore, from Figures 4.3 and 4.4, when the grinding depth is constant, increasing the workpiece speed can increase a_{gmax}. If the heat generated in the grinding zone can be effectively transferred, the grinding efficiency of nickel alloys can be significantly improved.

4.1.3 Effect of Workpiece Material Types

Grinding forces are a critical factor in machining processes, significantly influencing wheel wear, thermal effects, and workpiece surface quality, while also strongly influencing the behavior of material removal mechanisms. However, these measurable process parameters vary substantially depending on workpiece material characteristics and operating conditions.

In this investigation, since identical wheel specifications, dressing parameters, and grinding conditions were maintained across all tests, the initial wheel topography remained consistent for the four tested alloys. Consequently, variations in grinding forces primarily reflect differences in material properties, allowing comparative evaluation of their grindability.

The total grinding force comprises both normal and tangential components, which arise from two primary sources: chip formation forces and sliding friction forces. As established in grinding force models by researchers such as Li et al. [5] and Yao et al. [6], these force components can be mathematically described as follows:

$$\begin{cases} F_n' = F_{nc}' + F_{ns}' \\ F_t' = F_{tc}' + F_{ts}' \end{cases} \tag{4.9}$$

$$\begin{cases} F_{nc}' = K\dfrac{v_w}{v_s}a_p \\ F_{ns}' = \bar{\delta}\bar{p}A\left(c_1\right)^{\beta}\left(\dfrac{v_w}{v_s}\right)^{\alpha}\dfrac{d_e^{\frac{1-\alpha}{2}}}{1+\alpha}a_p^{\frac{1+\alpha}{2}} \end{cases} \tag{4.10}$$

$$\begin{cases} F_{tc}' = \dfrac{\pi K}{4\tan\gamma}\dfrac{v_w}{v_s}a_p \\ F_{ts}' = \mu\bar{\delta}\bar{p}A\left(c_1\right)^{\beta}\left(\dfrac{v_w}{v_s}\right)^{\alpha}\dfrac{d_e^{\frac{1-\alpha}{2}}}{1+\alpha}a_p^{\frac{1+\alpha}{2}} \end{cases} \tag{4.11}$$

where F_{nc}', F_{tc}', F_{ns}', and F_{ts}' are the normal and tangential chip formation force components and the normal and tangential friction force components, respectively; K is the force per grinding contact area; $\bar{\delta}$ is the tip area of the worn grain; $\bar{p}$ is the average contact pressure between the worn grain and the workpiece; A is the fraction of the wheel surface that has wear flats; α and β are exponential coefficients related to the grain distribution; c_1 is the static cutting edge density; γ is half the edge angle; μ is the coefficient of friction; and d_e is equivalent diameter of the grinding wheel. Due to the same state of grinding wheel prior to grinding, the exponential coefficients α and β can be equal. Eqs. (4.9)–(4.11) can be written as follows:

$$\begin{cases} F_n' = K_1\dfrac{v_w}{v_s}a_p + K_2\left(\dfrac{v_w}{v_s}\right)^{\alpha}\dfrac{D^{\frac{1-\alpha}{2}}}{1+\alpha}a_p^{\frac{1+\alpha}{2}} \\ F_t' = K_3\dfrac{v_w}{v_s}a_p + K_4\left(\dfrac{v_w}{v_s}\right)^{\alpha}\dfrac{D^{\frac{1-\alpha}{2}}}{1+\alpha}a_p^{\frac{1+\alpha}{2}} \end{cases} \tag{4.12}$$

where $K_1 = K$, $K_2 = \bar{\delta}\bar{p}A(c_1)^{\beta}$, $K_3 = \dfrac{\pi K}{4\tan\gamma}$, and $K_4 = \mu K_2$. For the same workpiece material, the coefficients K_1, K_2, K_3, and K_4 in Eq. (4.12) can be considered as constants when the grinding wheel is determined. Submitting the results of grinding force (Figure 4.3) into Eq. (4.12), these coefficients can be calculated as listed in Table 4.2.

It can be seen that the coefficients of K_1 and K_3 that are produced by the chip formation are very large. This indicates that the size effect in the micro-cutting process is very prominent [5]. Moreover, the values of K_1 and K_3 are found to increase in the order: GH4169, DZ408, K403, and DD6, and the observable difference of these coefficients shows that the correlation between the chip formation force and the workpiece material type is significant. Furthermore, the coefficients of K_2 and K_4 that are related to the friction force have relatively small changes and are increased in the following order: DD6, GH4169, DZ408, and K403. In particular, from Tables 4.1 and 4.2, it can be found that the changing trend of the workpiece material properties, e.g., hardness and elastic modulus, and the contact pressure $\overline{p}$ is generally consistent with that of the coefficients K_2 and K_4. All these factors ultimately determine the grindability of the alloys, but further investigation is needed to clarify their correlations.

Figure 4.5 presents the computational results of grinding force components under varying conditions, revealing distinct patterns in chip formation and friction forces. When maintaining a constant material removal rate, analysis shows friction forces dominate the grinding process, constituting over 80% of total forces, while chip formation contributes minimally (below 20%). This friction-dominated mechanism explains the observed high grinding forces.

Table 4.1 Main physical properties of the nickel-based superalloys.

Workpiece material	Density (g/cm^3)	Hardness HV (GPa)	Fracture toughness ($Mpa \cdot m^{0.5}$)	Elastic modulus (GPa)	Poisson ratio
GH4169	8.24	3.97	11.8	204	0.30
K403	8.10	4.10	12.1	208	0.27
DZ408	8.68	4.09	15.7	118	0.28
DD6	8.78	3.81	20.6	134	0.42

Table 4.2 Calculation results of the coefficients in Eq. (4-12).

					Average error	
Workpieces	K_1	K_2	K_3	K_4	F_n'	F_t'
GH4169	0.87×10^4	26.73	0.51×10^4	6.81	10.74%	11.58%
K403	2.43×10^4	35.52	1.44×10^4	7.34	3.13%	7.43%
DZ408	2.29×10^4	32.82	1.35×10^4	10.01	6.10%	2.23%
DD6	7.87×10^4	10.20	4.64×10^4	2.08	6.24%	11.17%

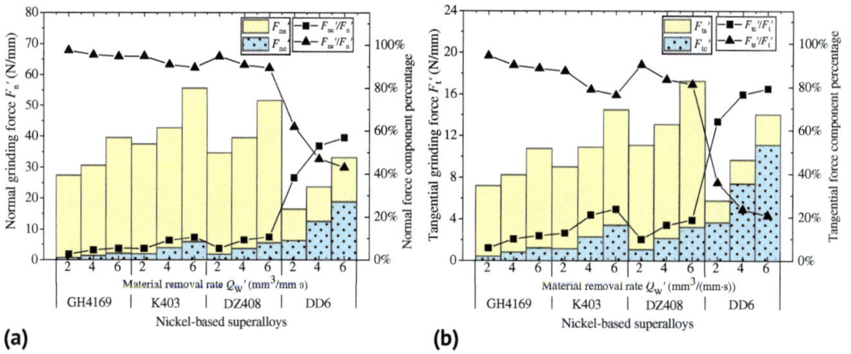

Figure 4.5 The grinding force components during grinding of different nickel alloys: (a) normal components and (b) tangential components.

Comparative analysis of DD6 alloy demonstrates unique force characteristics: normal chip formation (F_{nc}') approaches friction force (F_{ns}') in magnitude, while tangential chip formation (F_{tc}') exceeds friction (F_{ts}') by 2–3 times. This distinctive force distribution accounts for DD6's relatively lower grinding forces compared to other tested alloys.

The study further reveals an important trend with increasing material removal rate (Q_w'): as shown in Figure 4.5a, the proportion of normal chip formation force (F_{nc}') grows while friction (F_{ns}') diminishes. Corresponding tangential components (Figure 4.5b) exhibit parallel behavior. This phenomenon, consistently observed across all four tested alloys, likely stems from abrasive grain dynamics—either pull-out or fracture events that induce self-sharpening effects in alumina wheels. While beneficial for cutting performance, such grain behavior may accelerate wheel wear and reduce operational lifespan.

4.2 Grinding Force in Profile Grinding

All the profile grinding tests were conducted on the Profimat MT-408 grinding center as mentioned above (Figure 4.6a). Rectangular blocks of DD6 single-crystal alloy with the dimensions of 32 × 15 × 12 mm were employed as the workpiece materials. The workpiece was clamped by the jig, which was mounted on the machine worktable (Figure 4.6c). The MA wheels were used as grinding tools.

The experimental procedure examined two distinct grinding scenarios with systematically varied parameters including wheel speed (v_s), workpiece velocity (v_w), cutting depth (a_p), and cumulative grinding depth (h)—defined as the wheel's total vertical displacement (Figure 4.7). Tests were conducted on blade root specimens with 2.63 mm material allowance using creep-feed up-grinding (without spark-out), as outlined in Table 4.3. Each experimental case comprised six consecutive grinding passes, with wheel conditioning performed between passes using a 152.1 mm 45# diamond roller dresser (Figure 4.6b) operating at 20 m/s with −0.7 speed ratio, 1.3 mm/min feed rate, and 0.2 mm total dressing depth.

Figure 4.6 Experimental setups in profile grinding of DD6 alloy: (a) the grinding center; (b) diamond roller dresser; and (c) the DD6 workpiece.

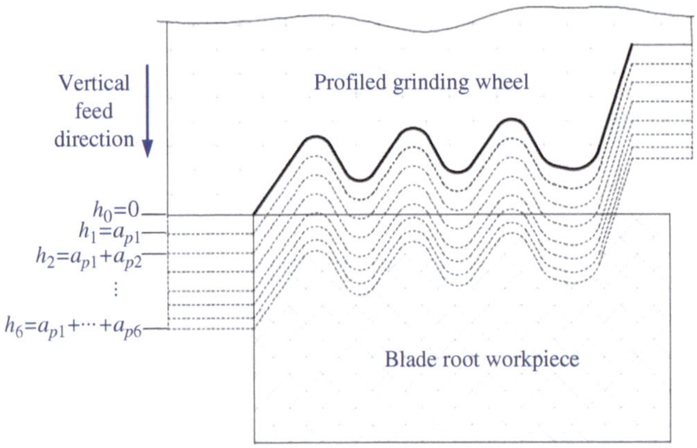

Figure 4.7 Schematic diagram of the profile grinding process in the vertical direction.

During the creep-feed profile grinding process, the grinding force components were measured using an HR-FP3407 three-component dynamometer. After grinding, residual stresses in selected regions of the blade root specimen were first measured via X-ray diffraction using an MG2000L goniometer with Mn Kα radiation (LXRD micro-area, Proto, Canada). Two measurements were taken for each selected region and averaged.

Subsequently, the specimen surface topography was examined using a Hitachi S3400 scanning electron microscope (SEM) (Japan), followed by sectioning perpendicular to the grinding direction. The specimen was then polished to achieve a surface roughness below Ra 0.1 μm and etched for approximately 30 seconds using a specific etchant solution (25 mL HNO_3 + 25 mL H_2O + 50 mL HCl).

Finally, subsurface microstructural characterization was performed using both an optical microscope (Leica DM3000, Germany) and the aforementioned SEM. Additionally, microhardness testing was conducted using an HXS-1000AK hardness tester with a Vickers indenter under a 25 g load applied for 15 seconds.

Table 4.3 Experimental arrangement for profile grinding.

Grinding cases	Grinding pass No.	Grinding parameters: Wheel speed v_s (m/s)	Workpiece speed v_w (mm/min)	Grinding depth a_p (mm)	Accumulated grinding depth h (mm)	The target root profile finished or not
I	1	25	240	1.50	1.50	No
	2	35	210	1.00	2.50	No
	3	35	300	0.13	2.63	Yes
	4	25	240	1.50	4.13	Yes
	5	35	210	1.00	5.13	Yes
	6	35	300	0.13	5.26	Yes
II	1	25	480	1.50	1.50	No
	2	35	420	1.00	2.50	No
	3	35	600	0.13	2.63	Yes
	4	25	480	1.50	4.13	Yes
	5	35	420	1.00	5.13	Yes
	6	35	600	0.13	5.26	Yes

4.2.1 Effect of Accumulated Grinding Depth

The evolution of grinding force components (tangential force F_t and normal force F_n) with increasing accumulated grinding depth (h) during creep-feed profile grinding is presented in Figure 4.8. Experimental results from Case I demonstrate a substantial increase in both F_t (118.8 N to 238.4 N) and F_n (287.5 N to 572.9N) as h progressed from 1.5mm (Pass 1) to 4.13mm (Pass 4). However, beyond 2.63mm (Pass 3), further depth increases to 5.26mm (Pass 6), resulting in only marginal force variations. Detailed analysis reveals an initial growth phase (Pass 1 to 2: $h = 1.5 - 2.5$mm) where forces increased to 144.1 N(F_t) and 424.5 N(F_n), followed by a subsequent reduction phase (Pass 4 to 5: $h = 4.13 - 5.13$mm), with forces decreasing to 145.2 N(F_t) and 361.5 N(F_n). Their force responses exhibited distinct behaviors. Parallel observations in Case II, which featured higher workpiece speeds, showed consistently greater force magnitudes, confirming that grinding forces are sensitive to both process parameters and accumulated material removal history.

4.2.2 Effect of Wheel–workpiece Contact Area

The grinding process fundamentally occurs within the wheel–workpiece contact zone, where material removal and surface formation simultaneously take place, making this interfacial region critically important. Research has demonstrated that grinding forces are predominantly determined by the undeformed chip thickness within this contact area, which governs the material removal mechanism [7, 8]. To better understand the complex interactions between grinding

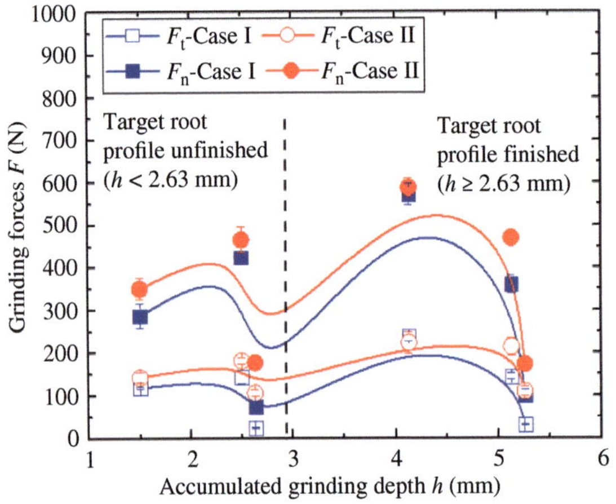

Figure 4.8 Effect of accumulated grinding depth on grinding forces.

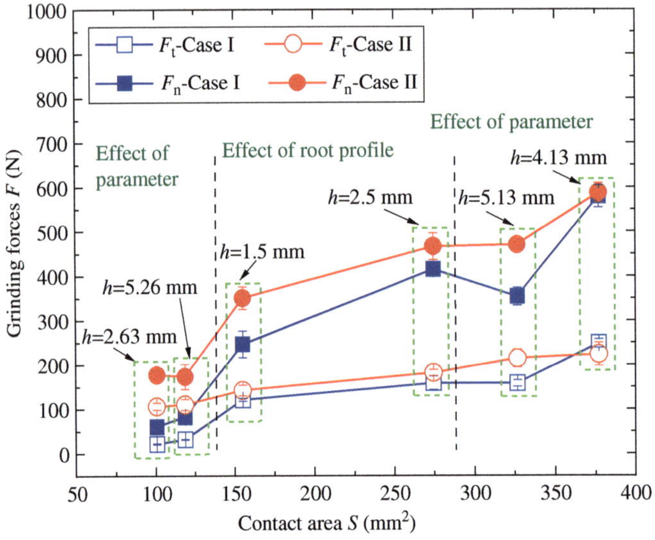

Figure 4.9 Effect of wheel–workpiece contact area on grinding forces.

forces, root profiles, and process parameters, Figure 4.9 presents the variation of grinding forces relative to the expanding wheel–workpiece contact area. The contact area (S) was precisely calculated through geometric analysis of the evolving root profile (Figure 4.7), employing the following computational approach:

$$S = l \cdot \sqrt{a_p d_s} \tag{4.13}$$

where l stands for the profile length of the root workpiece during grinding and d_s represents the wheel diameter. As can be seen in Figure 4.9, regardless of the specific grinding parameters utilized, an increase in the wheel–workpiece contact area generally led to a rise in grinding force, mainly because of the larger number of active abrasive grains engaged in cutting. Furthermore, the factors contributing to the variations in contact area and their subsequent influence on grinding force can be analyzed as follows:

1. Assuming a constant wheel diameter, Eq. (4.13) shows that a decrease in the root profile length l or the grinding depth a_p would result in a reduction of the wheel–workpiece contact area S. Apparently, when $h = 2.63$ mm or $h = 5.26$ mm, the root workpiece profile closely resembled the target profile and was highly similar (Table 4.3). At this moment, the small value of a_p led to a small contact area S (around 120 mm^2), which in turn caused a small grinding force F due to the diminished undeformed chip thickness. Therefore, during this period, it can be concluded that the grinding parameters were the key factors determining the variation of grinding force.
2. When (S) increased from 155 to 274 mm^2, an increase in grinding force was noticed in Figure 4.9, and the values of (h) ranged from 1.5 to 2.5 mm. At this time, according to Table 4.3, the target profile of the root workpiece had not been completed. However, despite the decrease in (a_p), both the contact area S and the grinding force F were increasing. Under these circumstances, the expansion of the root profile length l became the predominant factor affecting the grinding force.
3. The grinding force reached its maximum value when the contact area S increased to 377 mm^2, corresponding to $h = 4.13$ mm. As summarized in Table 4.3, the increase in a_p led to a large-undeformed chip thickness, while the root profile remained nearly unchanged. In such a situation, the increase in grinding force was mainly due to the large grinding parameters.

The experimental findings demonstrate the alternating interaction between workpiece root profile characteristics and process parameters on grinding force generation. Specifically, when maintaining a constant root profile geometry (quantified by accumulated grinding depth h), adopting aggressive grinding parameters (particularly increased depth of cut a_p) elevates grinding forces through the mechanism of greater undeformed chip thickness. Conversely, under fixed process parameters, reducing the root profile dimensions (thereby decreasing wheel–workpiece contact area S) lowers grinding forces by limiting the population of simultaneously engaged abrasive grains. These observations establish fundamental principles for force regulation in creep-feed grinding operations, suggesting two complementary control strategies: parameter optimization through depth of cut adjustment and geometric control via workpiece profile modification, with each approach offering distinct mechanisms for influencing the grinding force magnitude while maintaining process stability.

4.3 Grinding Force in Micro-grinding

The micro-grinding force is a crucial factor in the micro-grinding process. It stems from the elastic and plastic deformations, chip formation, and friction between the micro-grinding tool and the workpiece when they come into contact. Research on the micro-grinding force is valuable for understanding the grinding process. It is closely related to the integrity of the ground surface and acts as an essential index for evaluating the grindability of materials. Moreover, it has a significant influence on the wear of the micro-grinding tool, specific grinding energy, and the deformation of the process system. It can also lead to plastic deformation of metals, generate residual stresses, and consequently cause damage to the surface and subsurface. Such damages can seriously affect the fatigue performance and service life of components in practical applications. Additionally, the micro-grinding force can be used to calculate the grinding energy, which helps in understanding the micro-grinding mechanism. Furthermore, the micro-grinding force signal can be applied to monitor the wear of the micro-grinding tool, allowing for timely replacement. Hence, research on the micro-grinding force in the micro-grinding process holds great theoretical importance.

4.3.1 Micro-grinding Force

The grinding forces in micro-grinding originate from the elastic and plastic deformations of the micro-abrasive rod in contact with the workpiece, the formation of grinding chips, and the friction between the abrasive grains and the workpiece [9, 10]. Studying grinding forces is beneficial for grasping the grinding process. In grinding research, the grinding force is highly relevant to the integrity of the grinding surface and is one of the significant indicators for assessing the grindability of materials. At the same time, it has a considerable impact on the wear of the grinding rod, specific grinding energy, and the deformation of the process system [11–13]. Grinding forces can also result in plastic deformation and residual stresses in metals, which may cause surface and subsurface damage [14, 15]. These damages can have a severe impact on the fatigue performance and service life of the parts in practical applications. Besides, the grinding forces can be utilized to calculate the grinding energy and understand the grinding mechanism [16]. Moreover, the grinding force signal can be used to monitor the wear of the micro-grinding rods, enabling timely replacement. Therefore, the study of grinding force in the micro-grinding process is of high theoretical significance.

Park et al. [17] formulated an equation for the grinding force of a single abrasive grain by presuming that the shape of the abrasive grain on the surface layer of the micro-abrasive rod is hemispherical. The grinding force model is established as follows [18]:

$$\begin{cases} F_n = F_{nc} + F_{ns} \\ F_t = F_{tc} + F_{ts} \end{cases} \tag{4.14}$$

where F_n is the normal force, F_t is the tangential force, F_{nc} is the normal force aroused by the material deformation in the grinding process, and F_{ns} is the normal force aroused by friction force. F_{tc} is the tangential force aroused by the material deformation in the grinding process, and F_{ts} is the tangential force aroused by friction force.

To calculate the maximum undeformed grinding thickness and based on the actual shape of the abrasive grain on the micro-abrasive rod, the following two assumptions are made:

1. The abrasive grains are of the form of a circular table and have a taper γ.
2. All the protruding abrasive grains on the rod are involved in the grinding and the material is removed.

The measured and assumed shape of the abrasive grain on the surface of the bar is shown in Figure 4.10a and b.

The force condition of a single abrasive in micro-grinding is depicted in Figure 4.11. However, in the micro-scale grinding process, the diameter of the micro-abrasive rod is less than 1 mm, and even micron-level grinding rods have been developed overseas. In micro-scale grinding, the contact arc length between the grinding rod and the workpiece during grinding is reduced, and its size and the maximum undeformed cutting thickness of the grinding grain can even be of the same order of magnitude. Therefore, it is necessary to consider the influence of the size effect when establishing the micro-grinding process, as shown in Eqs. (4.15) and (4.16).

$$l_c = \frac{d_s \theta}{2} = \frac{d_s \arccos\left(1 - \frac{2a_p}{d_s}\right)}{2} \tag{4.15}$$

The average undeformed cutting thickness of the abrasive grains in the contact arc is then

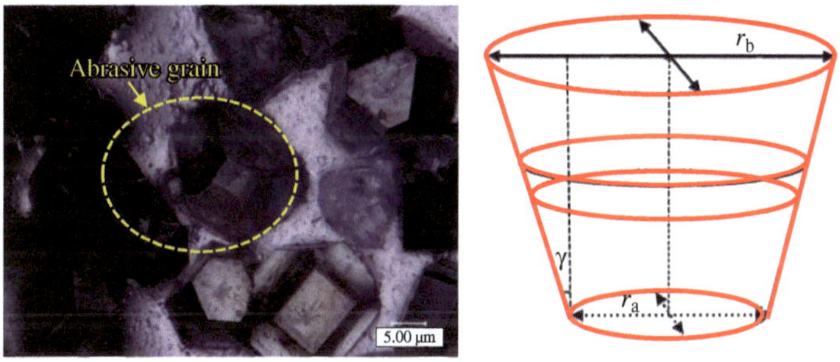

(a) (b)

Figure 4.10 Real shape and simplified model of abrasive: (a) the micro-grinding tool surface and (b) geometric model.

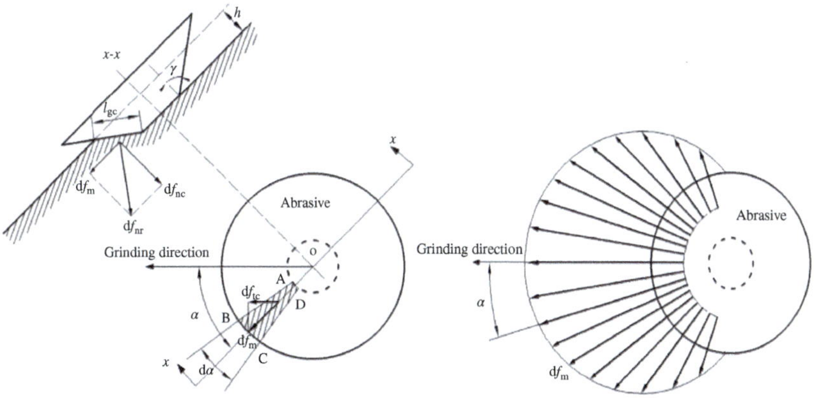

Figure 4.11 Mechanical interaction of the spherical abrasive in grinding.

$$E(h) = \frac{\sqrt{\pi^2 r_a^2 + \frac{4\pi a_p v_w \tan\gamma}{C l_c v_s}} - \pi r_a}{4\tan\gamma} \tag{4.16}$$

Thus, the deformed force on the single grain can be calculated as follows:

$$\begin{cases} \mathrm{d}f_{tc} = \frac{1}{2} k_s \left[\left(l_{gc} + \frac{\gamma_a}{\sin\gamma} \right)^2 \sin\gamma - \frac{\gamma_a^2}{\sin\gamma} \right] \cdot \cos^2\gamma \cos^2\alpha \, \mathrm{d}\alpha \\ \mathrm{d}f_{nc} = \frac{1}{2} k_s \left[\left(l_{gc} + \frac{\gamma_a}{\sin\gamma} \right)^2 \sin\gamma - \frac{\gamma_a^2}{\sin\gamma} \right] \cdot \cos\gamma \sin\gamma \cos\alpha \, \mathrm{d}\alpha \end{cases} \tag{4.17}$$

4.3.2 Experimental Validation of the Force Model

In order to verify the accuracy of micro-scale grinding force of single-crystal nickel alloy, five groups of micro-grinding parameters were selected to carry out grinding tests on the crystal face of nickel-based superalloy DD98(001), and F_n and F_t in the grinding process were measured. The top and bottom radii of a single abrasive particle were 4.55 and 9.15 μm measured by the VHX-1000E field ultra-depth 3D microscopic system. The initial edge morphology of micro-grinding rod is also focused and synthesized, and the exit edge height of the abrasive particle is measured to be 5.1 μm. Through calculation, the top cone angle of a single abrasive particle is about 42°. The cutting width is 1.5 mm, and the number of cutting edges per unit grinding rod area $C = 2209\,\mathrm{mm}^{-2}$. According to regression analysis, $k_s = 0.839$, $P_0 = 73.33$, and $\mu = 0.312$. Figures 4.12 and 4.13 display the comparisons between the theoretical and experimental values of the normal and tangential grinding forces, respectively. The relative errors between the theoretical and experimental values of the five groups of normal grinding forces were 6.9%, 8.93%, 11.28%, 13.24%, and 17.06%. The relative errors between the empirical and experimental values of the normal grinding

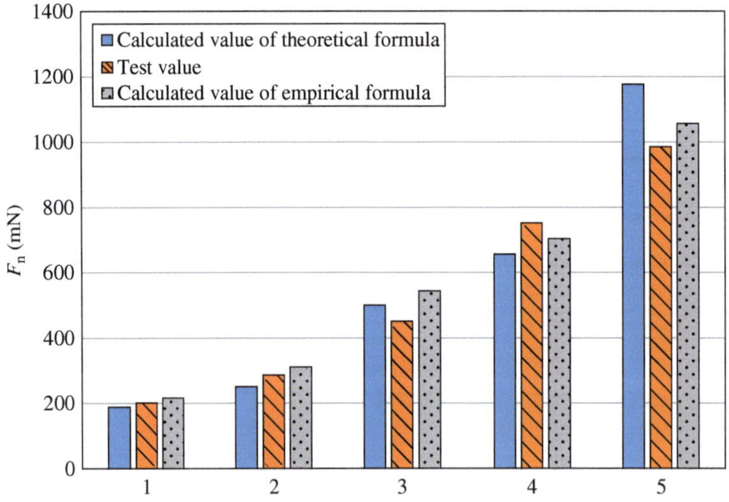

Figure 4.12 Comparison between theoretical and experimental values of the normal force.

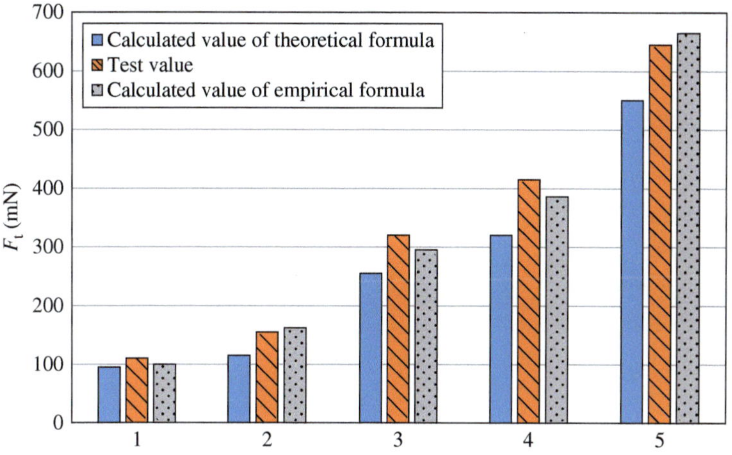

Figure 4.13 Comparison between theoretical and experimental values of the tangential force.

force were 7.88%, 6.53%, 19.31%, 8.52%, and 5.65%. For the tangential grinding forces, the relative errors between the theoretical and experimental values of the five tests were 25.4%, 14.84%, 22.67%, 22.28%, and 14.26%, while the relative errors between the empirical and experimental values were 16.1%, 5.44%, 8.39%, 10.17%, and 2.79%. The maximum error in the theoretical calculation of the normal grinding force was 17.06%, and the maximum error in the empirical value was 19.31%. The reasons for the error are as follows: firstly, in the process of deducing the grinding force model, the geometric contact length rather than the actual contact length between the workpiece and the grinding rod is used to calculate the dynamic abrasive particle number; secondly, in the process of micro-grinding, the abrasive particles on the micro-grinding rod in the grinding zone do not fully participate in grinding and produce grinding chips, and some abrasive particles only slip. Besides, micro-grinding environmental factors are not considered. As can be seen from Figures 4.12 and 4.13, the maximum error of this model is about 20%, which is not very large and is within the acceptable range. In addition, with the change of micro-scale grinding processing parameters, the calculated values of normal grinding force and tangential grinding force of this theoretical model and regression model are consistent with the measured values in the changing trend: it decreases with the increase of spindle speed. In conclusion, the theoretical model and regression model can accurately reflect the grinding force of nickel-based single-crystal superalloy micro-scale grinding, and can be used as an important theoretical reference in the machining process of nickel-based single-crystal superalloy micro-parts.

4.3.3 Influence of Parameters on Micro-grinding Force

A single-factor experimental design for micro-grinding DD98 was developed. The grinding method was down grinding, with a grinding width of $b = 1.5\,\text{mm}$. The experimental conditions are illustrated in Figure 4.14. The main objective of this

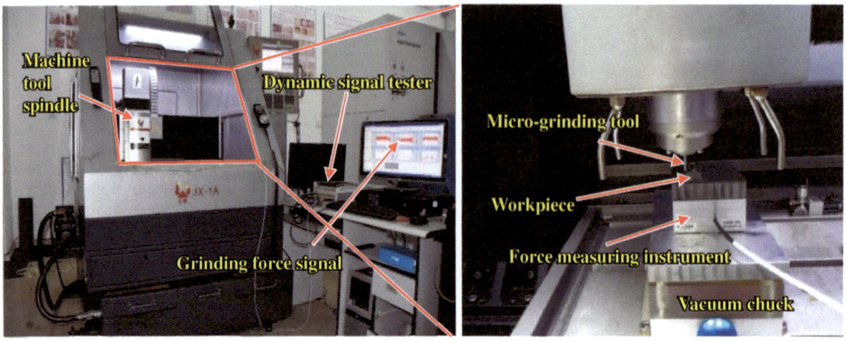

Figure 4.14 Experimental setup.

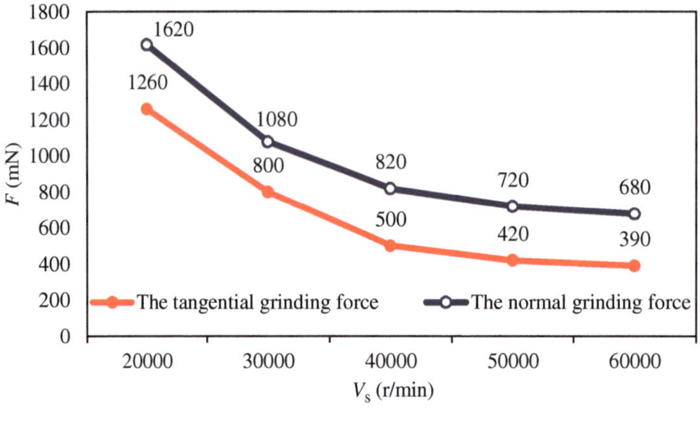

Figure 4.15 Effect of spindle speed on grinding force.

experiment was to explore the effects of spindle speed, feed rate, and grinding depth on the grinding forces.

As shown in Figure 4.15, when the spindle speed increases while other parameters remain constant, the normal and tangential forces of micro-grinding nickel-based single-crystal superalloy decrease, and the decreasing trend gradually levels off. The reason is that as the spindle speed increases, the number of abrasives involved in grinding per unit time rises, resulting in a thinner cutting thickness of a single abrasive, thinner chips, and reduced power consumption per unit time. However, the grinding arc length remains unchanged, leading to a decrease in the grinding force. Additionally, as the spindle speed continues to increase, the number of abrasives involved in plowing and rubbing increases, causing the decreasing trend to become more gradual.

Figure 4.16 demonstrates a clear correlation between feed rate and grinding forces under constant processing conditions. Both normal and tangential force components exhibit an upward trend with increasing feed rates, with the tangential force showing particularly pronounced growth that eventually surpasses the normal force. This phenomenon primarily stems from two interrelated mechanisms: firstly,

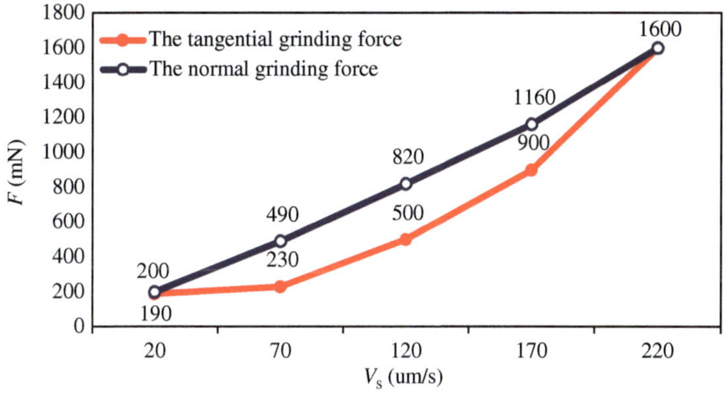

Figure 4.16 Effect of feed rate on grinding force.

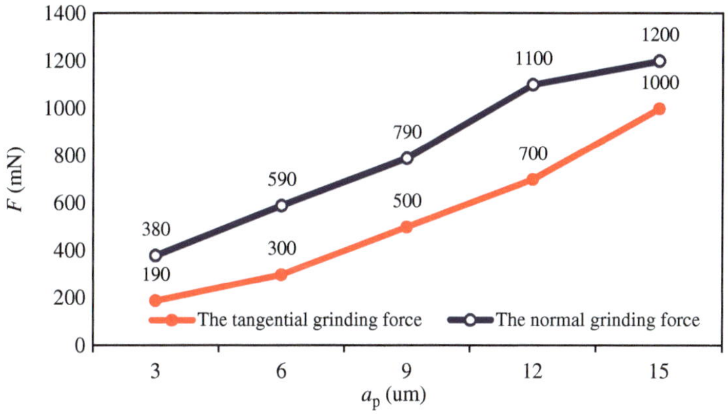

Figure 4.17 Effect of grinding depth on grinding force.

elevated feed rates lead to greater material removal volumes and increased chip thickness per abrasive grain, substantially raising energy consumption and resultant forces per unit time. Secondly, the unique characteristics of micro-grinding create a complex thermal-mechanical interaction where high-speed abrasive impacts generate concentrated heat accumulation in the shallow surface layer due to limited heat dissipation into the workpiece bulk. While the nickel-based single-crystal superalloy maintains its hardness through adiabatic shear effects, the combination of restricted chip evacuation and material adhesion to abrasive edges progressively dulls the grinding tool. This tool degradation, characterized by chips adhering to the sharp parts of the abrasives and clogging within inter-abrasive voids, amplifies grinding forces dramatically at higher feed rates.

Figure 4.17 clearly demonstrates a positive correlation between grinding depth and both normal and tangential force components. This phenomenon can be attributed to several interrelated factors that collectively influence the grinding mechanics. Firstly, the increased grinding depth leads to greater maximum undeformed chip thickness for individual abrasive grains. Secondly, the extended contact arc length between the grinding tool and workpiece surface results in more abrasive grains being simultaneously engaged in the cutting process. Furthermore, the enhanced material removal volume per abrasive grain under deeper cuts generates additional resistance. These combined effects—including thicker chips, prolonged contact, increased active grain count, and greater material displacement per grain—synergistically contribute to the observed elevation in grinding forces as depth of cut increases.

Experimental investigations on the anisotropic grinding behavior of (001)-oriented nickel-based single-crystal superalloys reveal significant crystallographic dependence of grinding forces. As illustrated in Figure 4.18 for grinding conditions ($v_s = 40000$ r/min, $v_w = 120$ µm/s, and $a_p = 9$ µm), both tangential and normal force components exhibit $\pi/2$-periodic variations relative to the [100] crystal direction in the (001) plane. The

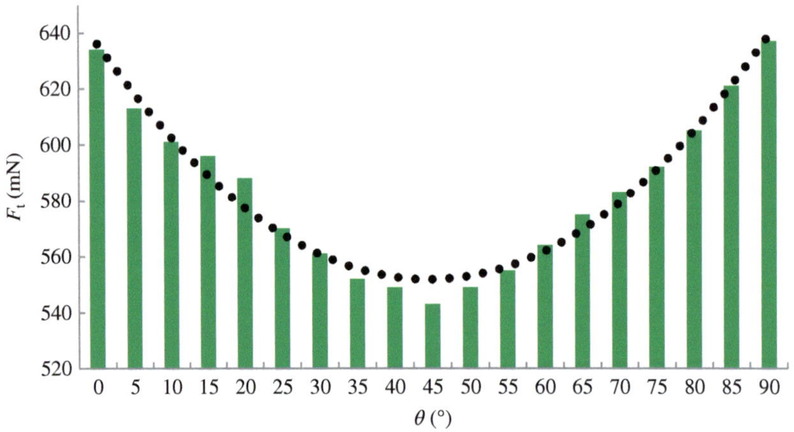

Figure 4.18 Tangential grinding force of grinding along θ with [100] in (001) crystal plane.

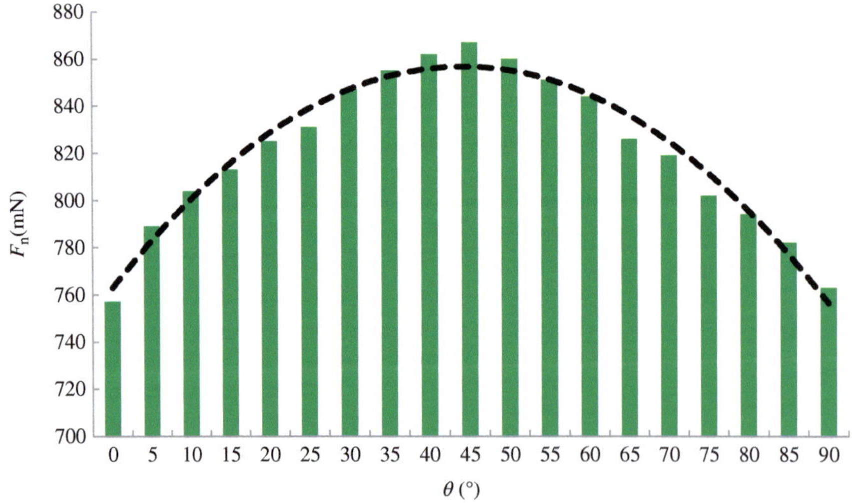

Figure 4.19 Normal grinding force of grinding along θ with [100] in (001) crystal plane.

minimum tangential forces occur at $\theta = 45°$ orientations (and equivalent 135°, 225°, and 315° positions), coinciding with directions of minimum shear modulus, where material shear deformation occurs most readily. Conversely, maximum tangential forces manifest along primary crystallographic axes (0°, 90°, 180°, and 270°) corresponding to peak shear modulus orientations. Normal forces demonstrate an inverse relationship, reaching maxima at 45° intervals where elastic modulus peaks enhance material rebound effects, while minima occur along primary crystal axes where reduced elastic modulus diminishes normal resistance (Figure 4.19). This periodic behavior stems from the intrinsic a typical regular "ten" dendrite architecture of the superalloy, which creates alternating zones of varying deformation resistance as grinding direction changes relative to the underlying crystal structure.

References

1 Wei, M., Klocke, F., Barth, S. et al. (2017). Detailed analysis and description of grinding wheel topographies. *Journal of Manufacturing Science and Engineering* 139 (5): 054502.

2 Liu, M.Z, Li, C.H., Zhang, Y.B. et al. (2023). Analysis of grinding mechanics and improved grinding force model based on randomized grain geometric characteristics. *Chinese Journal of Aeronautics* 36 (7): 160–193.

3 Ghosh, A. and Chattopadhyay, A.K. (2007). Experimental investigation on performance of touch-dressed single-layer brazed cBN wheels. *International Journal of Machine Tools and Manufacture* 47 (7–8): 1206–1213.

4 Qian, N., Ding, W.F., and Zhu, Y.J. (2018). Comparative investigation on grindability of K4125 and Inconel718 nickel-based superalloys. *The International Journal of Advanced Manufacturing Technology* 97: 1649–1661.

5 Li, L., Fu, J., and Peklenik, J. (1980). A study of grinding force mathematical model. *CIRP Annals-Manufacturing Technology* 29 (1): 245–249.

6 Yao, C.F., Wang, T., Xiao, W. et al. (2014). Experimental study on grinding force and grinding temperature of Aermet 100 steel in surface grinding. *Journal of Materials Processing Technology* 214 (11): 2191–2199.

7 Yang, M., Li, C.H., Zhang, Y.B. et al. (2017). Maximum undeformed equivalent chip thickness for ductile-brittle transition of zirconia ceramics under different lubrication conditions. *International Journal of Machine Tools and Manufacture* 122: 55–65.

8 Zhang, Y.B., Li, C.H., Ji, H.J. et al. (2017). Analysis of grinding mechanics and improved predictive force model based on material-removal and plastic-stacking mechanisms. *International Journal of Machine Tools and Manufacture* 122: 81–97.

9 Sun, Y., Su, Z.P., Jin, L.Y. et al. (2021). Modelling and analysis of micro-grinding surface characteristics and generated mechanism of hard brittle material machined by micro abrasive tools with helical chip pocket. *Journal of Materials Processing Technology* 297: 117242.

10 Cheng, J., Yin, G., Wen, Q. et al. (2015). Study on grinding force modelling and ductile regime propelling technology in micro drill-grinding of hard-brittle materials. *Journal of Materials Processing Technology* 223: 150–163.

11 Li, C., Wang, K.C., Piao, Y.C. et al. (2024). Surface micro-morphology model involved in grinding of GaN crystals driven by strain-rate and abrasive coupling effects. *International Journal of Machine Tools and Manufacture* 201: 104197.

12 Jamshidi, H., Gurtan, M., and Budak, E. (2019). Identification of active number of grits and its effects on mechanics and dynamics of abrasive processes. *Journal of Materials Processing Technology* 273: 116239.

13 Perveen, A., San, W.Y., and Rahman, M. (2021). Fabrication of different geometry cutting tools and their effect on the vertical micro-grinding of BK7 glass. *The International Journal of Advanced Manufacturing Technology* 61: 101–115.

14 Li, C., Hu, Y.X., Wei, Z.Z. et al. (2024). Damage evolution and removal behaviors of GaN crystals involved in double-grits grinding. *International Journal of Extreme Manufacturing* 6 (2): 025103.

15 Setti, D., Arrabiyeh, P.A., Kirsch, B. et al. (2020). Analytical and experimental investigations on the mechanisms of surface generation in micro grinding. *International Journal of Machine Tools and Manufacture* 149: 103489.

16 Piao Y C, Li C, Hu Y X, et al. (2024). Nanoindentation induced anisotropy of deformation and damage behaviors of MgF_2 crystals. *Journal of Materials Research and Technology* 28: 4615–4625.

17 Park, H.W. and Liang S.Y. (2008). Force modeling of micro-grinding incorporating crystallographic effects. *International Journal of Machine Tools and Manufacture* 48 (15): 1658–1667.

18 Gong, Y.D., Zhou, Y.G., Wen, X.L. et al. (2017). Experimental study on micro grinding force and subsurface microstructure of nickel-based single crystal superalloy in micro grinding. *Journal of Mechanical Science and Technology* 31 (7): 3397–3410.

5

Grinding Temperature Evaluation

5.1 Grinding Temperature in Surface Grinding

Temperature measurements were conducted using a thermocouple technique, which has been proven effective for high-depth grinding applications due to its fast response characteristics [1]. The experimental setup (Figure 5.1a) featured a specialized thermocouple configuration consisting of a 0.03 mm constantan foil electrically insulated by dual 0.01 mm mica layers, all precisely positioned between two sections of a split workpiece. During operation, the junction exposed to abrasive contact functioned as the thermal measurement point (hot junction), while the reference junction was temperature-stabilized in an ice bath. The complete thermal circuit was established when the grinding wheel traversed the workpiece surface, enabling signal acquisition through a high-speed 16-channel data acquisition system (NI USB-6211, USA) operating at 2 kHz, with subsequent signal processing performed in Labview software. System calibration involved comparative measurements against standard references under controlled induction heating conditions, resulting in the well-characterized voltage–temperature correlation shown in Figure 5.1b that shows excellent consistency with published data [2]. This calibration enabled precise temperature determination during nickel alloy grinding through mathematical conversion of the acquired thermoelectric potentials, employing an established thin-foil thermocouple methodology that has been documented in previous studies [3].

5.1.1 Effect of Grinding Speed

The impact of grinding speed v_s on the grinding temperature is shown in Figure 5.2a and b. Generally, the grinding temperature increased as v_s rose from 20 to 35 m/s. For the grinding wheel, in the case of GH4169, the BA wheel generated a grinding temperature ranging from 100.2 to 426.8 °C when v_s varied from 20 to 30 m/s. When v_s reached 35 m/s, the temperature increased to 755.3 °C. Under such circumstances, although no obvious surface defects were observed, the blackening of the side regions of the workpiece indicated the occurrence of slight oxidation. For DZ408, the BA wheel induced

Grinding of Single-Crystal Superalloys: Fundamentals and Technologies, First Edition.
Wenfeng Ding et al.
© 2026 Wiley-VCH GmbH. Published 2026 by Wiley-VCH GmbH.

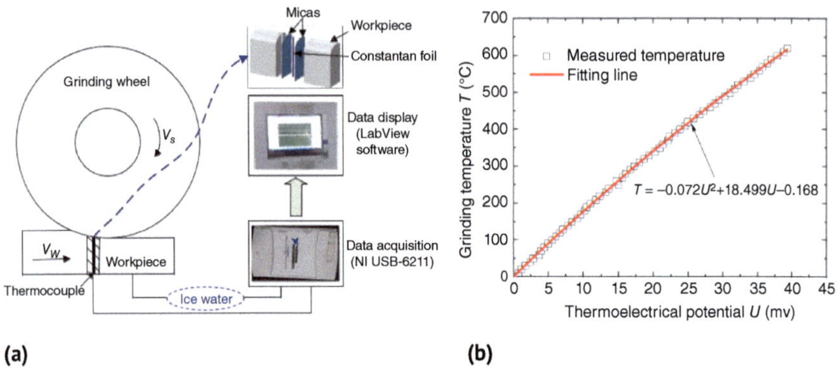

Figure 5.1 Grinding temperature measurement of DD6 alloy: (a) schematic illustration and (b) calibration curve.

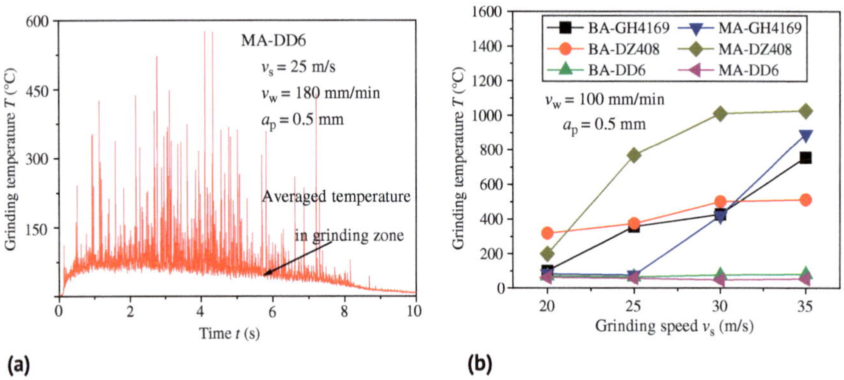

Figure 5.2 Effect of grinding speed on grinding temperature: (a) typical temperature signal and (b) temperature results.

temperatures from 320.1 to 512.7 °C when v_s ranged from 20 to 35 m/s. The temperature variation amplitude was smaller than that of GH4169, which was similar to the results obtained under shallow grinding conditions [4]. For DD6, the grinding temperatures of the BA wheel were relatively low (from 72.5 to 82.2 °C) when v_s increased from 20 to 35 m/s.

When the MA wheel was used, the grinding temperature of GH4169 was not significantly different from that of the BA wheel. However, the grinding temperature of DZ408 increased sharply from 200.9 to 1028.3 °C when using the MA wheel. Severe surface oxidation occurred on the DZ408 surface. This phenomenon was not observed when grinding DD6, whose temperature ranged from 55.4 to 65.3 °C. The DD6 surface was smooth and the texture was clear.

The above analysis reveals that grinding GH4169 and DZ408 with either the BA or MA wheel easily led to high temperatures (>1000 °C) even with relatively gentle parameters. In contrast, the temperature of DD6 remained at a low level (<100 °C) without grinding burn. These results also suggest that the mechanical properties of these superalloys have a crucial influence on the grinding temperature.

5.1.2 Effect of Workpiece Speed

When v_s was fixed at 25 mm/min and a_p was kept at 0.5 mm, the effect of workpiece speed v_w on grinding temperature was studied and the experimental results were shown in Figure 5.3. Significant rise in temperature was observed generally when v_w varied in a range of 80 – 240 mm/min. For grinding with BA wheel, the grinding temperature of GH4169 was changed from 282.1 to 850.4 °C as v_w increased from 80 to 150 mm/min. The temperature can reach 900 °C if v_w = 240 mm/min, at which the GH4169 surface was going dark. DZ408 had the higher grinding temperature of 338.8–988.8 °C than GH4169. However, the lowest temperature can be obtained when grinding of DD6, ranged from 59.6 to 85.6 °C with v_w from 80 to 180 mm/min, then it increased sharply to 786.2 °C when v_w became 240 mm/min. Oxidation was detected from the blackening of the side region of the DD6 workpiece.

For grinding with the MA wheel, the variation in grinding temperature was similar to that of the BA wheel. The difference was that for grinding DD6, v_w could be extended to 210 mm/min, at which the temperature was less than 100 °C. However, once v_w increased to 240 mm/min, the temperature of DD6 quickly rose to 790.9 °C and the grinding surface quality deteriorated.

The above results follow a similar trend, that is, the temperature initially remained at a very low level (i.e., <130 °C) and then increased remarkably with more aggressive parameters. This is the typical temperature change trend resulting from the transition between film and nucleate boiling in the grinding zone during the creep-feed grinding process.

5.1.3 Effect of Grinding Depth

Figure 5.4 shows the effect of grinding depth a_p on the grinding temperature. Undoubtedly, the temperature increased significantly when a_p changed from 0.2 to 1.5 mm. When a_p was 0.2 mm, the grinding temperature of all workpieces varied within the range of 60.1–85.1 °C. However, when using the BA wheel at a grinding depth of 0.5 mm, the grinding temperatures of GH4169 and DZ408 could reach 357.2 and 418.9 °C, respectively. When a_p was 1.2 mm, grinding with the BA wheel

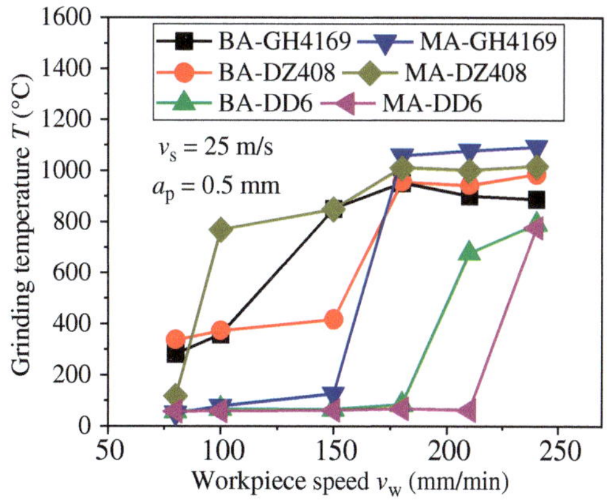

Figure 5.3 Effect of workpiece speed on grinding temperature.

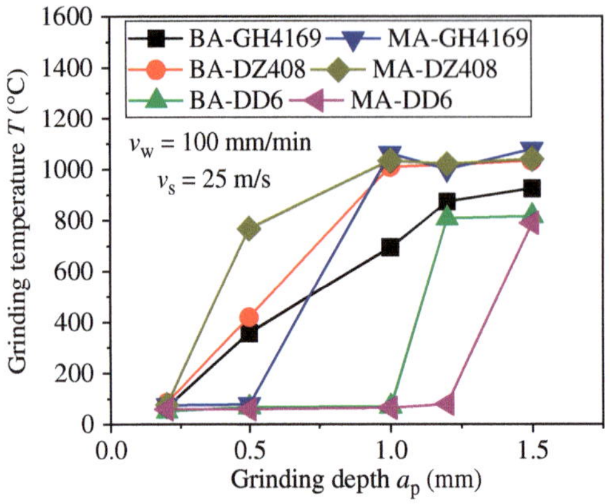

Figure 5.4 Effect of grinding depth on grinding temperature.

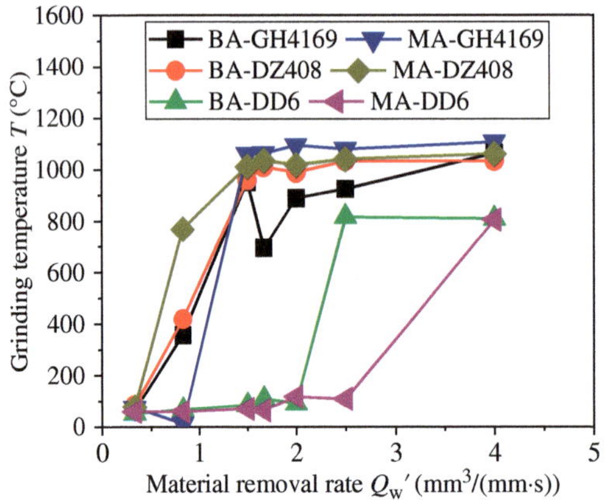

Figure 5.5 Effect of material removal rate on grinding temperature.

resulted in a high temperature (695.2–1000.3 °C) regardless of the workpiece. In contrast, when grinding DD6, the MA wheel had the lowest temperature (65.9 °C). When a_p was further increased to 1.5 mm, the temperatures of 789.7–1081 °C were generated by both grinding wheels. These results indicate that the grinding temperature rises rapidly when increasing the grinding depth using alumina abrasive wheels. Although aggressive parameters contribute to high-efficiency grinding, the risk of grinding burn due to high temperature must be faced. If the grinding heat cannot be dissipated effectively and timely, grinding burn cannot be avoided.

5.1.4 Effect of Material Removal Rate

The grinding speed was fixed at 25 m/s, and variations in workpiece speed and grinding depth were used to obtain different material removal rates (MRRs) Q_w'. The effect of Q_w' on the grinding temperature is shown in Figure 5.5. For GH4169 and DZ408 with the BA wheel, the temperatures ranged from 67.7 to 958.1 °C when Q_w' varied from 0.33 to 1.67 mm³/(mm · s). It increased rapidly to 1064.3 °C when Q_w' was 4 mm³/(mm · s), causing oxidation with a dark color in the side regions of the workpieces. For DD6, when using the BA wheel, the grinding temperature could be as low as 53.4–67.9 °C if Q_w' was less than 1.5 mm³/(mm · s). When grinding GH4169 and DZ408 with the MA wheel, the grinding temperature could ultimately reach as high as 900 °C if $Q_w' > 4$ mm³/(mm · s). Nevertheless, the MA wheel exhibited better performance than the BA wheel for Q_w' in the range of 0.33 – 3.5 mm³/(mm · s) as it produced lower temperatures of 60.1–87.7 °C.

As described earlier, the cooling of the grinding zone plays a vital role in determining the increase in grinding temperature. The cooling effectiveness is highly dependent on the physical state of the coolant in the grinding zone. If nucleate boiling occurs, it can enhance heat transfer by accelerating the coolant flow and bubble growth. If film boiling occurs, the heat transfer process may be inhibited by the continuous steam formation [5, 6]. Given the grinding wheel specifications and coolant parameters, the critical temperature for the transition from nucleate boiling to film boiling can be determined. For water-based coolants, the critical temperature is approximately 130 °C. Thus, when film boiling occurs at $Q_w' = 4$ mm³/(mm · s), the temperature in the grinding zone can reach around 800–1000 °C. The temperature rise of the workpiece surface can be 400–500 °C, which may not be sufficient to cause typical grinding burn on the surface, but the side regions of the workpiece turn black. It should be noted that under such high-temperature environments, the grinding wheel wear can be significantly increased when grinding difficult-to-cut nickel-based superalloys, ultimately degrading the wheel's service life, machining accuracy, and quality.

5.2 Grinding Temperature in Profile Grinding

The vitrified microcrystalline alumina wheel was precisely profiled to match the blade root's final cross-sectional geometry, as evidenced by Figure 5.6a, which shows the MA wheel condition along with pre-grinding and post-grinding blade root specimens. The experimental workpieces were fabricated from DD6 single-crystal nickel alloy in standardized rectangular blocks ($32 \times 15 \times 8$ mm^3). Temperature distribution analysis was conducted using constantan wire–workpiece semi-natural thermocouples positioned at strategic locations (A–D in Figure 5.7) to capture thermal variations across both peak and valley regions of the blade root profile (Figure 5.6b). Complete profile grinding methodology is detailed in Section 4.2, with temperature measurement specifications systematically presented in Table 5.1, including all critical operational parameters for thermal characterization.

The thermal characteristics during turbine blade fir-tree root profile grinding are presented in Figure 5.7, with representative temperature profiles for surface positions A and C shown in Figure 5.7a and b, respectively. Under parameter group I conditions, measured temperatures ranged from 68 to 208 °C, exhibiting consistent spatial variations where position B recorded the minimum values while position D showed peak temperatures. Thermal analysis reveals that root peak regions (positions A/D) consistently maintain approximately 20% higher temperatures compared to valley areas (positions B/C). Notably, parameter group II operations induced severe thermal spikes reaching 900 °C, indicating grinding burn initiation. The observed temperature distribution patterns demonstrate significant influence from root geometry, as established in prior research [7, 8], where reduced heat

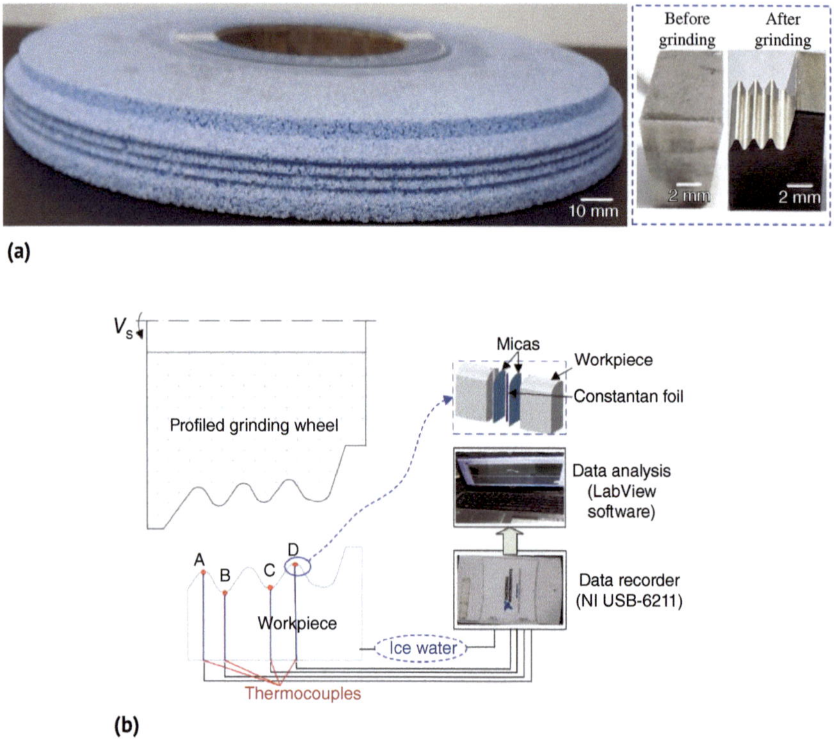

Figure 5.6 The temperature measurement in profile grinding of DD6 alloy: (a) profiled grinding wheel and the blade roots before and after grinding and (b) diagram of temperature measurement system.

Table 5.1 Profile grinding parameters for testing temperatures.

Grinding group No.	**I (without grinding burn)**			**II (with grinding burn)**		
Grinding step No.	1	2	3	1	2	3
Grinding speed v_s (m/s)	25	35	35	25	35	35
Workpiece speed v_w (mm/min)	240	210	300	600	600	600
Grinding depth a_p (mm)	1.5	1.0	0.13	1.5	1.0	0.13
Material removal rate Q_w' ($mm^3/(mm \cdot s)$)	6	3.6	0.65	15	10	1.3

conduction capacity in peak regions and restricted coolant flow due to complex profile features [9, 10] contribute to localized temperature elevation.

The experimental data (Figures 5.1 and 5.7) demonstrate that the maximum temperature rise consistently occurs during grinding step No. 2 at all measurement points. This thermal behavior stems from distinct material removal mechanisms: in step No. 1, only the grinding wheel's peak surfaces engage with the rectangular

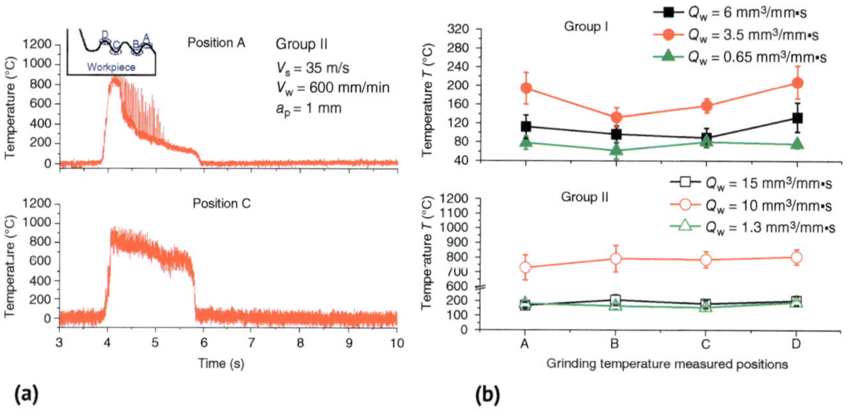

Figure 5.7 Grinding temperature distribution of blade root workpiece: (a) typical temperature signals at different positions and (b) temperature variations along the root profile.

workpiece, creating high specific grinding energy (Q_w') but limited material removal. As the process advances to step No. 2, the active cutting area expands significantly when forming the root profile geometry, resulting in substantially increased MRRs and consequent thermal loading. The subsequent temperature reduction in step No. 3 correlates with the application of conservative grinding parameters, particularly reduced depth of cut. These observations confirm that workpiece geometry critically influences grinding temperature and force in profile grinding operations.

5.3 Grinding Temperature in Micro-grinding

Grinding temperature is a crucial factor influencing the machining quality in micro-grinding. In the micro-grinding process of nickel-based single-crystal high-temperature alloys, the influence of grinding parameters on the grinding temperature is more pronounced. The grinding process is essentially a continuation of the abrasive grains sliding, ploughing, and cutting on the workpiece surface, and most of the consumed energy accumulates in the grinding zone, resulting in high temperatures. The temperature at the grinding point is the temperature at the contact location between the abrasive grains and the chips, which is the heat source of grinding. The temperature in the grinding zone is the average temperature in the contact area between the grinding rod and the workpiece material. In conventional macro-scale grinding, thermocouples are easy to fabricate, the grinding temperatures are higher, and the grinding signals are easier to collect. However, in the micro-grinding of nickel-based single-crystal high-temperature alloys, the small size of the grinding workpiece and tool leads to a small grinding area, making it difficult to make thermocouples. Therefore, the traditional grinding temperature measurement method cannot be fully applied to micro-grinding. To measure the temperature in the micro-grinding of nickel-based single-crystal high-temperature alloys, different thermocouple arrangement options, such as unipolar thermocouples, bipolar grindable thermocouples, and grindable bipolar thermocouples, were explored. Through numerous experiments, it was found that in the micro-grinding process of nickel-based single-crystal high-temperature alloys, unipolar thermocouples and bipolar grindable thermocouples have a lower success rate in signal acquisition. This may be due to the small grinding area of the workpiece in micro-grinding, requiring the use of an electric coupling with a smaller diameter and thickness. However, the grinding force during the grinding process is not large, making it difficult to form a circuit between the electric coupling and the workpiece, and between the electric coupling wires, resulting in difficult signal acquisition. The grindable bipolar thermocouple can collect the signal from the grinding area of the grinding rod, and the signal measured when the junction of the electric coupling wire is ground out is used as the temperature signal of the grinding area. A Nichrome–nickel–aluminum thermocouple wire with a diameter of 0.08 mm and a thick insulating mica sheet with a thickness of 0.01 mm are bonded together using epoxy resin. The fabricated artificial thermocouple temperature measurement specimen is shown in Figure 5.8.

To study the influence of grinding parameters on the grinding temperature in the micro-grinding of nickel-based single-crystal high-temperature alloys, the nickel-based single-crystal high-temperature alloy DD98 was used as the test material. The grinding direction was along the < 110 > crystal direction within the (001) crystal plane. A one-way experimental program was designed to examine the effects of v_s, v_w, and a_p on the temperature of the grinding zone.

5.3.1 Influence of Spindle Speed

Figure 5.9 presents the relationship between grinding temperature and spindle speed during micro-grinding of nickel-based single-crystal superalloys under constant parameters ($v_w = 120\ \mu m/s$, $a_p = 10\ \mu m$). The experimental results demonstrate a clear positive correlation between temperature and spindle speed (20,000–60,000 rpm), with distinct thermal characteristics observed across different speed ranges. The initial speed increase (20,000 – 30,000 rpm) produces a more pronounced temperature rise compared to higher speed ranges (30,000 – 60,000 rpm). This nonlinear behavior stems from competing thermal mechanisms: while higher speeds increase grain

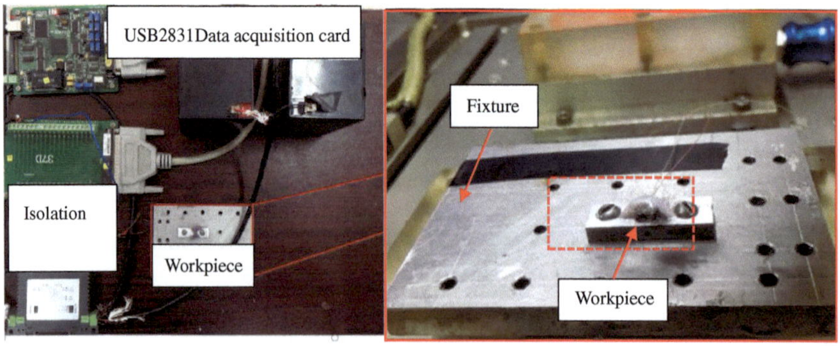

Figure 5.8 Specimen of measuring temperature.

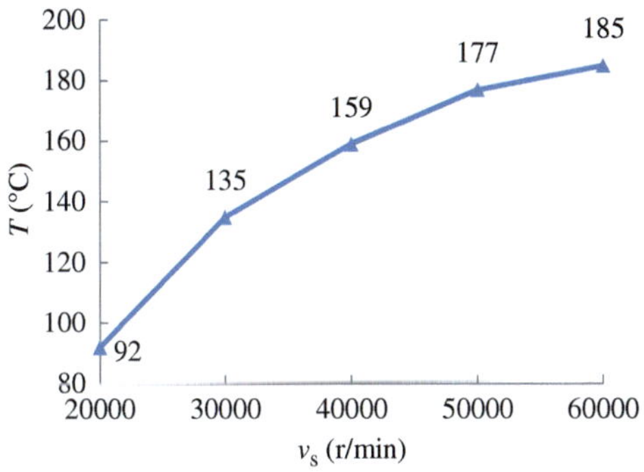

Figure 5.9 Influence of spindle speed on grinding temperature.

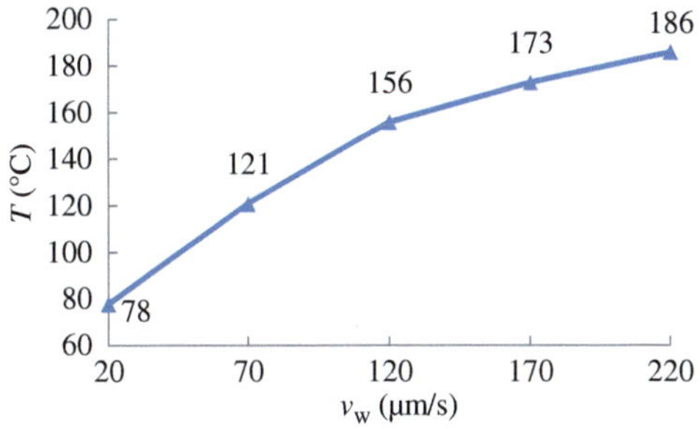

Figure 5.10 Influence of feeding rate on micro-grinding temperature.

participation rate and frictional heat generation, they also enhance chip evacuation efficiency and airflow cooling effects. The micro-grinding process exhibits unique thermal dynamics due to the constrained tool diameter, which limits achievable linear speeds despite high rotational rates. These findings suggest an optimal speed range where thermal accumulation balances with heat dissipation mechanisms, recommending against excessive spindle speeds to prevent thermal damage while considering the process's inherent speed limitations.

5.3.2 Influence of Feed Rate

Figure 5.10 presents the relationship between workpiece feed rate and grinding temperature in micro-grinding of nickel-based single-crystal superalloys under fixed parameters ($v_s = 40{,}000$ rpm, $a_p = 10$ μm). The experimental data demonstrate a nonlinear temperature response to increasing feed rate v_w (20 – 220 μm/s), characterized by two distinct regimes: an initial rapid temperature rise at lower feed rates (< 120 μm/s) followed by a moderated increase at higher velocities. This nonlinear behavior stems from competing thermal mechanisms—while elevated feed rates enhance heat generation through greater undeformed chip thickness (increasing specific energy requirements) and intensified frictional interactions between tool and workpiece, the accelerated movement of the heat source simultaneously improves convective cooling. The transition point at 120 μm/s marks the threshold where these opposing effects reach dynamic equilibrium, causing the temperature growth rate to stabilize despite continued feed rate increases.

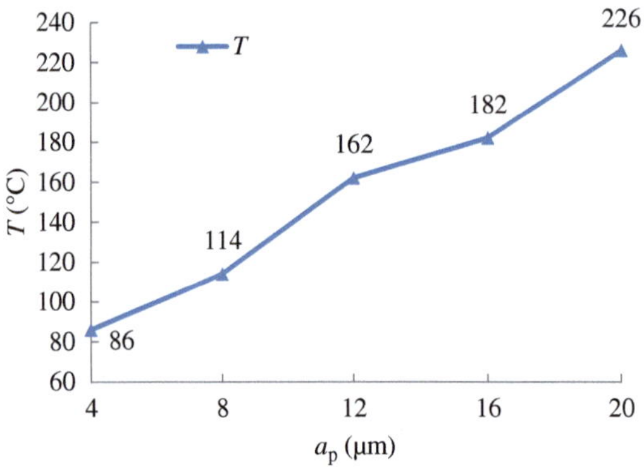

Figure 5.11 Influence of micro-grinding depth on grinding temperature.

5.3.3 Influence of Grinding Depth

The experimental results presented in Figure 5.11 demonstrate the correlation between grinding depth and temperature rise in the micro-grinding of nickel-based single-crystal superalloys under fixed parameters of wheel speed (v_s = 40,000 r/min) and workpiece feed rate (v_w = 120 μm/s), with grinding depths (a_p) varying from 4 to 20 μm. The observed temperature elevation in the grinding zone with increasing depth can be attributed to multiple interrelated factors: as the grinding depth becomes greater, each abrasive grain experiences increased undeformed chip thickness, leading to amplified tangential and normal grinding forces that elevate specific energy consumption; this is accompanied by heightened deformation resistance in the chips and intensified frictional interactions between abrasives and workpiece material, collectively converting more mechanical energy into thermal energy. Furthermore, deeper cuts extend the contact arc length between grinding tool and workpiece, which not only strengthens the ploughing effect but also reduces heat dissipation efficiency due to prolonged tool–workpiece engagement, resulting in progressive temperature accumulation within the grinding zone during the processing of these high-temperature alloys.

5.3.4 Influence of Crystal Orientation

To overcome the challenges in measuring grinding temperatures for microscale components, an ABAQUS-based thermal simulation model was developed to investigate temperature field evolution during micro-grinding of single-crystal superalloys. The model accounts for key factors, including process parameters and material anisotropy, by approximating the micro-grinding process as single-grit cutting with negative rake angles. Prior to simulation, to simulate the stress and temperature fields in the micro-grinding process of nickel-based single-crystal high-temperature alloys more realistically, detailed characterization of abrasive morphology was conducted using ultra-depth microscopy, revealing 500# CBN grains (≈ 20 μm) with distinct quadrangular prism shapes on the grinding rod surface (Figure 5.12), which were subsequently incorporated into the model. The workpiece material was defined with orthotropic anisotropy properties, with crystal orientation effects simulated through coordinate system rotation. Material removal was modeled using

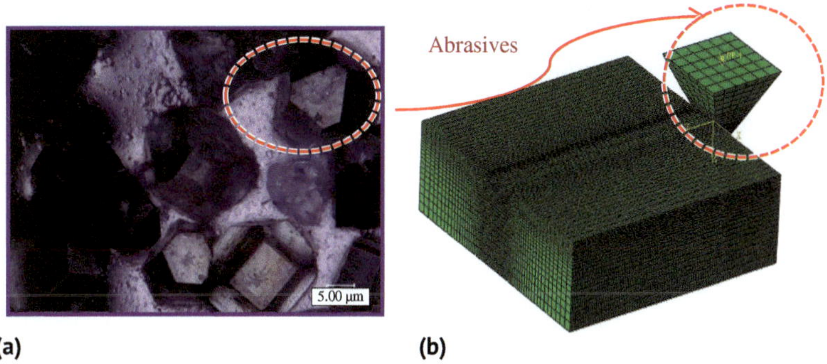

Figure 5.12 (a) Image of micro-grinding tool and abrasive and (b) simulation model.

ABAQUS's cumulative damage criterion, ultimately generating the representative single-grit cutting simulation shown in Figure 5.12 that captures both abrasive geometry and workpiece interaction dynamics.

Nickel-based single-crystal high-temperature alloys exhibit anisotropic properties during plastic deformation. In recent years, scholars have proposed several intrinsic models to describe the anisotropy of single-crystal materials, including the macroscopic image-only model represented by the Hill model and the crystal slip intrinsic model. The latter lacks an effective method for determining the latent hardening coefficients reflecting the interaction between slip systems, while the Hill model is widely used in practice due to its simplicity and convenience. Correct material modelling is the basis for realizing the simulation. In this chapter, the Hill model was chosen as the material model, which is commonly used to simulate orthotropic anisotropic material models where the tensile and compressive yield stresses are equal or not significantly different. When simulating and analyzing the anisotropy of a material, the coordinate system of the workpiece can be rotated in the material model setup to simulate the grinding process in different directions within the same crystal plane. Therefore, the Hill model is suitable for defining the material of nickel-based single-crystal high-temperature alloys in micro-grinding simulations. The abrasive grain material in the micro-grinding process is CBN, which has a much larger modulus of elasticity than nickel-based single-crystal high-temperature alloys and is therefore regarded as a rigid body. For orthotropic anisotropic materials, the elastic constitutive relations are obtained according to Hooke's law:

$$\{\varepsilon\} = [S]\{\sigma\} \tag{5.1}$$

[S] is the 6×6 order flexibility matrix, which is shown in Eq. (5.2) for an orthotropic anisotropic material nickel-based single-crystal high-temperature alloy with a face-centered cubic structure.

$$[S] = \begin{bmatrix} S_{11} & S_{12} & S_{13} & 0 & 0 & 0 \\ S_{12} & S_{11} & S_{22} & 0 & 0 & 0 \\ S_{12} & S_{12} & S_{11} & 0 & 0 & 0 \\ 0 & 0 & 0 & S_{44} & 0 & 0 \\ 0 & 0 & 0 & 0 & S_{44} & 0 \\ 0 & 0 & 0 & 0 & 0 & S_{44} \end{bmatrix} \tag{5.2}$$

where $S_{11} = \frac{1}{E}; S_{44} = \frac{1}{G}; S_{12} = -\frac{\mu}{E}$.

The yield criterion for the Hill material model can be expressed in Eq. (5-3).

$$\sigma_S^2 = A(\sigma_{22} - \sigma_{33})^2 + B(\sigma_{33} - \sigma_{11})^2 + C(\sigma_{11} - \sigma_{22})^2 + 3D\sigma_{23}^2 + 3E\sigma_{31}^2 + 3F\sigma_{12}^2 \quad (5.3)$$

where σ_S is the reference yield stress, generally taken as the yield stress in a certain principal axis direction, A, B, C, D, E, and F are six independent anisotropic parameters, which are related to the main direction of the material, and can be determined by unidirectional tensile tests and shear yield tests along the three main directions of the material. For single-crystal materials with FCC structure, the yield properties along the three main axes of the material, < 001 >, < 110 >, and < 111 >, are equal. From material symmetry: $A = B = C = 1$, $D = E = F$. Then, Eq. (5-2) is transformed into Eq. (5.3) to obtain a simplified expression for Hill's material model, as shown in Eq. (5.3).

In the friction model, there is significant friction between the abrasive grains and the workpiece in micro-grinding, and the heat generated by the friction increases the temperature in the grinding area between the abrasive grains and the workpiece. The heat generated by the friction also changes the stress–strain field of the workpiece. In the friction model for micro-grinding, substantial friction occurs between the abrasive grains and the workpiece. The heat produced by this friction raises the temperature in the grinding zone between them. This frictional heat causes alterations in the stress–strain field of the workpiece. Moreover, the coupling effect of stress and strain leads to changes in the plastic deformation of the micro-grinding subsurface. To obtain more ideal and realistic simulation outcomes, it is essential to accurately understand the type of frictional contact between the abrasive particles and the workpiece material.

The cutting separation criteria for metallic materials mainly consist of geometrical and physical separation criteria. Considering the actual situation in the micro-grinding process of nickel-based single-crystal high-temperature alloys, the grinding process can be simplified as the cutting process of the material with a negative rake angle cutting edge. In this chapter, the shear criterion within the physical separation criterion is applied in the grinding simulation modeling process.

Regarding the influence of different grain orientations on the simulated values of grinding temperature within the (001) grain surface, Figure 5.13 depicts the variation of the simulated micro-grinding temperature within the (001) crystal surface with θ. Here, θ represents the angle between the grinding direction within the (001) crystal plane and the [100] crystal direction. Figure 5.14a–g displays the micro-grinding temperature fields of the abrasive grains in the (001) crystal plane along $\theta = 0°$, 15°, 30°, 45°, 60°, 75°, and 90° directions, respectively. As shown in Figures 5.13 and 5.14, within the range of $\theta = 0° - 45°$, the simulated grinding temperature increases as θ increases. In the range of $\theta = 45° - 90°$, the simulated micro-grinding temperature decreases as θ increases. The temperature value at $\theta = 0°$ is close to that at $\theta = 90°$; the temperature value at $\theta = 15°$ is close to that at $\theta = 75°$; the temperature value at $\theta = 30°$ is close to that at $\theta = 60°$. The maximum temperature simulation value occurs at $\theta = 45°$. Due to the symmetry of the face-centered cubic structure of the nickel-based single-crystal high-temperature alloy, it can be observed

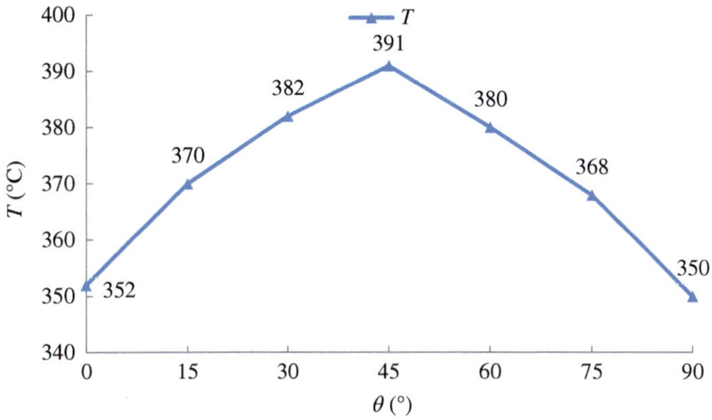

Figure 5.13 Changing of micro-grinding temperature with the crystal orientation in (001) plan.

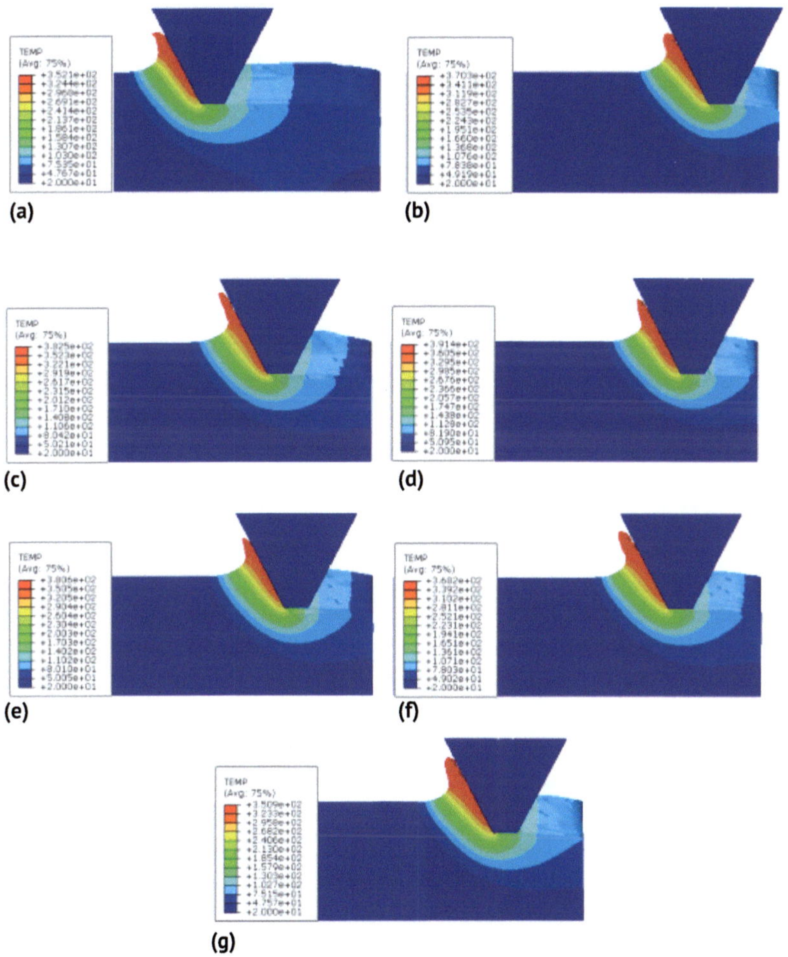

Figure 5.14 Micro-grinding temperature simulation with different crystal orientations [11]: (a) $\theta = 0°$; (b) $\theta = 15°$; (c) $\theta = 30°$; (d) $\theta = 45°$; (e) $\theta = 60°$; (f) $\theta = 75°$; and (g) $\theta = 90°$.

that during grinding within the crystal plane of the nickel-based single-crystal high-temperature alloy (001), lower grinding temperatures are obtained when grinding along $\theta = 0°, 90°, 180°$, and $270°$. Higher grinding temperatures are present when grinding in $\theta = 45°, 135°, 225°$, and $315°$ directions. The reason is that within the range of $\theta = 0° - 45°$, the modulus of elasticity of the material increases with θ, the normal grinding force in the micro-grinding process increases, and the friction between the abrasive grain and the workpiece material in the micro-grinding process is enhanced, resulting in increased grinding heat generation and a higher micro-grinding temperature. In the range of $\theta = 45° - 90°$, as θ increases, the modulus of elasticity of the material decreases, the friction between the abrasive particles and the workpiece material decreases, the heat generated by the friction decreases, and the micro-grinding temperature becomes lower. It can be seen that the variation of the micro-grinding temperature values during the micro-grinding of nickel-based single-crystal high-temperature alloys is consistent with the trend of the elastic modulus of the material.

References

1 Nie, Z.G., Wang, G., Liu, D.H. et al. (2018). A statistical model of equivalent grinding heat source based on random distributed grains. *ASME Journal of Manufacturing Science and Engineering* 140: 051016.

2 Yao, C.F., Jin, C.Q., Huang, X.C. et al. (2013). Research on surface integrity of grinding inconel 718. *International Journal of Advanced Manufacturing Technology* 65: 1019–1030.

3 Lefebvre, A., Vieville, P., Lipinske, P. et al. (2006). Numerical analysis of grinding temperature measurement by the foil/workpiece thermocouple method. *International Journal of Machine Tools and Manufacture* 46: 1716–1726.

4 Sinha, M.K., Setti, D., Ghosh, S. et al. (2016). An investigation on surface burn during grinding of Inconel 718. *Journal of Manufacturing Process* 21: 124–133.

5 Guo, C. and Malkin, S. (1994). Analytical and experimental investigation of burnout in creep-feed grinding. *CIRP Annals-Manufacturing Technology* 43: 283–286.

6 Li, H.N. and Axinte, D. (2018). On the inverse design of discontinuous abrasive surface to lower friction-induced temperature in grinding: an example of engineered abrasive tools. *International Journal of Machine Tools and Manufacture* 132: 50–63.

7 Jin, T., Yi, J., and Li, P. (2017). Temperature distributions in form grinding of involute gears. *International Journal of Advanced Manufacturing Technology* 88 (9–12): 2609–2620.

8 Brinksmeier, E., Aurich, J., Govekar, E. et al. (2006). Advances in modelling and simulation of grinding processes. *CIRP Annals-Manufacturing Technology* 55: 667–696.

9 Wen, J., Zhou, W.H., Tang, J.Y. et al. (2024). Integrated novel modeling approaches for non-uniform heat source and temperature field in gear profile grinding. *The International Journal of Advanced Manufacturing Technology* 130 (9): 4397–4414.

10 Rowe, W. (2017). Temperatures in grinding-A review. *ASME Journal of Manufacturing Science and Engineering* 139: 121001.
11 Zhou, Y.G. (2017). Study on the Technological Theory and Key Technology in Micro Grinding Nickel-based Single Crystal Superalloy. Northeastern University.

6

Grinding Wheel Wear Evaluation

6.1 Grinding Wheel Wear in Surface Grinding

The grinding force ratio was employed to evaluate the grinding wheel wear. The representative topographies of the worn wheel surfaces were analyzed using scanning electron microscopy (SEM). In particular, the worn surfaces of the abrasive wheels at different material removal rates were inspected online with an optical microscope at a magnification of × 100. The wheel radial wear was measured by the replica method [1, 2].

6.1.1 Wheel Radial Wear and Grinding Ratio

Figure 6.1 illustrates the notable differences in both radial wheel wear (ΔR) and wear volume (V_s) of alumina abrasive wheels when processing different materials at varying material removal rates (Q_w'). As shown in Figure 6.1a, during BA grinding with Q_w' increasing from 0.83 to 8 mm^3/(mm · s), the DD6 alloy caused the most severe wheel wear, with Vs ranging from 10.6 to 34.9 mm^3/mm. Under the same conditions, GH4169 and DZ408 alloys resulted in wear volumes of 10–28.5 mm^3/mm and 11.8–32.6 mm^3/mm, respectively. In contrast, MA grinding demonstrated significantly lower wear across all materials, with corresponding V_s values of 7.2–20.4 mm^3/mm (DD6), 10.6–20.2 mm^3/mm (GH4169), and 11.8–29.2 mm^3/mm (DZ408) for the same Q_w' range. These results clearly indicate that BA grinding consistently produces greater wheel wear than MA grinding, as Q_w' increases, irrespective of the workpiece material. Additionally, the severity of wheel wear during creep-feed grinding followed a consistent progression, with DD6 causing the highest wear, followed by GH4169 and then DZ408.

The wear progression of alumina wheels exhibits three characteristic phases during creep-feed grinding of nickel-based superalloys, as illustrated in Figure 6.1a and b. Initially, an accelerated wear phase occurs as abrasive grains become conditioned, followed by an extended steady-state period demonstrating stable wear rates. When material removal rates exceed 6 mm^3/(mm · s), a sharp transition to rapid wear becomes evident. This triphasic wear behavior correlates strongly with applied grinding loads, as confirmed by previous studies [3, 4], where excessive

Grinding of Single-Crystal Superalloys: Fundamentals and Technologies, First Edition.
Wenfeng Ding et al.
© 2026 Wiley-VCH GmbH. Published 2026 by Wiley-VCH GmbH.

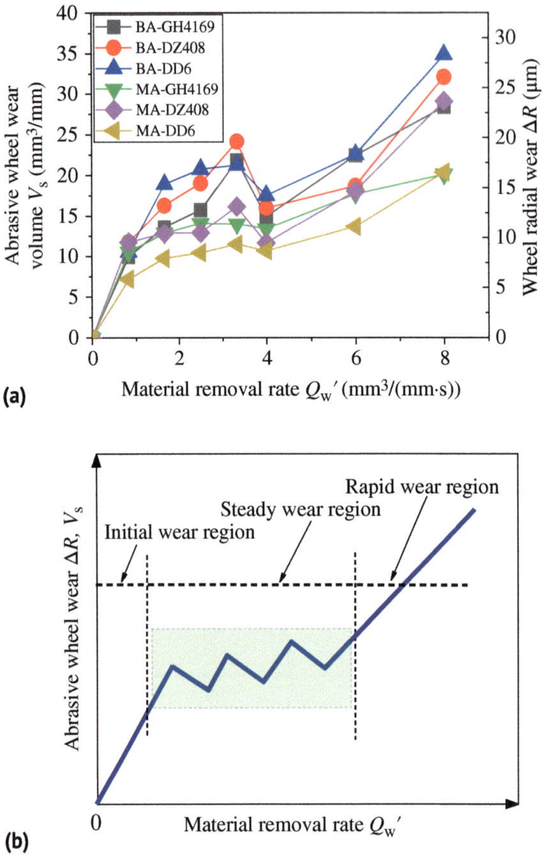

Figure 6.1 Wear of alumina abrasive wheels in the creep-feed grinding process: (a) effect of material removal rate on wheel wear volume and wheel radial wear and (b) schematic illustration for tool wear of alumina abrasive wheels.

mechanical stresses accelerate wheel degradation. The findings emphasize the critical need for periodic wheel dressing to maintain optimal cutting performance, particularly when processing high-strength superalloys under aggressive creep feed conditions. Proper dressing intervals help preserve wheel sharpness and prevent premature transition to the rapid wear regime, thereby ensuring consistent grinding quality and process efficiency.

Furthermore, the grinding ratio is an important parameter to characterize the wear resistance of abrasive wheel [5]. If the abrasive wheels and grinding conditions are maintained the same, the grinding ratio G would be controlled by the workpiece material properties. The grinding ratio G can be expressed as follows:

$$G = \frac{V_w}{V_s} \tag{6.1}$$

$$V_s = \pi d_s \cdot \Delta R \tag{6.2}$$

$$V_w = Q_w' \cdot \Delta t = a_p v_w \Delta t = a_p l \tag{6.3}$$

where V_s is the worn volume of an abrasive wheel per width, d_s is the abrasive wheel diameter, ΔR is the wheel radial wear, V_w is the material removal volume of the workpiece per grinding width, Q_w' is the material removal rate per grinding width, Δt is the grinding duration, and l is the length of the workpiece.

The relationship between grinding ratio (G) and material removal rate (Q_w') is presented in Figure 6.2, revealing significant performance differences between wheel types. Microcrystalline alumina (MA) wheels demonstrated superior efficiency, achieving a maximum G value of 3.8 when processing DD6 alloy, whereas bonded abrasive (BA) wheels showed minimum effectiveness $(G = 1.7)$ with DZ408 alloy. The data indicates a critical threshold at $Q_w' = 3.3\ \text{mm}^3/(\text{mm} \cdot \text{s})$, marking a transition in wear behavior: below this value, G increases proportionally with Q_w', mirroring the trends observed for radial wear (ΔR) and wear volume (V_s) in Figure 6.1a; beyond this point, further increases in removal rate lead to substantial G reduction. This inflection point represents an optimal balance between material removal efficiency and wheel degradation, suggesting that maximum grinding performance occurs at intermediate Q_w' values before excessive wear mechanisms dominate at higher removal rates.

6.1.2 Volumetric Wheel Wear Rate and Wear Topographic Evolution

To further analyze the wear behavior of the abrasive wheels, the wheel wear rate was introduced to evaluate the wear resistance of the abrasive wheels. According to Eqs. (6-1)–(6-3), the grinding ratio G can be written as:

$$G = \frac{V_w}{V_s} = \frac{a_p v_w \Delta t}{\pi d_s \Delta R} = \frac{1}{\pi d_s} \cdot \frac{Q_w'}{r_g} \tag{6.4}$$

where r_g is the wheel wear rate, which is related to the material removal rate Q_w' and grinding ratio G. Based on Eq. (6–4), the wheel wear rate r_g has been calculated and shown in Figure 6.3. It is found that with the increase of Q_w', the increasing rate of wheel wear rate r_g becomes higher; while it reaches a constant (about 5×10^{-4} mm/s) with Q_w' in the range 1.67–3.33 $\text{mm}^3/(\text{mm} \cdot \text{s})$. It can be found again that the wear region changed from the initial wear region to the steady wear region,

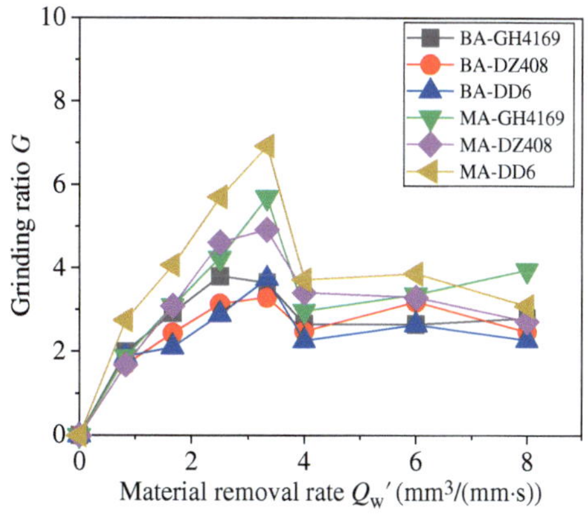

Figure 6.2 Effect of material removal rate on the grinding ratio.

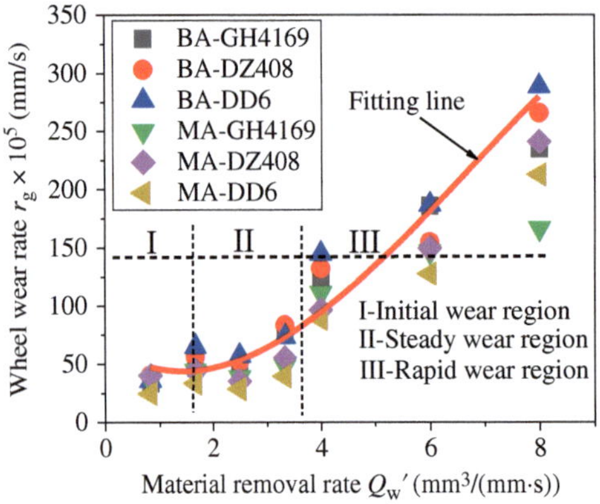

Figure 6.3 Wheel volumetric wear rate evolution.

and then to the rapid wear region. Furthermore, as can be observed from Figure 6.3, the MA wheel had a lower value of r_g than the BA wheel, suggesting a higher wear resistance of MA grains.

Optical microscopy analysis of wheel wear morphologies (Figure 6.4) reveals distinct material-dependent degradation patterns. Both BA and MA wheels exhibited substantial workpiece material adhesion when processing GH4169 and DZ408 alloys, with adhesion areas expanding significantly as Q_w' increased from 0.83 (Test No. 1) to 6 $mm^3/(mm \cdot s)$ (Test No. 7). Comparative evaluation shows MA wheels developed more pronounced wear flats than BA counterparts under identical conditions. Most notably, DD6 alloy grinding induced severe clogging phenomena in both wheel types, demonstrating this material's unique interaction with alumina abrasives. These observations highlight how wheel wear mechanisms vary considerably depending on both workpiece composition and operating parameters, with adhesion and clogging representing primary failure modes that progress with accumulated material removal.

The evolution of wheel wear morphology under varying process parameters reveals several key trends, as illustrated in Figure 6.4. Initial speed increases from 25 m/s (Test No. 1) to 35 m/s (Test No. 2) produced distinct wear patterns between wheel types despite identical workpiece materials. Progressive depth of cut augmentation (Tests No. 2 – No. 4, 0.5–1.5 mm) consistently enhanced grain-workpiece adhesion across all materials. While elevated workpiece speeds (100 → 240 mm/min, Tests No. 4 → No. 7) mainly expanded adhesion areas for GH4169 and DZ408 alloys, DD6 alloy exhibited a remarkable wear mechanism transition from severe clogging to adhesion-dominated wear. These observations demonstrate that wheel degradation modes in creep-feed grinding are governed by complex interactions between material properties (particularly for DD6 alloy) and process parameters, as documented in Refs. [6, 7]. The findings underscore the importance of parameter optimization based on specific workpiece-wheel combinations to control wear progression.

To gain deeper insights into the wear mechanisms of abrasive wheels, SEM was employed to examine the wheel surfaces, with representative images presented in Figure 6.5. Analysis of Figure 6.5a–d reveals that BA wheels processing GH4169 and DZ408 alloys exhibited significant attritious wear characteristics along with pronounced microfractures on the cutting edges. Under increasingly aggressive

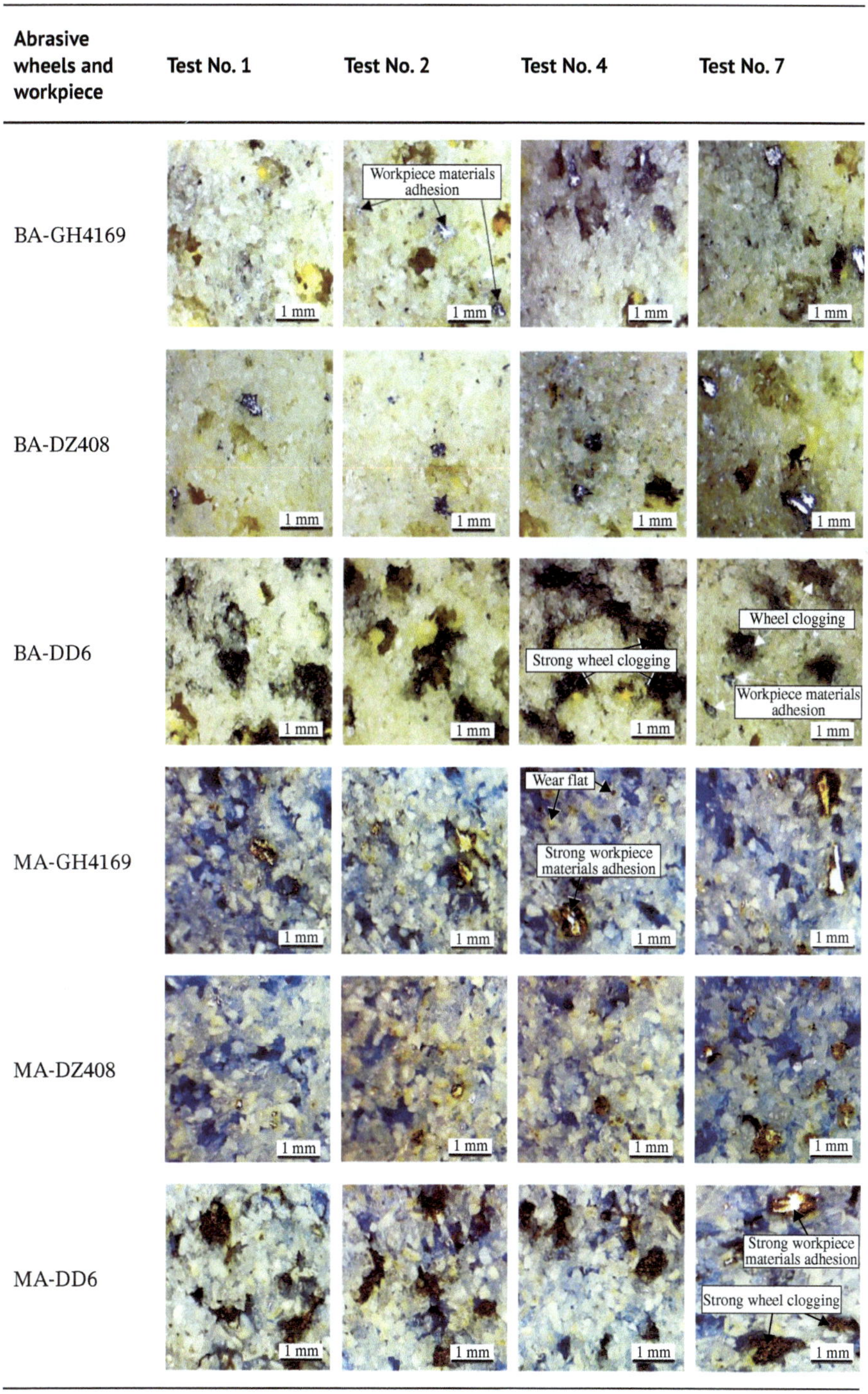

Figure 6.4 Wear topography evolution of different alumina abrasive wheels.

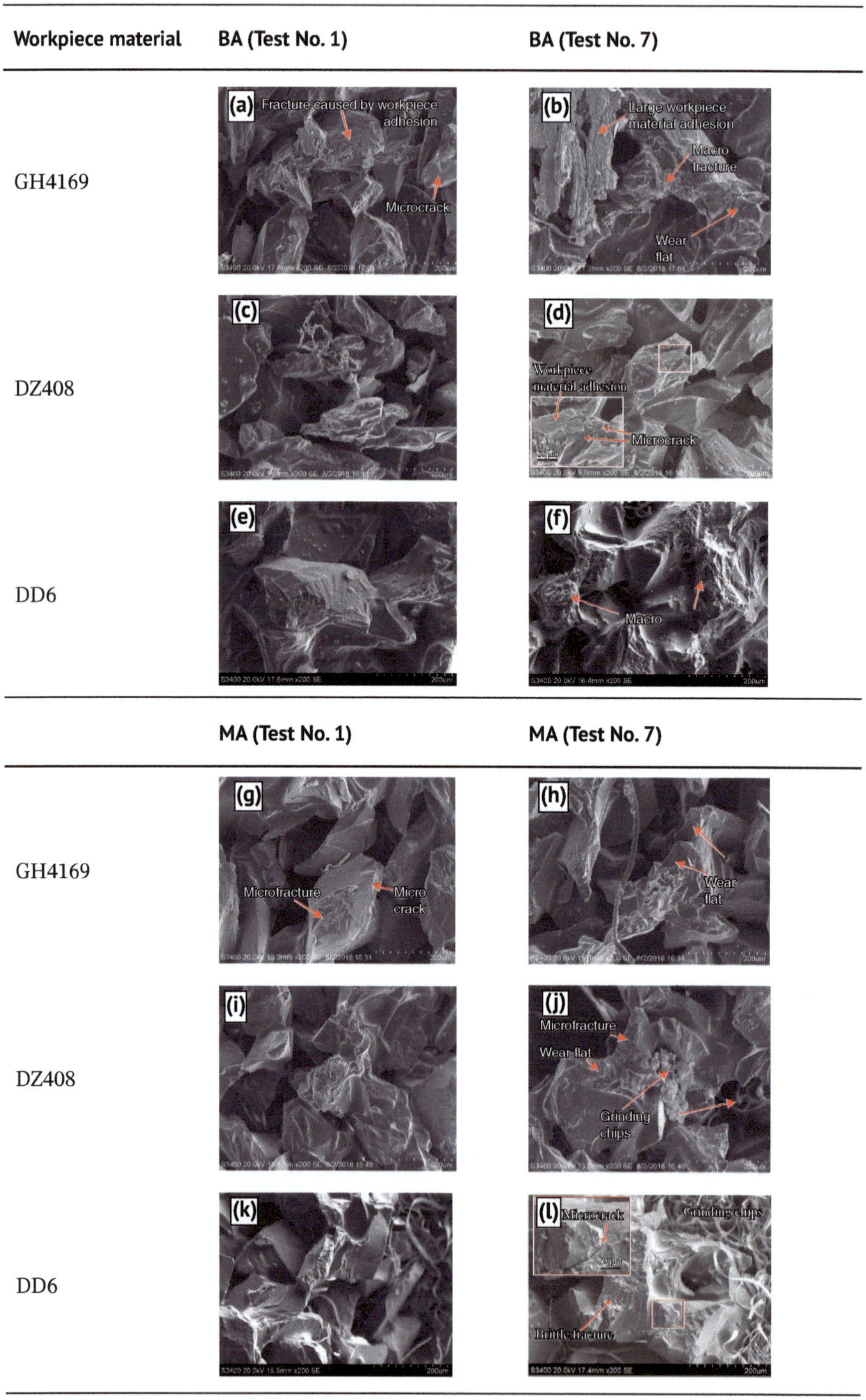

Figure 6.5 Scanning electron microscopy images of abrasive wheel wear topographies (a) BA-GH4169 (Test No. 1); (b) BA-GH4169 (Test No. 7); (c) BA-DZ408 (Test No. 1); (d) BA-DZ408 (Test No. 7); (e) BA-DD6 (Test No. 1); (f) BA-DD6 (Test No. 7); (g) MA-GH4169 (Test No. 1); (h) MA-GH4169 (Test No. 7); (i) MA-DZ408 (Test No. 1); (j) MA-DZ408 (Test No. 7); (k) MA-DD6 (Test No. 1); and (l) MA-DD6 (Test No. 7).

grinding parameters, these wheels demonstrated recurring and intensified occurrences of workpiece material adhesion, wear flat formation, and grain fracture. Particularly in Test No. 1 and No. 7 involving BA wheels grinding GH4169 alloy (Figure 6.5a–b), the observed cutting-edge fractures primarily resulted from substantial material adhesion. Previous studies [8] have confirmed that intense chemical bonding interactions between wheel and workpiece materials under high grinding loads can form strong welded junctions. When these interfacial bonds surpass the shear strength of the workpiece material, they may induce further adhesion, and if the bonding force becomes sufficiently large, it can lead to either abrasive grain fracture or bond bridge failure, ultimately causing particle dislodgement. This cyclic particle detachment at the wheel–workpiece interface can significantly compromise surface integrity. Comparable microfracture patterns were also identified in MA wheels during processing of GH4169 and DZ408 alloys, as evidenced in Figure 6.5g–j.

The wear characteristics during DD6 alloy grinding exhibited marked transitions with increasing material removal rates $\left(Q_{\mathrm{w}}' = 0.83 \rightarrow 6\ \mathrm{mm}^3/(\mathrm{mm} \cdot \mathrm{s})\right)$, as evidenced in Figure 6.5e–l. BA wheel surfaces progressed from minor micro-striations (Test No. 1, Figure 6.5e) to severe macrofractures with complete cutting edge deterioration (Test No. 7, Figure 6.5f). Conversely, MA wheels maintained predominantly micro-fracture patterns (Figure 6.5k and l), demonstrating superior performance attributed to their enhanced ductility and hardness properties. This comparative analysis reveals two distinct wear progression pathways: BA wheels undergo catastrophic failure through macroscopic grain fracture, while MA wheels exhibit controlled microfracture wear modes.

The SEM micrographs in Figure 6.5a, d, g, and l reveal extensive micro-cracking across all alumina wheel types, significantly accelerating grain fracture and consequently increasing radial wear rates (Figures 6.1 and 6.3). While these fractures reduce wheel longevity, they may beneficially maintain cutting efficiency through self-sharpening effects by continuously exposing fresh abrasive edges. This phenomenon suggests that during deep-cut creep-feed grinding, the predominant fracture mechanism stems from intense mechanical stresses at the grain-workpiece interface, despite existing research [9] proposing potential chemical interactions between alumina abrasives and nickel-based alloys.

6.1.3 Active Cutting Edge Density and Quantitative Analysis

Figure 6.6 presents the quantitative analysis of active cutting edge density (C) on wheel surfaces, determined through optical evaluation of grains exhibiting either workpiece adhesion or wear flats. Both BA and MA wheels maintained comparable active edge densities ranging from 6–10 mm^{-2}, with a modest reduction observed at higher material removal rates. This limited decline in cutting-edge density despite increasing grinding loads can be attributed to the unique self-sharpening behavior of alumina wheels described in Section 3.3, where continuous generation of fresh cutting edges compensates for edge attrition. The findings reveal an important characteristic of superalloy grinding: while conventional wear mechanics predict decreasing active edges under elevated loads, the dynamic nature of alumina wheel

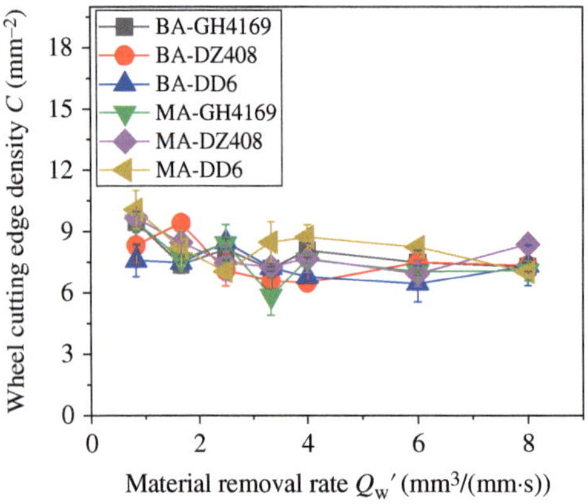

Figure 6.6 Effect of material removal rate on the wheel cutting edge density.

wear enables sustained cutting edge availability through progressive grain fracture and renewal mechanisms.

Grinding force serves as an effective indicator for predicting abrasive wheel wear patterns, with the normal force component being particularly significant due to its direct influence on grain-workpiece interactions and material removal mechanisms. This investigation establishes a comprehensive correlation between workpiece material properties, normal grinding forces, and operational parameters. Previous research [10, 11] has confirmed that grain dislodgement constitutes the primary wear mechanism when machining nickel-based superalloys, prompting detailed examination of wheel degradation behavior under creep-feed conditions.

According to the grinding force model proposed by Jahanmir et al. [12], the normal grinding force per abrasive grain F_{ngrain} in grinding can be expressed as:

$$F_{\text{ngrain}} = \beta \cdot \left(\frac{K_{\text{IC}}^{1/2} H^{9/10}}{E_{\text{w}}^{2/5}} \right) \left(\frac{v_{\text{w}}}{v_{\text{s}}} \right)^{3/4} a_{\text{p}}^{11/12} \cdot d_{\text{s}}^{1/12} \tag{6.5}$$

$$F_{\text{n}} = N_{\text{d}} F_{\text{ngrain}} = b l_{\text{c}} C F_{\text{ngrain}} \tag{6.6}$$

where β is the coefficient related to the wheel wear severity, E_{w} is the elastic modulus of the workpiece, N_{d} is the number of wheel active cutting edges, b is the width of the abrasive wheel, and the wheel–workpiece contact length is $l_{\text{c}} = (a_{\text{p}} \cdot d_{\text{s}})^{0.5}$.

Then, the coefficient related to the wheel wear severity β has a form as follows:

$$\beta = \frac{F_{\text{n}}}{bC\sqrt{a_{\text{p}} d_{\text{s}}}} \cdot \left[\left(\frac{K_{\text{IC}}{}^{1/2} H^{9/10}}{E_{\text{w}}{}^{2/5}} \right) \left(\frac{v_{\text{w}}}{v_{\text{s}}} \right)^{3/4} a_{\text{p}}{}^{11/12} \cdot d_{\text{s}}{}^{1/12} \right]^{-1} \tag{6.7}$$

The calculation of coefficient β through Eq. (6-7) incorporates experimental data from Tables 6.1 and 6.3 along with results from Figure 6.6, with the outcomes presented in Figure 6.7. Analysis reveals an inverse relationship between β and material removal rate, where higher Q_{w}' values correspond to lower β measurements. This behavior manifests in three distinct phases: an initial rapid decline during the wheel's first contact with the workpiece, followed by stabilization during steady-state operation, and finally a progressive reduction under high removal rates. While

Table 6.1 The experimental results of the normal grinding force.

	The normal grinding force F_n (N)					
Test No.	**BA-GH4169**	**BA-DZ408**	**BA-DD6**	**MA-GH4169**	**MA-DZ408**	**MA-DD6**
2	159.46	248.68	70.28	201.80	304.25	182.01
3	211.86	403.90	102.22	348.37	459.49	295.43
4	241.74	463.40	123.81	438.79	647.13	355.06
5	295.96	566.45	168.12	477.91	725.92	417.00
6	345.55	508.90	260.76	440.43	795.43	546.04
7	407.20	742.91	558.54	634.35	973.60	602.10
8	413.20	873.37	605.16	827.18	1024.00	696.79

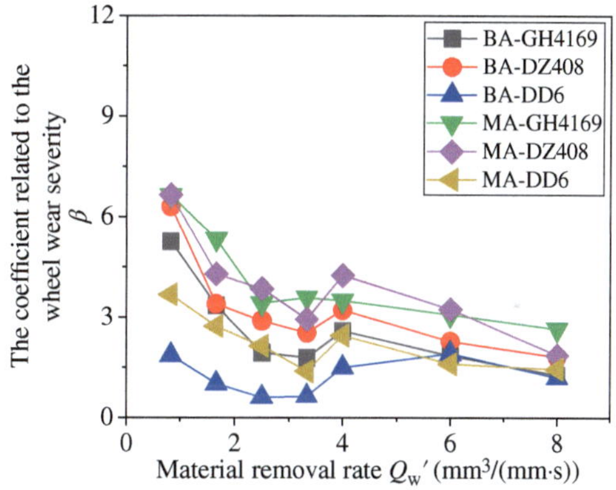

Figure 6.7 Alumina abrasive wheel wear behavior evaluation.

conventional expectations suggest β should rise with increased grinding loads, the observed decrease stems from fracture-induced improvements in wheel–workpiece contact conditions during creep-feed grinding. This phenomenon becomes particularly evident when Q_w' exceeds 4 $mm^3/(mm \cdot s)$, where enhanced interface conditions from grain fractures outweigh the effects of elevated mechanical loads.

The wheel wear model described in Eq. (6–7) currently does not account for radial wear resulting from grain and bond fractures, instead relying primarily on workpiece material properties and grinding parameters for calculating β. While higher grinding loads induce more frequent grain and bond fractures, the wheel's topography remains relatively stable due to continuous self-sharpening effects. However, this self-sharpening process simultaneously contributes to increased radial wear. A key challenge moving forward involves establishing stronger correlations between tool wear patterns, the material properties of both workpiece and abrasive grains, and grinding process parameters to develop more comprehensive wear prediction models.

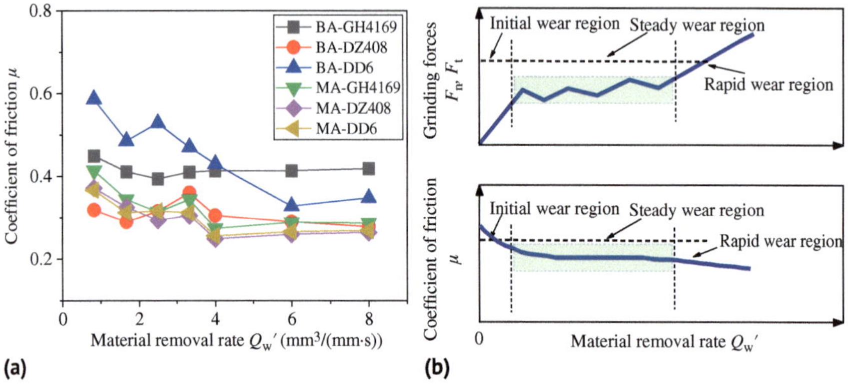

Figure 6.8 Effect of wear on grinding force ratio and illustrations of wheel wear states: (a) coefficient of friction and (b) schematic models.

6.1.4 Effect of Wheel Wear on Grinding Force Ratio

As illustrated in Figure 6.8, while alterations in wheel topography modify the contact conditions between abrasive wheels and workpiece materials, they exert minimal influence on cutting edge density. At material removal rates (Q_w') below 3.3 mm^3/(mm · s), both alumina wheels primarily exhibited wear mechanisms dominated by attritious wear, material adhesion, and wheel clogging. However, when Q_w' surpassed 4 mm^3/(mm · s), more severe fracture phenomena emerged, including both grain fracture and bond bridge fracture. This transition in wear mechanisms directly affected the friction characteristics, with the coefficient of friction ($\mu = F_t/F_n$) demonstrating three distinct phases: an initial rapid decline in the early wear stage, followed by relative stability during steady-state wear, and finally a gradual reduction in the accelerated wear phase, as depicted in Figure 6.8a. The schematic representation in Figure 6.8b further elucidates these characteristic variations in both friction coefficient and grinding forces throughout different wear stages.

It has been reported that an increase in the material removal rate leads to larger grinding loads, aggravated tool wear, and an extended force ratio. However, in the present study, the observed trend of the force ratio with respect to Q_w' was the opposite. This can be attributed to the self-sharpening effect of the alumina abrasive wheel due to grain pull-off and fracture during heavy-load grinding. The rapid wear of the wheel may be partially offset by the self-sharpening effect. This is consistent with the observations of Fathallah et al. [6] and Magham et al. [13] who investigated the wear behavior of alumina wheels under different grinding conditions.

6.2 Grinding Wheel Wear in Profile Grinding

6.2.1 Profiled Grinding Wheel Wear Variables

The experimental program comprised three distinct grinding groups (totaling nine individual tests) investigating parameter variations, including wheel speed (v_s), workpiece velocity (v_w), cutting depth (a_p), and accumulated grinding depth (h)—the latter representing vertical material removal as defined in Section 3.1.2. Critical experimental controls ensured consistent initial conditions: h was carefully maintained across groups to guarantee identical DD6 blade root profile starting geometry. All tests employed standardized wheel dressing protocols using a diamond roller dresser operating at 20 m/s with −0.7 speed ratio, 1.3 mm/min infeed rate, and 0.2 mm total dressing allowance. The complete parameter matrix is systematically presented in Table 6.2, providing full documentation of the controlled variables for each experimental condition.

The wear characteristics of the MA wheel profile were evaluated through a precise measurement methodology employing graphite replicas (Figure 6.9). A 5-mm-thick graphite impression was first created to capture the post-grinding wheel profile, which was subsequently scanned using a Videocheck IP400 3D optical

Table 6.2 Grinding parameters.

Grinding group No.	Grinding test No.	Wheel speed v_s (m/s)	Workpiece speed v_w (mm/min)	Grinding depth a_p(mm)	Accumulated grinding depth h(mm)	Wheel diameter (approximately) d_s(mm)
i	1	30	100	1.0	3.0	373.6
	2	30	150	1.0	3.0	369.2
	3	30	210	1.0	3.0	364.8
ii	4	25	100	1.0	4.0	372.6
	5	30	100	1.0	4.0	368.2
	6	35	100	1.0	4.0	363.8
iii	7	30	100	0.4	4.4	371.6
	8	30	100	1.0	5.0	367.2
	9	30	100	1.5	5.5	362.8

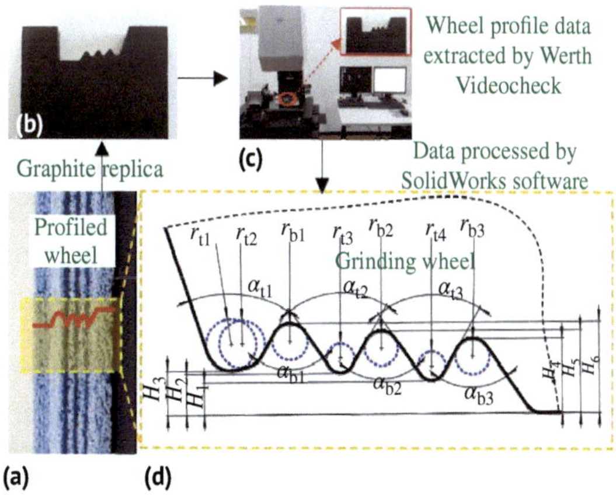

Figure 6.9 Measuring process of the profile wear of grinding wheel: (a) the wheel profile after grinding; (b) profile duplication; (c) profile scanning; and (d) profile measurement.

Table 6.3 Initial values of the shape and dimensions of the profiled grinding wheel after dressing.

	H_1	H_2	H_3	H_4	H_5	H_6
Heights (mm)	**0.88**	**1.11**	**1.18**	**2.04**	**2.26**	**2.50**
Convex radii (mm)	r_{t1}	r_{t2}	r_{t3}	r_{t4}	—	—
	0.675	0.600	0.350	0.350	—	—
Concave radii (mm)	r_{b1}	r_{b2}	r_{b3}	—	—	—
	0.475	0.475	0.475	—	—	—
Convex angles (°)	α_{t1}	α_{t2}	α_{t3}	—	—	—
	48.0	65.0	65.0	—	—	—
Concave angles (°)	α_{b1}	α_{b2}	α_{b3}	—	—	—
	65.0	65.0	65.0	—	—	—

measurement system. The acquired profile data underwent processing and analysis through SolidWorks software to determine dimensional changes. Initial baseline measurements of the freshly dressed wheel, including profile heights, radii, and angles, were recorded in Table 6.3 to establish reference values. Comparative analysis between pre-grinding and post-grinding measurements enabled accurate quantification of profile wear. Additional wheel topography examinations were conducted using a Hirox KH7700 optical microscope following completion of selected test runs to validate the replica-based measurements and provide supplementary wear pattern information.

6.2.2 Effect of Grinding Parameters on Wheel Wear Variable

Figure 6.10 presents the correlation between grinding parameters and wheel profile wear characteristics (height (Figure 6.10a), radius (Figure 6.10b), and angular deviations (Figure 6.10c)). Under constant conditions ($v_w = 100$ mm/min, $a_p = 1$ mm, $h = 4$ mm), increasing wheel speed from 25 to 35 m/s produced wear variations of 12–53 μm (height), 15–73 μm (radius), and 0.16–1.45° (angle), with optimal wear resistance observed at $v_s = 30$ m/s. When examining workpiece speed effects ($v_w = 100$–210 mm/min at $v_s = 30$ m/s, $a_p = 1$ mm, $h = 3$ mm), wear progressed from 11–36 μm (height), 13–89 μm (radius), and 0.11–1.91° (angle). Similarly, depth of cut variations ($a_p = 0.4$–1.5 mm at $v_s = 30$ m/s, $v_w = 100$ mm/min) demonstrated proportional wear increases ranging from 9–61 μm (height), 10–92 μm (radius), and 0.21–1.92° (angle). The data establishes clear monotonic relationships where both elevated workpiece speeds and cutting depths accelerate wheel profile degradation.

Analysis of the wheel wear patterns in Figure 6.10 reveals distinct spatial variations across different profile features. The height measurements demonstrate significantly greater wear in the H_1–H_3 regions compared to H_4–H_6, while convex radii

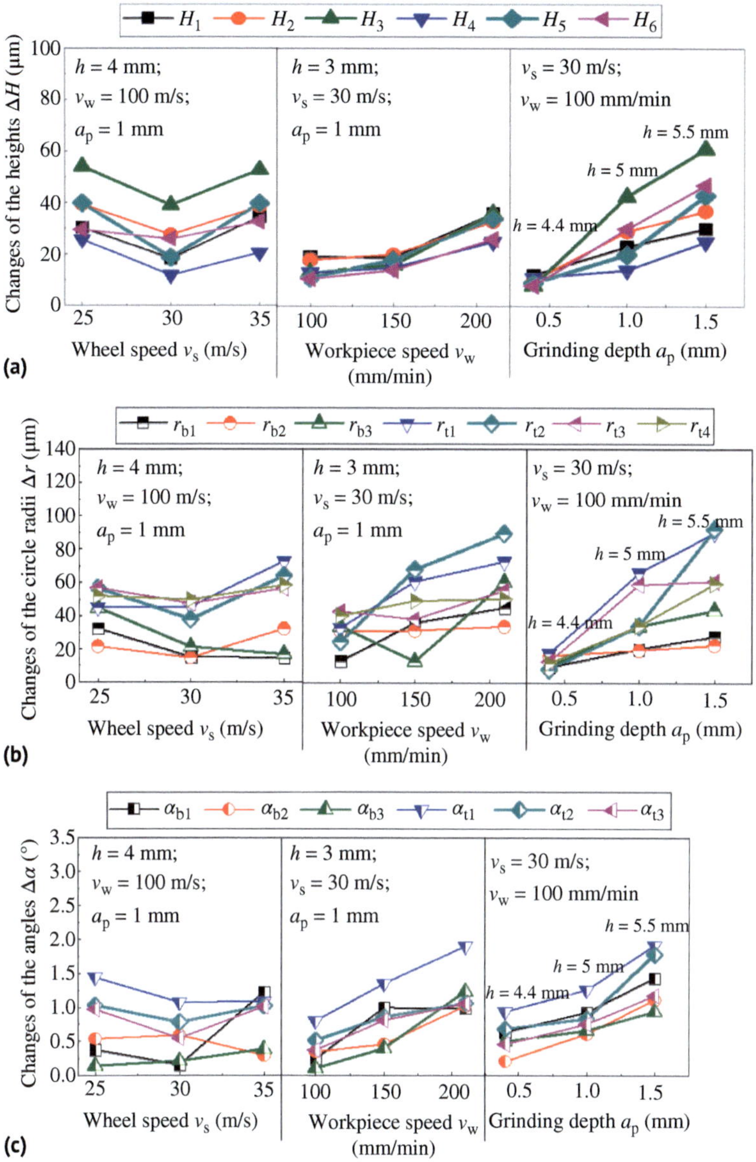

Figure 6.10 Effects of the grinding parameters on changes of the wheel profile wear in (a) heights, (b) circle radii, and (c) angles.

$(r_{t1}-r_{t4})$ exhibit faster degradation than concave radii $(r_{b1}-r_{b3})$. Angular measurements further show that smaller angles (α_{t1}) experience accelerated wear relative to larger angles $(\alpha_{t2}, \alpha_{t3})$. These observations confirm substantial nonuniform wear distribution along the wheel profile under identical grinding conditions. Notably, peak regions consistently demonstrate more severe wear than valleys, with average increases of 21.7% in profile heights, 42.8% in circle radii, and 34.5% in angular measurements across all tested parameters.

The study highlights the inherent complexity in wheel profile wear characterization during creep-feed grinding operations, where geometric parameters, including heights, radii, and angles, exhibit strong coupling effects. As evidenced in prior research [14], these dimensional parameters demonstrate significant interdependence—for instance, radius wear directly influences both height reduction and angular variations. The nonuniform thermal conditions across the contoured wheel surface, combined with varying mechanical loads at different profile locations, create a multifaceted wear mechanism that leads to distinct degradation patterns even for nominally identical geometric features $(r_{t3}/r_{t4}, r_{b2}/r_{b3})$ under equivalent process conditions. The experimental results presented in Figure 6.10 provide clear validation of this spatial wear variation phenomenon, confirming that different sections of the profiled wheel experience substantially different wear rates despite operating under the same nominal grinding parameters.

6.2.3 Grinding Ratio and Wheel Wear Topography

The grinding ratio is crucial for evaluating the wear resistance of a grinding wheel. Due to the difficulty in directly measuring the worn volume of the wheel, in this study, the average value of the heights of wheel profile wear, namely H_1, H_2, H_3, H_4, H_5, and H_6, was used to represent the wheel radial wear. Then, the grinding ratio G can be expressed as follows:

$$G = \frac{V_w}{V_s} \tag{6.8}$$

$$V_w = l_w \cdot l \cdot a_p \tag{6.9}$$

$$V_s = \pi d_s \cdot \Delta R \cdot l \tag{6.10}$$

$$\Delta R = \frac{1}{6}\left(\sum_{n=6}^{n=1} H_n\right) \tag{6.11}$$

$$G = \frac{l_w \cdot a_p}{\pi d_s \cdot \frac{1}{6}\left(\sum_{n=6}^{n=1} H_n\right)} \tag{6.12}$$

where V_w is the material removal volume of the workpiece, V_s is the worn volume of the grinding wheel, l_w is the workpiece length, l is the wheel–workpiece contact length in the normal section, and ΔR is the wheel radial wear.

The influence of process parameters on grinding performance is quantitatively presented in Figure 6.11, revealing significant variations in grinding ratio (G) across different operating conditions. Optimal machining efficiency ($G \approx 2$) was achieved at intermediate parameters ($v_s = 30$ m/s, $v_w = 100$ mm/min, $a_p = 1$ mm, $h = 3$ mm), while the most unfavorable combination ($v_s = 25$ m/s with otherwise identical parameters) yielded substantially reduced efficiency ($G \approx 0.75$), highlighting the considerable challenges in nickel-based superalloy machining and rapid wheel

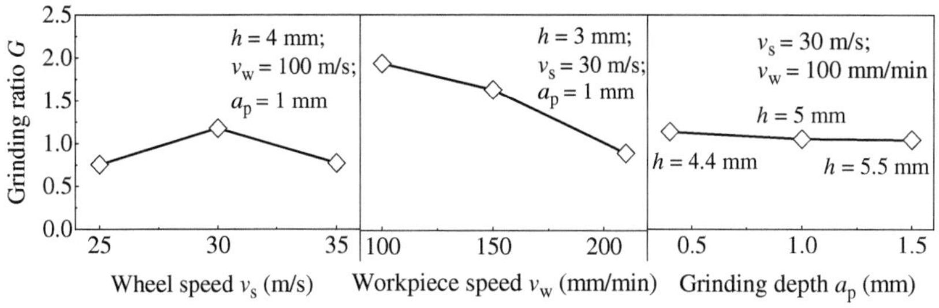

Figure 6.11 Effects of the grinding parameters on the grinding ratio.

degradation characteristic of creep-feed profile grinding. Parameter sensitivity analysis demonstrates that workpiece speed exerts the most pronounced effect on grinding ratio, whereas wheel speed and depth of cut show relatively moderate influences.

According to Eq. (6.12), in this study, the grinding ratio G is closely related to the grinding depth a_p and the wheel radial wear ΔR. If the grinding depth remains the same, the grinding ratio can be controlled by the wheel radial wear. Generally, increasing the wheel speed tends to result in lower grinding forces, better surface quality, and longer wheel life due to the decreasing undeformed chip thickness. Here, the undeformed chip thickness a_{gmax} is expressed as:

$$a_{gmax} = 2\lambda \cdot \frac{v_w}{v_s}\sqrt{\frac{a_p}{d_s}} \tag{6.13}$$

where λ is the distance between two adjacent active grains. However, fast wheel speed also leads to an increase in the sliding length for an individual grain in the wheel–workpiece contact zone. It may lead to more wheel wear. The sliding length is defined as the total distance travelled by an individual grain when in contact with the workpiece during grinding, and the calculation can be given by

$$l_r = \frac{6000}{\pi d_s} \cdot v_s \cdot \frac{l_w}{v_w} \cdot l_c \tag{6.14}$$

$$l_c = \sqrt{a_p \cdot d_s} \tag{6.15}$$

where l_r is the sliding length for an individual grain and l_c is the arc length of the wheel–workpiece contact zone. Moreover, when the workpiece speed is increased, the opposite trends can be observed. That is, increasing the workpiece speed accelerates the wheel wear due to the increasing undeformed chip thickness (Eq. (6.13)), while providing a shorter sliding length for an individual grain, reducing the wheel wear (Eq. (6.14)). Therefore, the final wear state of the wheel is determined by the competing dominance between the accelerating and reducing effects of the changes in both the wheel speed and the workpiece speed. Based on the analysis and the results in Figure 6.11, it is inferred that the effect of the wheel speed on the decreasing grinding ratio is mainly due to the increasing sliding length of the grain, while the effect of the workpiece speed is mainly due to the increasing undeformed chip thickness in the current creep-feed profile grinding process.

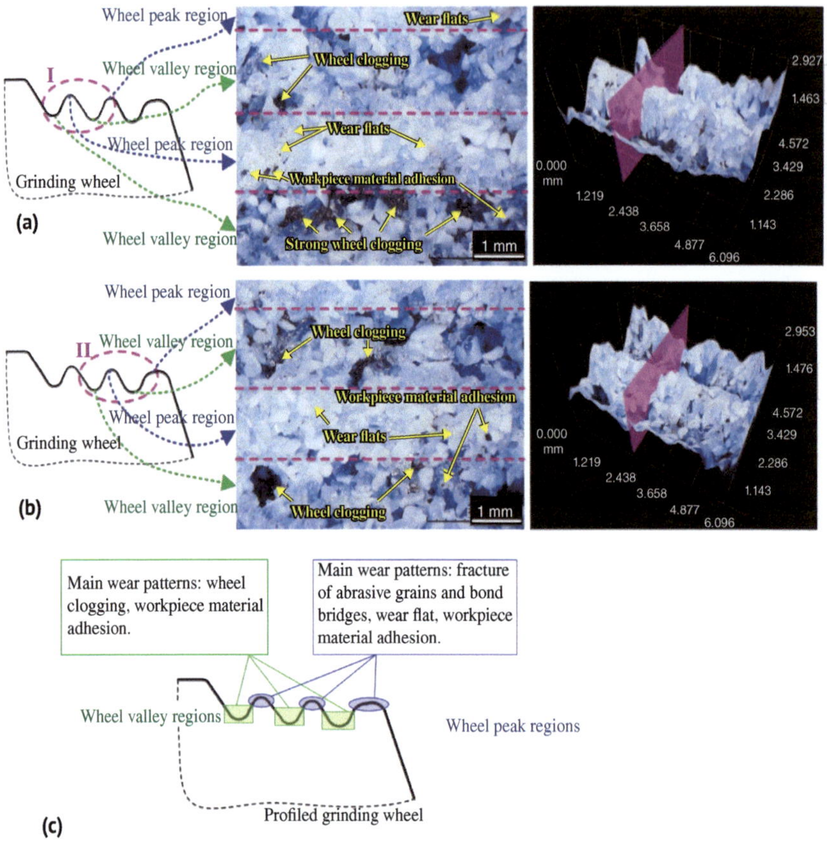

Figure 6.12 Typical wear topography on the profiled wheel surface at (a) position I and (b) position II, coupled with (c) the main wheel wear patterns.

The post-grinding wheel morphology analysis in Figure 6.12 reveals distinct wear characteristics across different wheel regions. Peak areas exhibit extensive wear flats and workpiece material adhesion (Figure 6.12a), while valley regions demonstrate severe clogging phenomena—observations consistent across multiple tests (Figure 6.12b). These findings align with prior research [15] identifying clogging as the predominant wear mechanism in nickel-based superalloy grinding, though the current study shows this can transition to grain/bond fractures and wear flats under elevated mechanical loads. The differential wear patterns between peak and valley regions indicate substantially higher stress concentrations in peak areas during profile grinding operations. When correlated with the significant profile wear dimensions reported in Figure 6.10, the observed morphology strongly suggests the occurrence of severe fracture mechanisms, including both grain fragmentation and bond bridge failure, particularly in high-contact regions of the wheel profile.

The wear mechanisms of profiled wheels exhibit distinct zonal characteristics during creep-feed grinding of single-crystal nickel-based superalloy blade roots, as systematically categorized in Figure 6.12c. Regional analysis reveals fundamentally different dominant wear behaviors: wheel peak regions primarily suffer from wear flat formation, material adhesion, and combined grain/bond bridge fractures, whereas valley regions are predominantly affected by severe clogging phenomena accompanied by workpiece material buildup.

6.2.4 Analytical Model for Predicting Wheel Wear

Rigney [16] reported that the wear volume of materials is proportional to the sliding distance and the normal load applied to the materials. Previous studies have shown that among the components of grinding force, the frictional force generated by the wear flat rubbing of abrasive grains on the machined surface usually plays a dominant role, especially when using alumina wheels to grind difficult-to-machine materials such as nickel-based superalloys [17]. Therefore, in this study, it is assumed that the wear of abrasive grains is completely caused by rubbing, and the abrasive grains do not break during contact with the workpiece but form a wear flat area. Then, the wear volume V_g of a single grain after sliding a distance l_r can be expressed as:

$$V_g \propto F_{nsg} l_r \tag{6.16}$$

$$F_{nsg} = \bar{p}\bar{\delta} \tag{6.17}$$

$$V_g = \bar{\delta}\Delta r_s = C_1 \bar{p}\bar{\delta} l_r \tag{6.18}$$

where F_{nsg} is the normal load that the individual grain endures during grinding, C_1 is a constant, Δr_s is wheel wear during the sliding distance l_r, and $\bar{p}$ is the constant average contact strength between the workpiece and the grain wear flat. Thus, the wheel wear can be expressed as:

$$\Delta r_s = C_1 \bar{p} l_r \tag{6.19}$$

During the grinding process, due to the nonconformity between the grinding wheel and the tribological trajectory of an individual abrasive grain, it is found that the average contact strength is significantly affected by the workpiece speed rather than the grinding depth. In this scenario, interference may occur between the workpiece surface and the grain wear flat. To quantify such interference, Hahn [18] proposed the curvature difference Λ between the grinding wheel and the trochoidal path, which can be expressed as:

$$\Lambda = \frac{1 + 2\left(\frac{v_s}{v_w}\right)}{r_s\left(1 + \frac{v_s}{v_w}\right)^2} \tag{6.20}$$

where r_s is the wheel radius. According to Eq. (6.20), the curvature difference is likely to be highly dependent on the workpiece speed even though the large grinding depth is usually used in the creep-feed grinding process. Furthermore, Malkin [19] even demonstrated experimentally that the average contact strength $\bar{p}$ depended on the curvature difference and could be linearly related to the curvature difference in

Table 6.4 The calculated coefficients in Eq. (6.22).

Wheel profile dimensions	C_4 (mm)	$C_5 \times 10^7$
H_1	6.32	1.05
H_2	9.74	−12.5
H_3	7.79	−9.99
H_4	3.49	7.85
H_5	8.76	−11.2
r_{t1}	37.7	−92.9
r_{t2}	25.4	−54.5
r_{t3}	60.9	−168
r_{t4}	43.1	−99.1
r_{b1}	2.41	9.41
r_{b2}	14.5	−16.1
r_{b3}	26.8	−64.3

general if the friction coefficient between the wear flat and the workpiece surface remained constant. Thus, the following relationship can be established:

$$\bar{p} = C_2\Lambda + C_3 \tag{6.21}$$

where C_2 and C_3 are constants.

Combining Eqs. (6.19)–(6.21), the final equation for the wheel wear has the following form:

$$\Delta r_s = \left[C_4 \cdot \frac{1 + 2\left(\frac{v_s}{v_w}\right)}{r_s\left(1 + \frac{v_s}{v_w}\right)^2} + C_5\right] \cdot l_r \tag{6.22}$$

where $C_4 = C_1C_2$, and $C_5 = C_1C_3$. This equation can be used to assess numerically a particular grinding wheel wear. Substituting the results of Figure 6.10a and b to Eq. (6.22), Δr_s can be calculated quantitatively.

Table 6.4 presents the calculated coefficients (C_4 and C_5) in Eq. (6.22) for wheel profile wear parameters including heights, convex radii, and concave radii. Figure 6.13 compares modeled and experimental results of wheel profile wear, showing general consistency at low wear levels (less than 30 μm for profile heights in Figure 6.13a and less than 50 μm for profile radii in Figure 6.13b), but significant discrepancies at higher wear where modeled values are notably smaller than experimental observations, reflecting weakened trend matching; this is attributed to creep-feed grinding with high material removal rates inducing increased mechanical loads that shift the main wear mechanism from attritious wear (wear flat) to grain and bond bridge fractures, which are unaccounted for in Eq. (6.22), though Figure 6.13c shows an average error of 17.3% when only attritious wear is considered, indicating the model's partial acceptability.

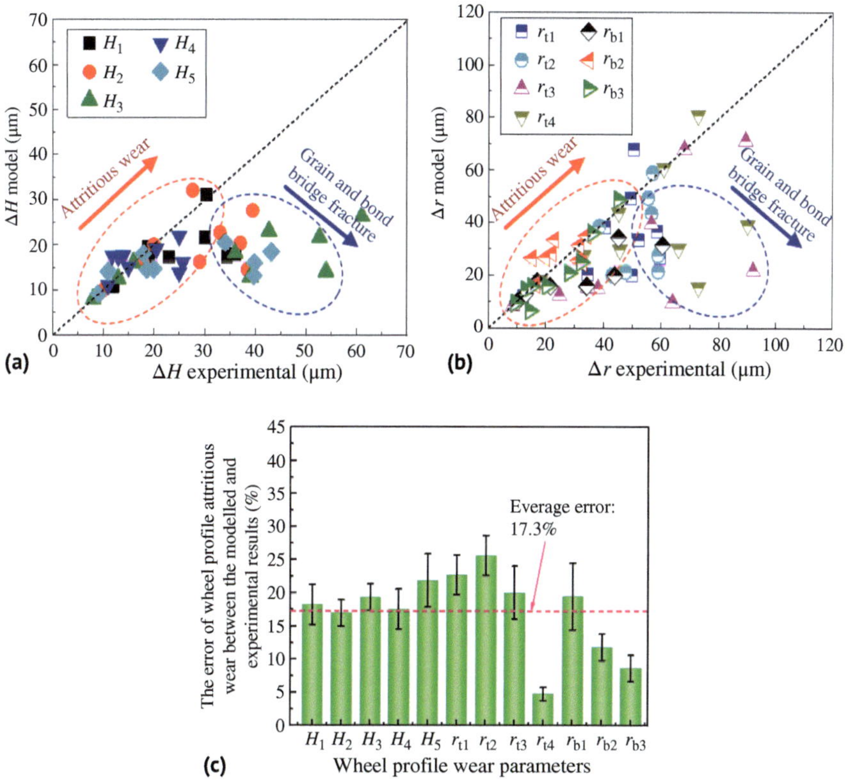

Figure 6.13 Comparison of the wheel profile wear obtained from the model with the experimental results in (a) the heights ΔH, (b) the circle radii Δr, and (c) the calculated error.

Additionally, as depicted in Figure 6.12, the uneven profile wear across the wheel surface means that the shape and dimensional accuracy of the blade root workpiece are typically governed by the wheel's most prone-to-wear regions. Using the current model to predict wheel wear can roughly indicate the appropriate timing for wheel dressing before these vulnerable areas reach critical designed values under varying grinding conditions. Furthermore, Eq. (6.22) incorporates key grinding parameters in creep feed profile grinding of nickel-based superalloys—wheel speed, workpiece speed, and sliding length—enabling the selection of optimal parameters prioritizing either grinding efficiency or surface quality.

6.3 Grinding Wheel Wear in Micro-grinding

6.3.1 Effect of Grinding Parameters on Wheel Wear

Figure 6.14 shows the diameter change of the grinding tool under different material removal volumes and spindle speeds during the micro-grinding of nickel-based single-crystal superalloys. In the experiment, the diameter of the grinding tool with different material removal volumes was measured using a VHX-1000E microscope, as depicted in Figure 6.15. This figure also shows the abrasive wear on the grinding tool surface for different material removal volumes.

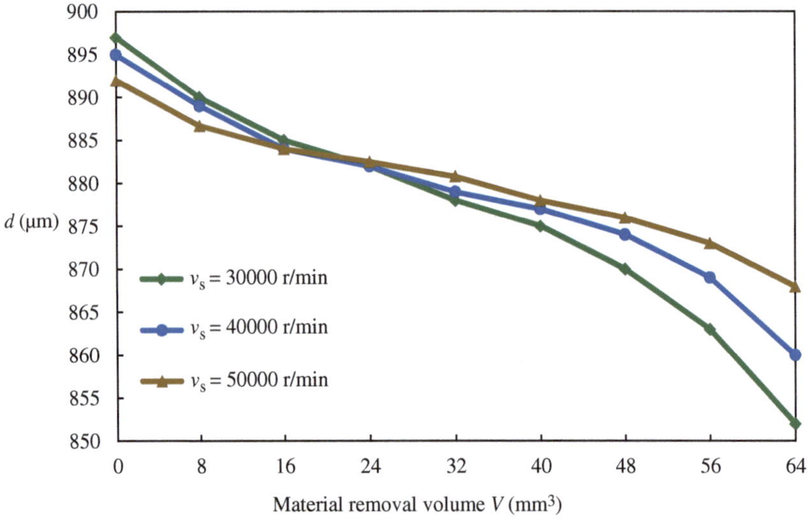

Figure 6.14 Influence of spindle speed on the radial wear of micro-grinding tool.

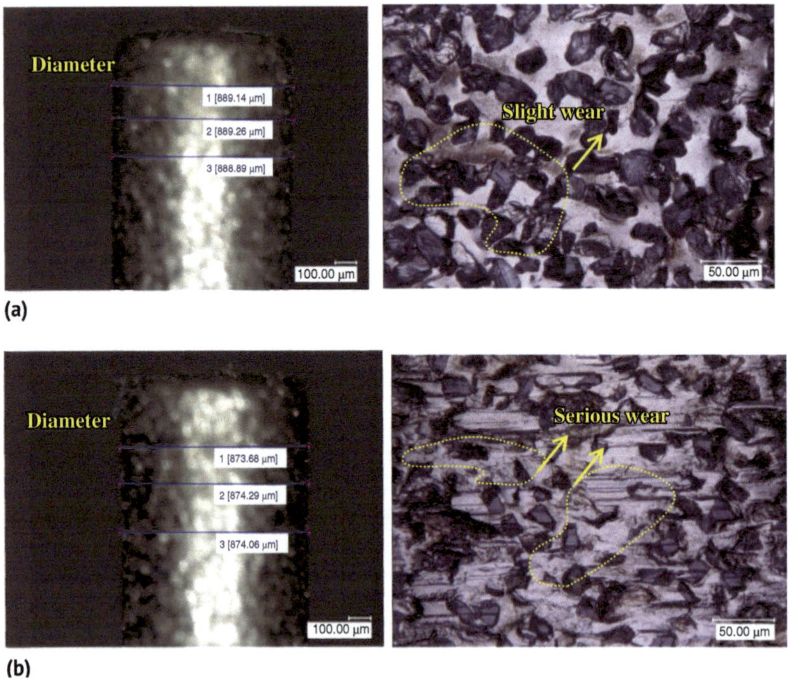

Figure 6.15 Changes of the wear pattern of the grinding tool with the material removal volume: (a) $V = 8\,\text{mm}^3$ and (b) $V = 48\,\text{mm}^3$.

Figure 6.15 demonstrates a three-stage evolution pattern in grinding tool diameter (d, μm) relative to material removal volume. The initial phase shows rapid diameter reduction, followed by a period of gradual decline, and concluding with another accelerated decrease. This characteristic wear progression can be explained by several mechanisms: during the initial wear stage, the nonuniform height distribution of protruding abrasives causes those with greater protrusion and weaker bonding to experience premature detachment and fracture under grinding forces, leading to swift diameter reduction. The subsequent steady wear phase features predominantly frictional wear mechanisms between abrasives and workpiece, manifesting as gradual abrasive attrition and micro-crack formation that produce more stable diameter changes. In the final accelerated wear stage, cumulative abrasive wear significantly increases grinding forces and temperatures, triggering rapid abrasive deterioration. This critical phase is marked by accelerated diameter reduction, intensified wear rates, massive abrasive particle loss, and eventual tool failure, completing the tool's service lifecycle.

The experimental results in Figure 6.14 reveal a clear relationship between spindle speed and micro-grinding tool wear characteristics. As the spindle speed increases, the radial wear of the micro-grinding tool shows a decreasing trend. Specifically, at a spindle speed of 40, 000 r/min, the tool enters the accelerated wear phase after removing 35 mm^3 of material. Comparatively, at lower speeds (30, 000 r/min), this rapid wear stage occurs earlier, while higher speeds (50, 000 r/min) delay the onset of accelerated wear. This phenomenon can be attributed to several key factors: higher spindle speeds increase the number of active abrasives per unit time while simultaneously reducing the cutting thickness and grinding force experienced by individual abrasive grains. Consequently, this leads to decreased wear rates of both the abrasive particles and the electroplated bonding layer [20].

Figure 6.16 shows the grinding process of three grinding tools at different feeding rates. The diameter $d(\mu m)$ of the grinding tool changes with the volume of material removed. With the increase in material removal volume, the diameter of the grinding tools at different feeding rates decreases rapidly first, then steadily, and finally rapidly.

As can be seen from Figure 6.16, as the feeding rate increases, the variation in the diameter of the micro-grinding tool increases. When $v_w = 170\,\mu m/s$, the diameter of the micro-grinding tool decreased rapidly, and the tool soon entered the rapid wear period. This is mainly because as the feeding rate increases, the material removal

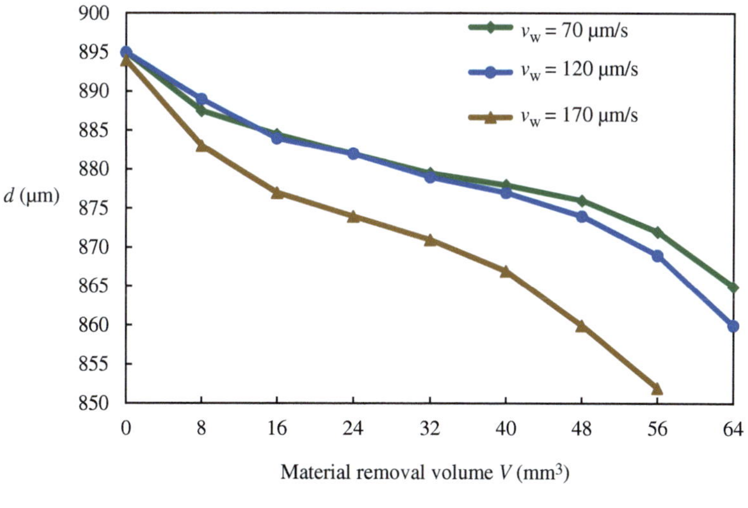

Figure 6.16 Influence of feeding rate on the radial wear of micro-grinding tool.

volume per unit time in micro-grinding and the cutting thickness of a single abrasive increase. This leads to an increase in the grinding force and grinding energy on the abrasives. The large grinding force and high machining temperature accelerate the wear of the grinding tool.

The micro-grinding process of single-crystal superalloys presents unique challenges due to their viscoplastic properties. At elevated feed rates, the material's plastic deformation characteristics hinder effective chip evacuation, causing workpiece material to accumulate around the abrasive grains and electroplated bonding layer. This material accumulation leads to progressive tool clogging, which significantly impacts the grinding process by elevating both grinding forces and temperatures—factors that accelerate tool degradation. Furthermore, high feed rates induce process vibrations that generate intermittent spikes in grinding forces, exacerbating tool wear. These combined effects result in accelerated deterioration of both the abrasive particles and the supporting electroplated matrix, ultimately compromising tool performance and lifespan.

Figure 6.17 shows the change of the grinding tool's diameter with material removal volume under different a_p during micro-grinding of superalloy. With the increase in material removal volume, the diameter of the grinding tool at different grinding depths first decreases rapidly, then decreases steadily, and finally decreases rapidly.

As shown in Figure 6.17, with the increase of grinding depth, the variation in the diameter of the micro-grinding tool becomes more significant, the tool wear rate accelerates, and the tool more readily enters the rapid wear phase. This is primarily due to the fact that a larger grinding depth increases the cutting thickness of individual abrasives and extends the contact arc length between the grinding tool and the workpiece, making it difficult for coolant to penetrate the grinding zone while increasing tangential and normal forces during grinding. These factors enhance the deformation resistance of chips and the friction between abrasives and the workpiece, generating more heat. Additionally, as a high-temperature-resistant material with poor thermal conductivity, the superalloy DD98 causes the heat generated during grinding to accumulate in the grinding area, potentially accelerating the wear of the micro-grinding tool.

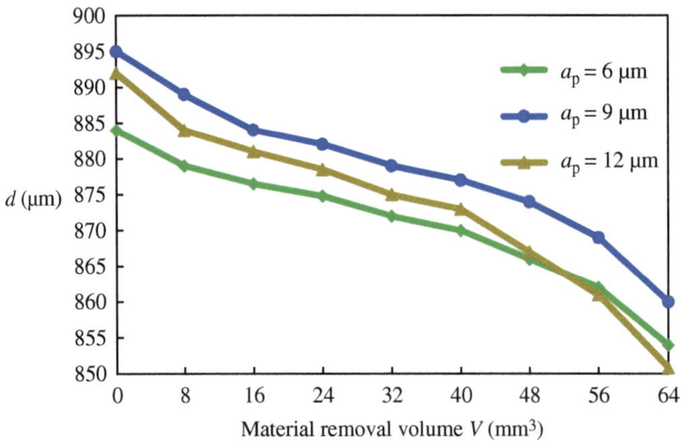

Figure 6.17 Influence of grinding depth on the radial wear of micro-grinding tool.

6.3.2 Effect of Wheel Wear on Grinding Force and Temperature

Figure 6.18 shows the relationship between the wear of the micro-grinding rod and the grinding forces. It can be observed from Figure 6.18 that as the removal volume of nickel-based single-crystal superalloy material in micro-grinding increases, the wear condition of the grinding rod surface intensifies, and both the normal grinding force and the tangential grinding force increase. The normal grinding force increases significantly, with a slow increase in the initial wear stage and a faster increase later.

The reasons are as follows: in the initial wear stage of the grinding rod, the surface grinding particles are sharp with good cutting performance, allowing a large number of particles to easily form chips during material cutting while fewer particles are involved in friction and ploughing; moreover, the small contact area between the micro-grinding rod and the single-crystal alloy results in relatively low grinding force on the rod. As the removal volume of the single-crystal alloy increases, some sharp abrasive particles undergo microfracture or fall off, leading to increased friction and contact area between the grinding rod and the single-crystal alloy, which in turn causes an increase in grinding force.

As can be seen from the grinding force signal in Figure 6.18, when the grinding rod is in the early wear stage, the amplitude of the change in the normal grinding force and tangential grinding force signals is small, and the vibration is minor. When the grinding rod is severely worn, the vibration of the normal and tangential grinding force signals is large. The grinding force signals can reflect the wear condition of the grinding rod to some extent.

Figure 6.19 shows the change in grinding temperature with the wear of the micro-grinding rod. The temperature values in the figure were measured using a thermocouple. As can be seen from Figure 6.19, as the wear of the micro-grinding rod intensifies, the contact area between the abrasive particle and the workpiece to be ground increases; the friction and extrusion between the abrasive particle and the workpiece are enhanced, and the heat generated during the grinding process increases.

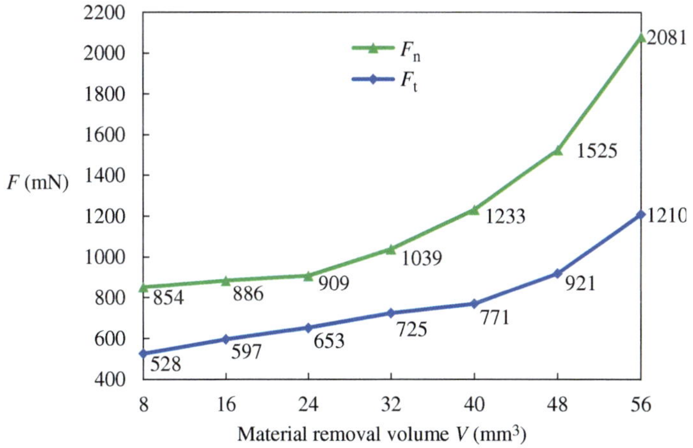

Figure 6.18 Micro-grinding force variation with material removal volume.

When the material removal volume is below 40 mm^3, the grinding temperature changes slowly. When the material removal volume increases to more than 40 mm^3, the grinding temperature shows a rapid rising trend. The reasons are as follows: in the initial wear stage of the grinding rod, the cutting edge of the grinding particle on the surface is sharp, and its cutting performance is good. The grinding heat generated is small, and the grinding temperature increases slowly. When the wear of the micro-grinding rod is severe, the abrasive particles on the surface are micro-broken and form wear platforms. The abrasive particles on the surface of the rod slip on the workpiece surface, and the number of ploughing actions increases. The friction between the grinding rod and the workpiece is enhanced, generating more heat, and the grinding temperature increases rapidly.

6.3.3 Effect of Wheel Wear on Surface Integrity

Figure 6.20 shows the change of surface roughness value of nickel-based single-crystal superalloy micro-scale grinding with material removal volume (V). As can be seen from Figure 6.20, except for a small decrease in surface roughness

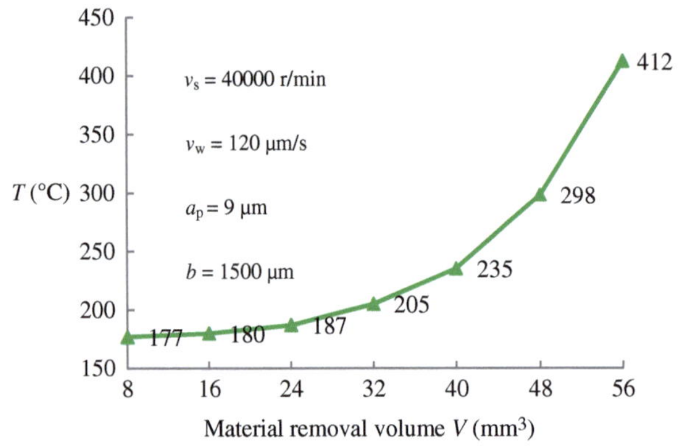

Figure 6.19 Grinding temperature variation with removal volume.

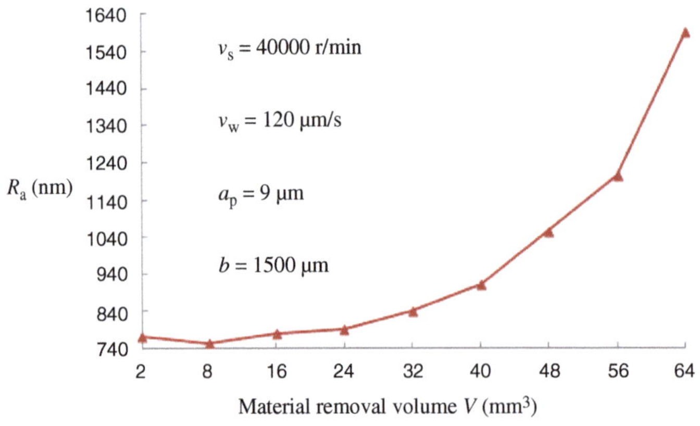

Figure 6.20 Surface roughness variation with removal volume (adapted from [9]).

at the beginning with an increase in material removal volume, surface roughness in micro-grinding shows an upward trend with increasing material removal volume. The reasons are as follows: in the production process of the new grinding rod, some abrasive particles may have a high protruding height, and the number of abrasive particles involved in the initial grinding stage is small, which will leave relatively deep scratches on the grinding surface. With the increase of the material removal volume, some abrasive particles with a high protruding height involved in the initial grinding stage will fall off due to the weak strength of the electroplating layer bonded with the grinding rod. The number of dynamically effective abrasive particles involved in grinding on the micro-grinding rod increases, and the surface roughness decreases first. However, as the material removal volume continues to increase, the friction between the abrasive particles and the workpiece surface is enhanced; the wear degree of the micro-grinding rod is increased; the wear and falling off of the abrasive particles appear; the micro-grinding rod becomes dull; the grinding force and heat increase; the grinding surface quality deteriorates; and the roughness value increases.

Figure 6.21 shows the surface morphology of the micro-grinding rod and the corresponding grinding surface morphology under different material removal volumes (V). With the aggravation of rod wear, the roughness value of the grinding surface increases, and the increasing trend continues to intensify. When the wear of the grinding rod is severe, the contact area between the grinding rod and the workpiece increases, and the material becomes difficult to remove. The strong extrusion between the workpiece and the grinding wheel particle and the high grinding temperature causes serious creasing on the surface of the workpiece; plastic sliding and

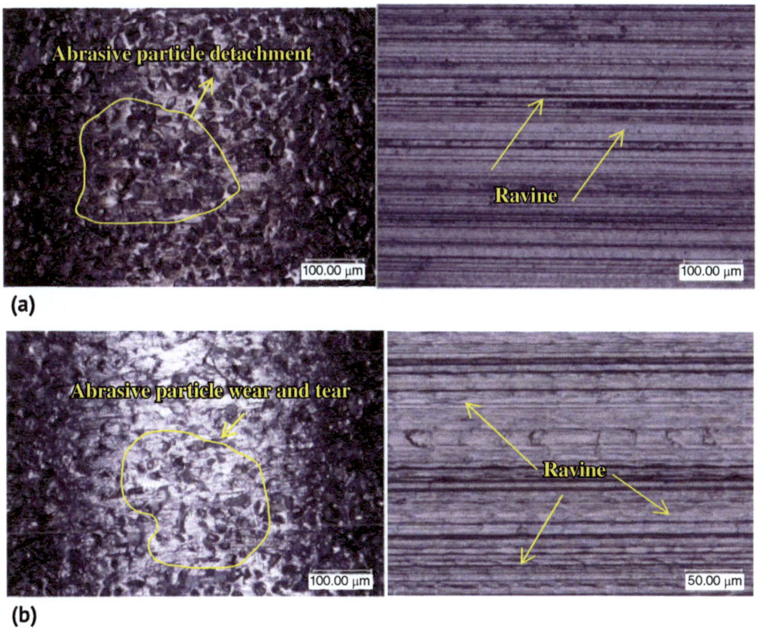

Figure 6.21 Topography of micro-grinding tool and corresponding grinding surface: (a) $V = 25\,\text{mm}^3$ and (b) $V = 45\,\text{mm}^3$.

vibration marks appear on the grinding surface materials, and the grinding texture is blurred, as shown in Figure 6.21a and b. At the same time, with the wear of the abrasive particles and the coating, the abrasive particles become dull, the shape of the abrasive chips becomes large, and a wide gully appears on the grinding surface. When the wear of the grinding rod is small, the grinding particle is sharp, the grinding surface quality is better, the grinding chip shape is small, and the grinding surface gully is narrow. Figure 6.21a and b shows the three-dimensional (3D) morphology of the grinding surface with $V = 25\,\text{mm}^3$ and $V = 45\,\text{mm}^3$, and the surface roughness is 881 nm and 1680 nm ($v_s = 30\,\text{kr/min}$, $a_p = 10\,\mu\text{m}$, $v_w = 100\,\mu\text{m/s}$), respectively. The gullies in 3D morphology and cross section of the surface in Figure 6.21b are deeper and wider than those in Figure 6.21a. The height distribution range of surface topography in Figure 6.21b is wider than that in Figure 6.21a, whereas the difference of surface topography height in Figure 6.21a is smaller; the height distribution is relatively concentrated, and the grinding quality of the ground surface is better. This is mainly because: as the wear of the grinding rod increases, the abrasive particles become dull, the friction and extrusion between the abrasive particles and the workpiece surface increase, the grinding performance of the abrasive particles decreases, the abrasive rod particles and the electroplating layer wear and fall off, the cutting thickness of a single abrasive particle increases, and grinding surface quality deteriorates. The reasons for the influence of grinding rod wear on the grinding surface topography will be analyzed as follows: in the micro-scale grinding process of nickel-based single-crystal superalloy, the main wear forms of micro-grinding rod include electroplating wear, abrasive wear, abrasive fall off, and electroplating fall off, as shown in Figure 6.22a and b. In the early grinding stage, although some abrasive particles fall off, the end wear of electroplating and abrasive particles is the main wear form of micro-grinding rod, which can be seen in almost any micro-grinding rod topography. As the grinding continues, the CBN grinding rod and the workpiece produce intense friction. Under the combined action of friction and friction heat, the abrasive tip and the coating begins to wear. In severe cases, it will lead to detachment of some CBN grinding particles and the coating, affecting the material removal during the grinding process. In the later wear stage, due to the severe wear of the end face, the grinding force and grinding heat suffered by the grinding particles during the grinding process are intensified, and the grinding surface materials are difficult to remove. Vibration marks appear on the surface of the skipping tool, and abrasive particles on the surface of the grinding rod appear to fall off across a large area.

The experimental results depicted in Figures 6.23 and 6.24 demonstrate a significant relationship between grinding tool wear and subsurface damage evolution during micro-grinding of nickel-based single-crystal superalloys. As material removal volume accumulates, the grinding rod undergoes progressive wear characterized by abrasive particle dulling and partial detachment, leading to heightened grinding forces and thermal effects. This tool degradation directly correlates with an accelerated expansion of the subsurface plastic deformation zone—the more severe the tool wear, the more rapidly the damage layer grows. This phenomenon occurs because worn abrasives require greater cutting forces, while generating increased

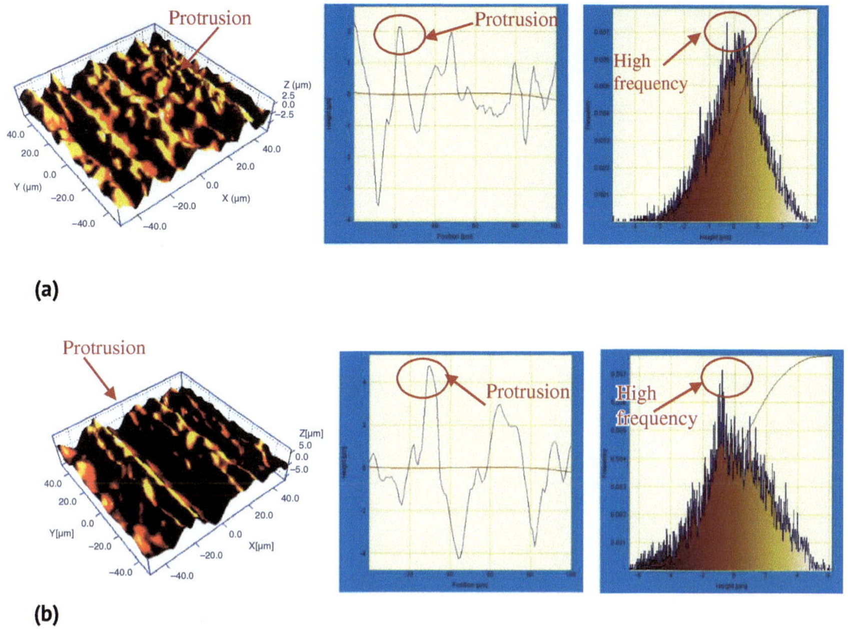

Figure 6.22 3D grinding surface profile and section profile under different wear of grinding wheel: (a) $V = 25\,\text{mm}^3$ and (b) $V = 45\,\text{mm}^3$.

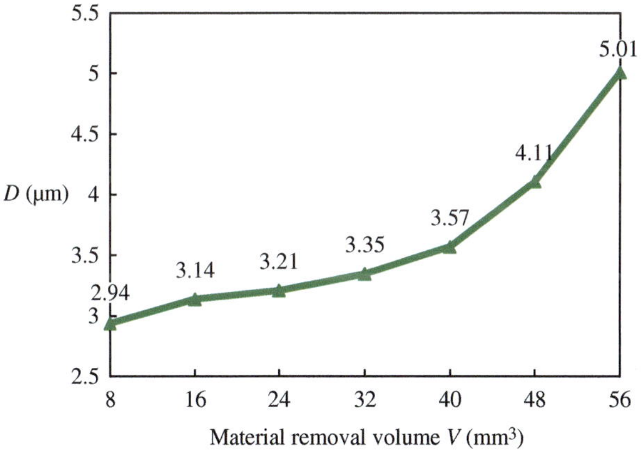

Figure 6.23 Thickness of subsurface deformation layer variation with removal.

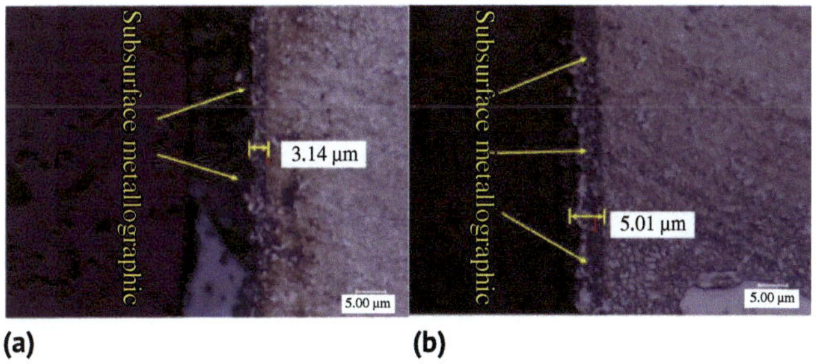

Figure 6.24 Subsurface metallographic morphology with different wear conditions of micro-grinding tool: (a) $V = 16\,\text{mm}^3$ and (b) $V = 56\,\text{mm}^3$.

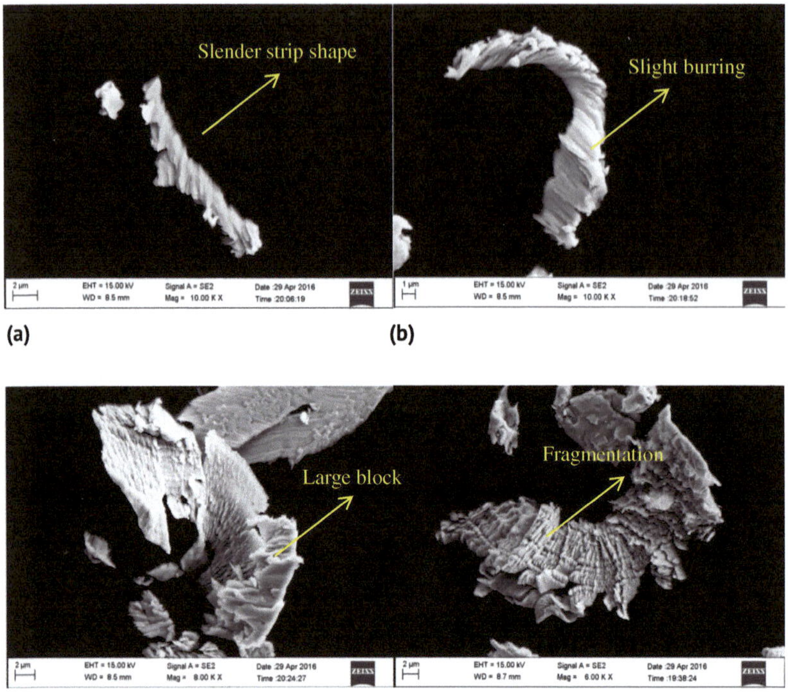

(a) (b) (c) (d)

Figure 6.25 Chip morphology with different wear conditions of micro-grinding tool: (a) $V = 8\,\text{mm}^3$; (b) $V = 16\,\text{mm}^3$; (c) $V = 40\,\text{mm}^3$; and (d) $V = 48\,\text{mm}^3$.

frictional heat, both of which contribute to deeper material deformation beneath the machined surface. Of particular concern is that such subsurface damage may trigger recrystallization during high-temperature operation, potentially compromising the component's performance. These findings highlight the critical need for timely tool replacement to maintain optimal machining conditions and preserve the structural integrity of single-crystal superalloy components.

Figure 6.25a–d shows the chip morphology under different wear conditions during micro-scale grinding of nickel-based single-crystal superalloy. From the point of view of the morphology of grinding chips, the shape characteristics of their grinding particles are also significantly different. As can be seen from Figure 6.25, with the intensification of wear of the grinding rod, the width and thickness of the grinding chips become wider and thicker. When the wear of the grinding rod is light, the grinding chip is shaped into a slender strip with a slight rough edge on the side of the grinding chip, as shown in Figure 6.25a–c. The grinding rod CBN abrasive particles appear dull and partially shed. With the aggravation of the wear of the micro-grinding rod, part of the abrasive particles on the surface of the grinding rod appear to fall off and wear, which increases the contact width between the abrasive particles and the workpiece surface. The phenomenon of the abrasive particles sliding and ploughing on the grinding surface becomes more severe, resulting in a wider width of the abrasive chips and a gradually thicker thickness. At this time, the abrasive chips are removed into large chunks on the grinding surface of the material. In addition, due to the dull abrasive particles during the grinding process, the workpiece material undergoes large deformation, the abrasive chips are stressed unevenly, and large cracks and fragmentation appear at the edges of the abrasive chips.

6.3.4 Wheel Wear Monitoring Based on Acoustic Emission

The effects of spindle speed, workpiece feed speed, grinding depth, and the wear state of the micro-grinding rod on the acoustic emission signal during micro-grinding were investigated. In traditional metal material grinding processes, the common approach is to observe the sparks or vibration signals generated when the grinding wheel contacts the workpiece [21–23]. However, in the micro-grinding process, due to the small size of the grinding tool and workpiece, no sparks are produced. Moreover, the vibration signal has a poor signal-to-noise ratio and low credibility. This poses significant challenges for monitoring the tool during micro-part processing. The tool process is difficult to observe with the naked eye, and it is extremely time-consuming and labor-intensive, seriously affecting the processing efficiency [24–26].

The acoustic emission sensor features high sensitivity and strong anti-interference ability. When the grinding tool contacts the workpiece, the friction, ploughing, and cutting between the surface particles of the grinding rod and the single-crystal superalloy material cause the grinding zone to emit distinct acoustic emission signals. These signals are detected by the acoustic emission sensor, enabling the determination of the contact between the grinding rod and the workpiece. Figure 6.26 shows the acoustic emission signal of the grinding rod during the micro-grinding process, from idling to cutting into the single-crystal alloy material and then to the stable grinding stage.

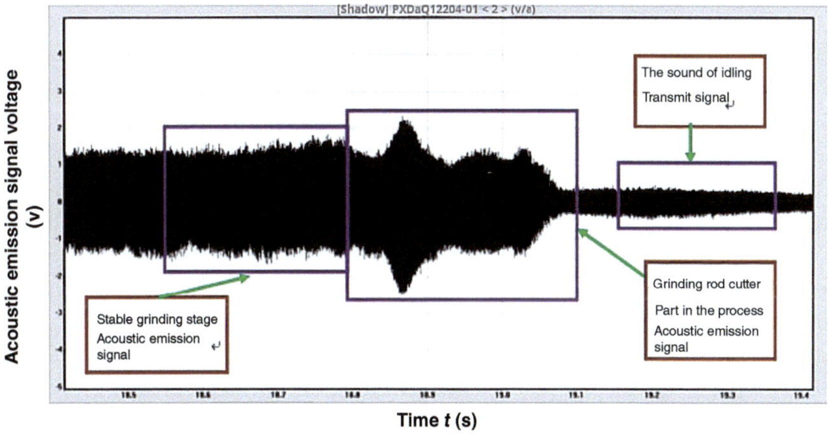

Figure 6.26 Acoustic emission signal of the grinding tool cutting into the workpiece.

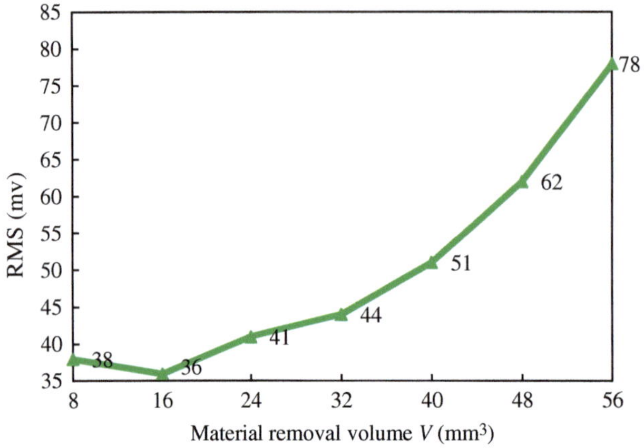

Figure 6.27 Influence of the grinding tool's wear on the root mean square (RMS) of acoustic emission.

The acoustic emission signals of the grinding rod under different wear conditions were collected. Figure 6.27 shows the change in the acoustic emission signal intensity with the wear of the grinding rod during the grinding of nickel-based single-crystal superalloys. As can be seen from the figure, the root-mean-square (RMS) value of the acoustic emission signals is large in the initial grinding stage of the micro-grinding rod. As the grinding process progresses, the RMS value of the acoustic emission signals decreases slightly and tends to flatten. After a while, as the wear of the grinding rod increases, the RMS value of the acoustic emission signals begins to rise. When the material removal volume is less than $32\,mm^3$, the acoustic emission signal intensity does not change significantly. When the material removal volume exceeds $32\,mm^3$, the acoustic emission signal intensity increases with the increase in material removal volume.

This phenomenon can be attributed to several interrelated factors: as the grinding rod's abrasive particles undergo progressive wear and clogging, their cutting efficiency diminishes while sliding and ploughing effects become more dominant during workpiece interaction. This transition in material removal mechanism causes substantial increases in both normal and tangential grinding forces, while simultaneously elevating process temperatures through enhanced frictional heating. The consequent intensification of elastic–plastic contact between tool and workpiece generates more energetic stress waves, manifesting as stronger acoustic emission signals. Experimental evidence demonstrates a consistent and quantifiable relationship between specific acoustic emission parameters (including root mean square values and energy content) and the grinding rod's wear progression during the processing of single-crystal superalloys.

Figure 6.28 presents the acoustic emission signal characteristics corresponding to various wear stages of the grinding rod during micro-grinding of single-crystal superalloys. The experimental data reveals a progressive intensification of acoustic emission activity that directly correlates with increasing material

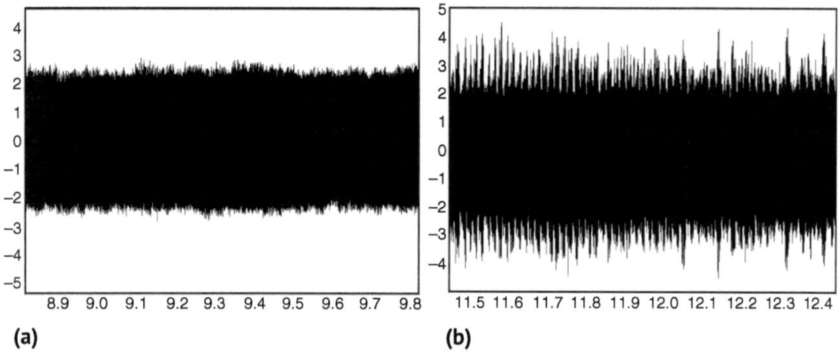

Figure 6.28 Acoustic emission signal with different wear patterns of the grinding tool: (a) $V = 16\,mm^3$ and (b) $V = 56\,mm^3$.

removal volume and consequent tool wear development. Three distinct signal evolution phases can be identified: first, a stable initial phase with relatively low amplitude; second, a transitional phase exhibiting abrupt signal amplification following tool edge dulling; and third, a chaotic final phase characterized by high-amplitude fluctuations and frequent spike occurrences during severe wear. Comparative analysis shows that the signal amplitude under severe wear conditions typically exceeds that of mild wear by approximately 40–60%, while the signal's root mean square (RMS) value increases by 2–3 times. These measurable changes in signal parameters, combined with the evident transition from ordered to disordered waveform patterns, provide robust indicators for wear state classification.

The RMS value of the acoustic emission signal can only reflect the intensity of the signal, and signals from different acoustic emission sources may have the same RMS value. To analyze the differences in acoustic emission signals, this chapter presents a spectrum analysis of the acoustic emission signals during the micro-grinding process, as shown in Figure 6.29.

As can be seen from Figure 6.29, the centroid of the acoustic reflection signal spectrum is not distinct, and the acoustic emission signals contain a relatively rich range of different frequencies. Most of these signals have a frequency distribution between 0 and 60 kHz. Some frequency images in the spectrum diagram exhibit large spikes, indicating that the energy equivalent of the acoustic emission signal at these frequencies is relatively high.

Based on the spectrum analysis of acoustic emission signals during the micro-scale grinding of nickel-based single-crystal superalloys, it can be observed that the acoustic emission signals in the frequency bands of 5, 20–35, and 50–60 kHz have relatively concentrated energy. Moreover, as the spindle speed increases, the energy in the 5 and 20–35 kHz bands decreases significantly, as shown in Figure 6.29a and b. The signal energy in the 50–60 kHz frequency band is relatively small and remains basically unchanged.

Similarly, through the spectral analysis of acoustic emission signals under different v_w and different a_p, it can be seen that with the increase of v_w and a_p, the energy

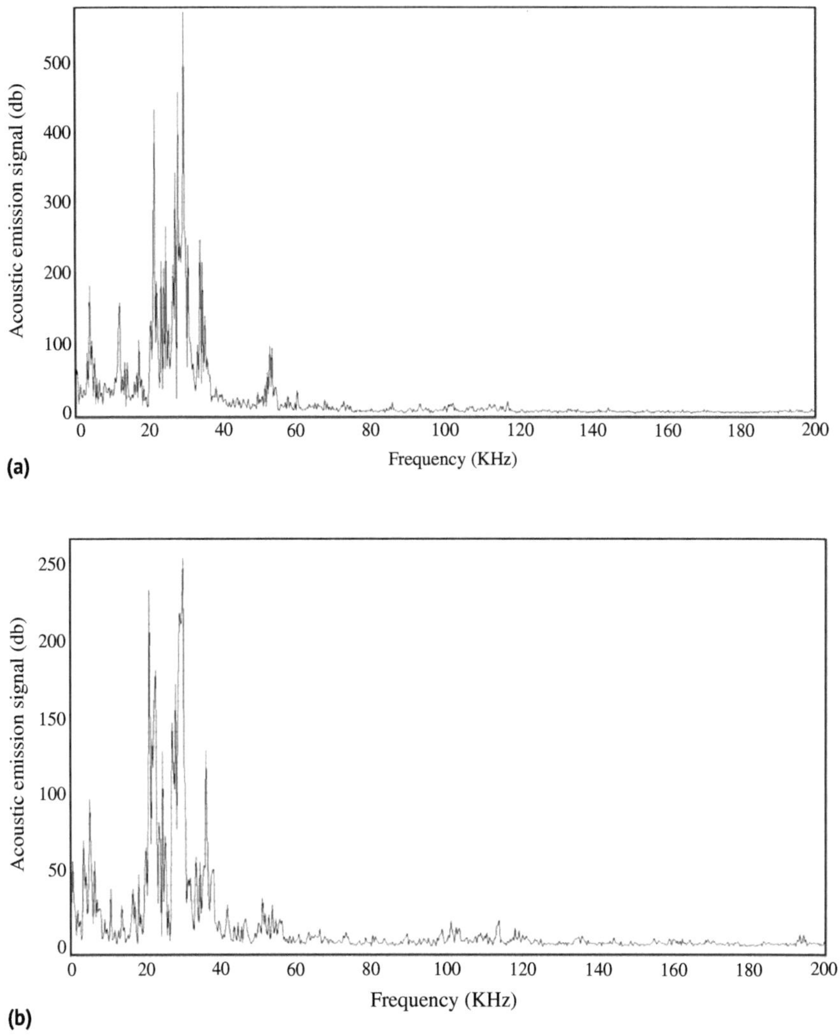

Figure 6.29 Spectrum map of acoustic emission signal with different spindle speeds: (a) $v_s = 20\,\text{Kr/min}$, $v_w = 120\,\mu\text{m/s}$, $a_p = 9\,\mu\text{m}$ and (b) $v_s = 60\,\text{Kr/min}$, $v_w = 120\,\mu\text{m/s}$, $a_p = 9\,\mu\text{m}$.

equivalent of 5 and 20–35 kHz bands increases greatly, while the energy equivalent of 50–60 kHz bands does not change basically.

Figure 6.30 shows the acoustic emission signal spectrum of the micro-grinding rod under different wear states. The grinding parameters are $v_s = 40\,\text{Kr/min}$, $v_w = 120\ \mu\text{m/s}$, $a_p = 9\,\mu\text{m}$. By comparing Figure 6.30a and b, it can be seen that when the grinding rod is worn, the energy equivalent of the acoustic emission signal in the frequency band of 5 and 20–35 kHz increases sharply. Moreover, there is a very high spike around 55 kHz.

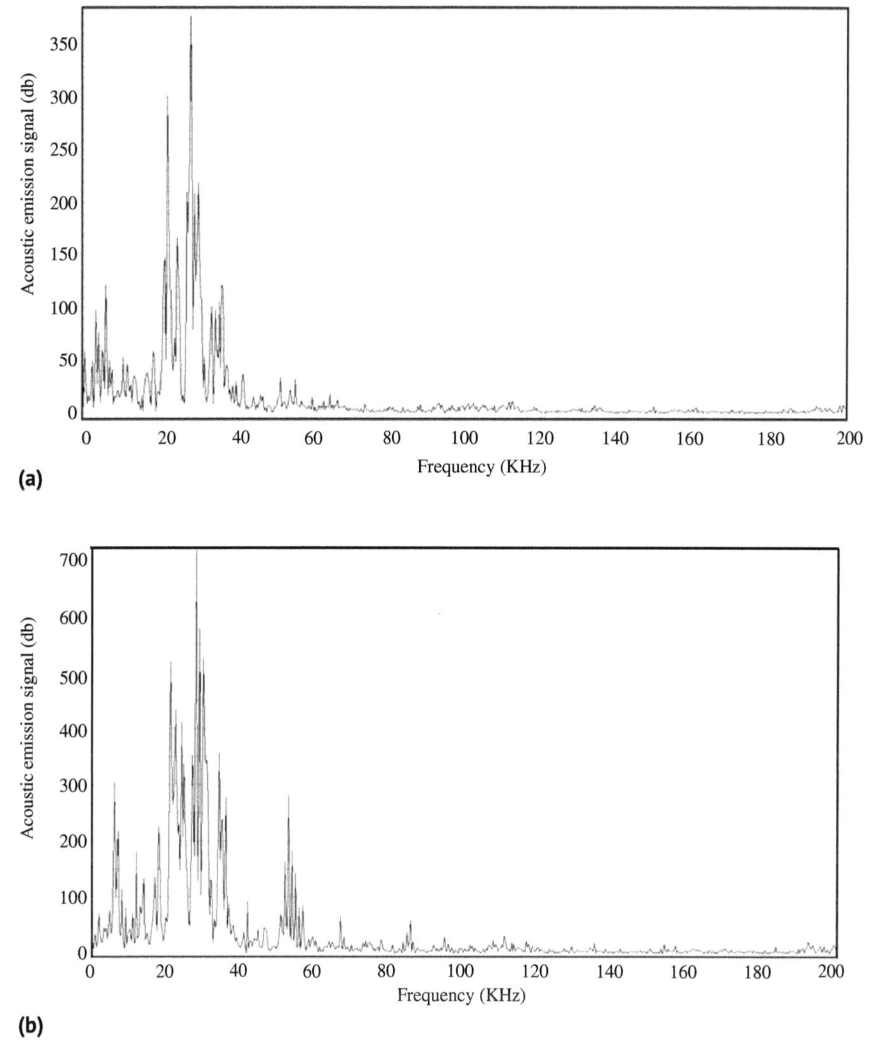

Figure 6.30 Spectrum map of acoustic emission signal with the wear pattern of the grinding tool: (a) $V = 16\,mm^3$ and (b) $V = 48\,mm^3$.

Therefore, the wear state of the micro-grinding rod can be judged by the change in the energy equivalent of the spectrum diagram. Through the above analysis, it can be concluded that the influence of grinding process parameters on the intensity of the acoustic emission signal is consistent with the influence trend of these parameters on the energy equivalent value of the acoustic emission signal in the 5 and 20–35 KHz frequency bands. The acoustic emission signal spectrum can reflect the energy release during the micro-grinding process, which is of great significance for studying the properties of acoustic emission signal sources. Observing the acoustic emission signal spectrogram is also a very effective method for monitoring the micro-grinding process.

References

1 Dyakonov, A.A. and Ardashev, D.V. (2017). Prediction of blunting area of abrasive grains on a grinding wheel. *Journal of Manufacturing Science and Engineering* 139 (12): 121004.

2 Dai, C.W., Ding, W.F., Xu, J.H. et al. (2017). Influence of grain wear on material removal behavior during grinding nickel-based superalloy with a single diamond grain. *International Journal of Machine Tools and Manufacture* 113: 49–58.

3 Nadolny, K. (2015). Wear phenomena of grinding wheels with sol–gel alumina abrasive grains and glass–ceramic vitrified bond during internal cylindrical traverse grinding of 100Cr6 steel. *The International Journal of Advanced Manufacturing Technology* 77: 83–98.

4 Cui, X., Li, C.H., Zhang, Y.B. et al. (2023). Comparative assessment of force, temperature, and wheel wear in sustainable grinding aerospace alloy using biolubricant. *Frontiers of Mechanical Engineering* 18 (1): 3.

5 Rao, X.S., Zhang, F.H., and Xu, J.B. (2023). Wheel wear behavior and its influence on grinding performance in electrical discharge diamond grinding of reaction-bonded silicon carbide. *Journal of Manufacturing Processes* 85: 904–914.

6 Fathallah, B.B., Fredj, N.B., Sidhom, H. et al. (2009). Effects of abrasive type cooling mode and peripheral grinding wheel speed on the AISI D2 steel ground surface integrity. *International Journal of Machine Tools and Manufacture* 49 (3–4): 261–272.

7 Xu, D., Zan, S., Liao, Z. et al. (2023). Investigation of surface integrity in grinding of nickel based superalloy under different cooling conditions. *Journal of Manufacturing Processes* 107: 240–251.

8 Zhou, Y.G., Gong, Y.D., Cai, M. et al. (2017). Study on surface quality and subsurface recrystallization of nickel-based single-crystal superalloy in micro-grinding. *International Journal of Advanced Manufacturing Technology* 90: 1747–1768.

9 Zhou, Y.G., Gong, Y.D., Zhu, Z.X. et al. (2016). Modelling and optimization of surface roughness from microgrinding of nickel-based single crystal superalloy using the response methodology and genetic algorithm. *International Journal of Advanced Manufacturing Technology* 85: 2607–2622.

10 Agarwal, S. and Rao, P.V. (2013). Predictive modeling of force and power based on a new analytical undeformed chip thickness model in ceramic grinding. *International Journal of Machine Tools and Manufacture* 65: 68–78.

11 Öpöz, T.T. and Chen, X. (2012). Experimental investigation of material removal mechanism in single grit grinding[J]. *International Journal of Machine Tools and Manufacture* 63: 32–40.

12 Xu, H., Jahanmir, S., and Ives, L.K. (1998). Mechanisms of material removal in abrasive machining of ceramics. *InterCeram: International Ceramic Review* 47: 380–385.

13 Magham, H.S.R., Vijayaraghavan, L., Sankaran, S. et al. (2023). Grindability studies of thermomechanically processed advanced high strength steel using sol-gel and fused alumina grain-based grinding wheels. *Proceedings of the*

Institution of Mechanical Engineers, Part B: Journal of Engineering Manufacture 237 (13): 1970–1985.
14 Malkin, S. and Guo, C. (2007). Thermal analysis of grinding. *CIRP Annals* 56 (2): 760–782.
15 Gu, Y.L., Li, H.N., Du, B.C. et al. (2019). Towards the understanding of creep-feed deep grinding of DD6 nickel-based single-crystal superalloy. *International Journal of Advanced Manufacturing Technology* 100: 445–455.
16 Rigney, D.A. (1988). Sliding wear of metals. *Annual Reviews of Materials Science* 18: 141–163.
17 Li, H.N., Yu, T.B., Wang, Z.X. et al. (2017). Detailed modeling of cutting forces in grinding process considering variable stages of grain-workpiece micro interactions. *International Journal of Mechanical Sciences* 126: 319–339.
18 Hahn, R.S. (1955). The effect of wheel-work conformity in precision grinding. *Transactions of the American Society of Mechanical Engineers* 77 (8): 1325–1329.
19 Malkin, S. (1976). Selection of operating parameters in surface grinding of steels. *Journal of Engineering for Industry -Transactions of the ASME* 98: 56–62.
20 Zhou, Y.G. (2017). Study on the Technological Theory and Key Technology in Micro Grinding Nickel-based Single Crystal Superalloy. Northeastern University.
21 Li, C., Zhang, F.H., Meng, B.B. et al. (2017). Material removal mechanism and grinding force modelling of ultrasonic vibration assisted grinding for SiC ceramics. *Ceramics International* 43: 2981–2993.
22 Yang, Z.C., Zhu, L.D., Lin, B. et al. (2019). The grinding force modeling and experimental study of ZrO2 ceramic materials in ultrasonic vibration assisted grinding. *Ceramics International* 45 (7): 8873–8889.
23 Macerol, N., Franca, L.F.P., and Krajnik, P. (2020). Effect of the grit shape on the performance of vitrified-bonded CBN grinding wheel. *Journal of Materials Processing Technology* 277: 116453.
24 Perveen, A., Rahman, M., and Wong, Y.S. (2014). Modeling and simulation of cutting forces generated during vertical micro-grinding. *The International Journal of Advanced Manufacturing Technology* 71: 1539–1548.
25 Zhao, M., Ji, X., Li, B.Z. et al. (2017). Investigation on the influence of material crystallographic orientation on grinding force in the micro-grinding of single-crystal copper with single grit. *The International Journal of Advanced Manufacturing Technology* 90: 3347–3355.
26 Zhang, J.H., Li, H., Zhang, M.L. et al. (2017). Study on force modeling considering size effect in ultrasonic-assisted micro-end grinding of silica glass and Al2O3 ceramic. *The International Journal of Advanced Manufacturing Technology* 89: 1173–1192.

Part III

Surface Integrity by High-efficiency Grinding

Surface integrity determines the service performance of single-crystal nickel alloy components. Due to the severe interaction between the workpiece and grinding wheel, strong thermo-mechanical effects are induced in the single-crystal nickel alloy, which influence the ground surface integrity significantly and pose great challenges for controlling the grinding quality. Under such circumstances, further understanding the variations and the underlying mechanism of surface integrity of single-crystal nickel alloy in the grinding process is of great importance, which could provide effective guidance for optimizing grinding parameters and improving grinding quality. This part mainly concerns the grinding-induced surface integrity of single-crystal nickel alloy and the fretting wear behavior of the machined surface. Because a large grinding depth is usually used for machining single-crystal nickel alloy, surface grinding and profile grinding can be treated as high-efficiency grinding methods, compared to the micro-grinding method.

7

Surface and Subsurface Microstructures in High-efficiency Grinding

7.1 Surface Microstructure and Surface Roughness in Surface Grinding

Figure 7.1 exhibits scanning electron microscope (SEM) images that depict the surfaces of workpieces subjected to various grinding parameters. Several surface imperfections were observed: prominent grooves and overlaps, smeared workpiece materials, scratches, ladder-like indentations, and redeposited materials.

The diverse types of abrasive grains and their associated wear mechanisms imparted distinct imperfections on the workpiece surfaces. Specifically, Figure 7.1a illustrates a brown alumina (BA) grinding wheel that produced deep grooves, approximately 5–20 micrometers (μm) in width, on the ground surface of the GH4169 alloy, some of which were oriented oblique to the grinding direction. Material extraction was also evident. For the DZ408 workpiece (Figure 7.1b), scratching resulted from abrasive grain blunting, accompanied by the formation of redeposited materials due to the adherence of hard particles. The DZ408 surface also exhibited micro-ladder-like indentations and numerous micro-cutting streaks, attributed to abrupt attrition or abrasive grain fracture. In Figure 7.1c, notable smearing was observed on the ground surface of the DD6 alloy, caused by extensive plastic deformation of the workpiece material; the groove containing smeared material was angled relative to the grinding direction.

Compared to BA grinding, surfaces exposed to microcrystalline alumina (MA) grinding wheels consistently exhibited deep grooves, as demonstrated in Figure 7.1d–f. Notably, Figure 7.1d shows that the removal of a section of GH4169 material led to the formation of micro-cavities and folded materials on the surface. For DZ408 material (Figure 7.1e), plowing by hard particles from fractured or displaced abrasive grains created short, prominent grooves. Additionally, Figure 7.1f illustrates black spots on the DD6 ground surface, resulting from material oxidation during grinding.

Surface imperfections in grinding are mainly governed by the wear characteristics of the abrasive wheels. The orientation of these defects often deviates from the grinding direction due to the fracturing or displacement of particles within the abrasive grains. BA abrasive grains, due to their lower strength, are more prone to

Grinding of Single-Crystal Superalloys: Fundamentals and Technologies, First Edition.
Wenfeng Ding et al.
© 2026 Wiley-VCH GmbH. Published 2026 by Wiley-VCH GmbH.

fracture under high grinding loads. In contrast, MA abrasive grains exhibit superior ductility and hardness, which help prevent brittle fractures during grinding. However, this self-sharpening effect diminishes as creep-feed grinding continues, leading to bond bridge fractures between the MA grains.

Additionally, alumina abrasive wheels used in grinding nickel-based superalloys are often designed with high porosity to improve chip storage capacity [1, 2]. When abrasive grains fracture or displace, the resulting hard particles can move into the wheel's pores or the spaces between grains, causing irregular cutting directions and leading to surface imperfections that do not align with the intended grinding pattern. Moreover, the removal and redeposition of particles from both the workpiece and abrasive grains can lead to defects such as smeared material, deep grooves, and overlaps. These hard particles can also cause dragging and tearing defects in subsequent grinding passes [3–5].

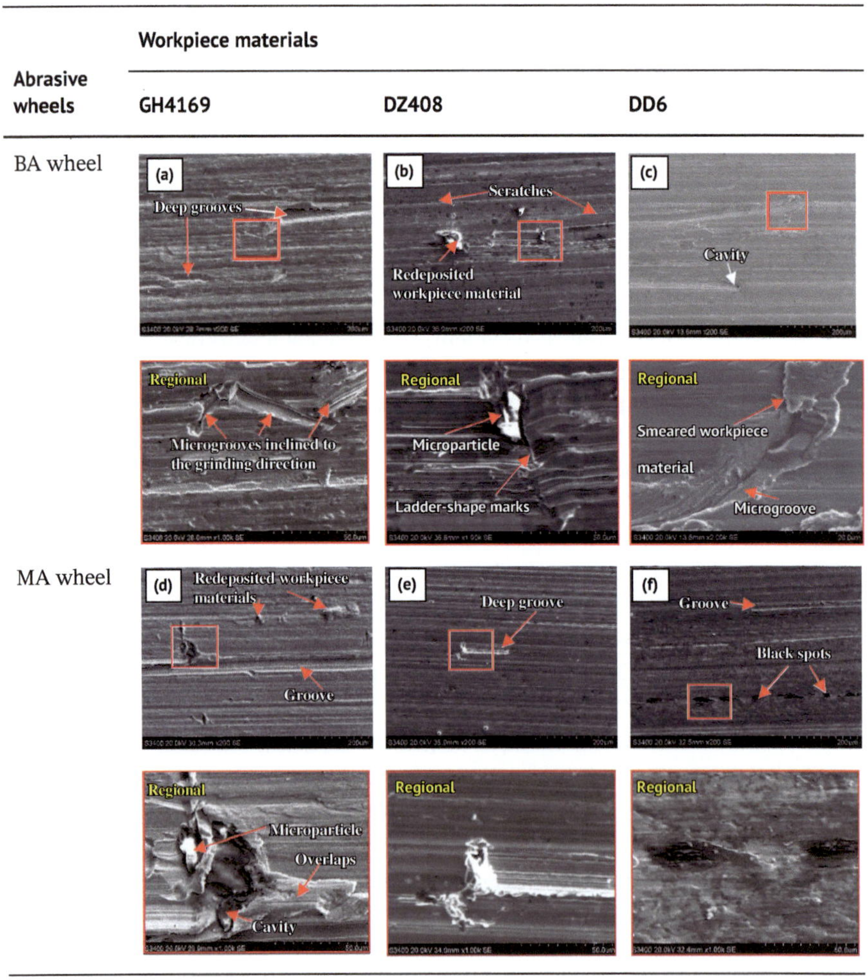

Figure 7.1 Representative surface defects induced by high-efficiency surface grinding: (a) BA-GH4169; (b) BA-DZ408; (c) BA-DD6; (d) MA-GH4169; (e) MA-DZ408; and (f) MA-DD6.

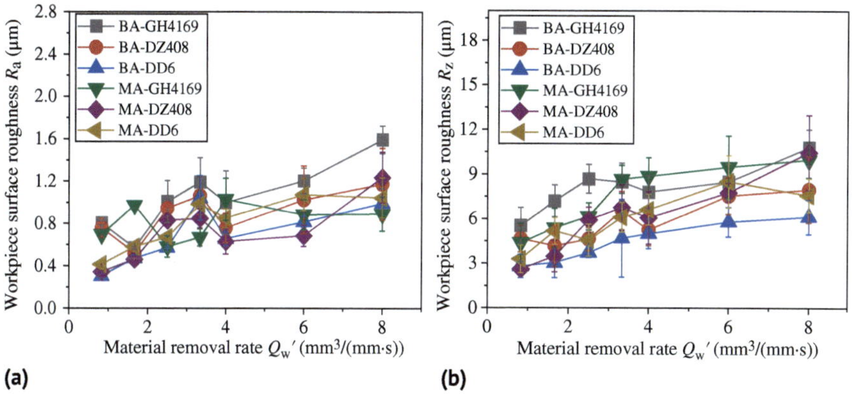

Figure 7.2 The ground surface roughness: (a) surface roughness R_a and (b) surface roughness R_z.

Figure 7.2 displays the roughness characteristics of the ground surface. The GH4169 alloy exhibited the highest roughness ($R_a = 1.6$ μm) when processed using a BA wheel, whereas the DD6 alloy achieved the lowest roughness ($R_a = 0.27$ μm) with the same wheel type. Analysis revealed that with the MA wheel, frequent microfractures of abrasive grains occur, leading to the formation of randomly protruding new cutting edges on the wheel's surface, thereby increasing the surface roughness. In Figure 7.2a, a notable increase in both R_a and R_z values is observed as the material removal rate (MRR) increases. A slight variation in surface roughness is evident when the MRR is below 4 $mm^3/(mm \cdot s)$. Studies indicate that abrasive grain fractures are more prevalent at higher MRRs, with the resultant broken particles introducing additional defects into the ground surfaces, as depicted in Figure 7.1. Under these circumstances, the measured errors exhibit significant variability (Figure 7.2b), primarily attributed to surface imperfections such as deep grooves and cavities, which contribute to an increase in the R_z value of surface roughness.

7.2 Subsurface Microstructure in Surface Grinding

Figure 7.3 presents cross-sectional images of the ground surface and subsurface for various superalloys processed using an alumina wheel. Generally, it is observed that a high MRR ($Q_w' = 6$ $mm^3/(mm \cdot s)$) leads to the formation of numerous surface and subsurface defects, including cavities, cracks, and material deformation, regardless of the workpiece material (Figure 7.3b, d, and f). Conversely, samples machined at a moderate MRR ($Q_w' = 2$ $mm^3/(mm \cdot s)$) exhibit smoother surfaces or slight material deformation (Figure 7.3a, c, and e). However, the manifestation of these defects varies based on the workpiece material. During the grinding of GH4169 alloy, Figure 7.3a reveals a material deformation layer approximately 3 μm thick, while Figure 7.3b shows an arc-shaped crack, approximately 5 μm long, running perpendicular to the machined surface. This crack is comparable to those reported by Herbert et al. [5] in their study on the drilling of Inconel 718, where surface cracks of approximately 4 μm were detected. In the grinding of DZ408 and DD6 alloys, as depicted in

Figure 7.3d and f, respectively, bending of the γ and γ' phase grids is evident, accompanied by severe material deformation extending to depths of about 10 μm, 6 μm, and 5 μm, respectively. Notably, Figure 7.3d shows cavities on the machined surface resulting from workpiece material fracture, accompanied by microcracks. Similar damage is observed in Figure 7.3f during the grinding of DD6 alloy, with cracks approximately 2 μm long running parallel to the machined surface. In contrast, no cracks are detected when the MRR is 2 $mm^3/(mm \cdot s)$, as expected due to the workpiece materials' ability to withstand lower grinding loads [6, 7].

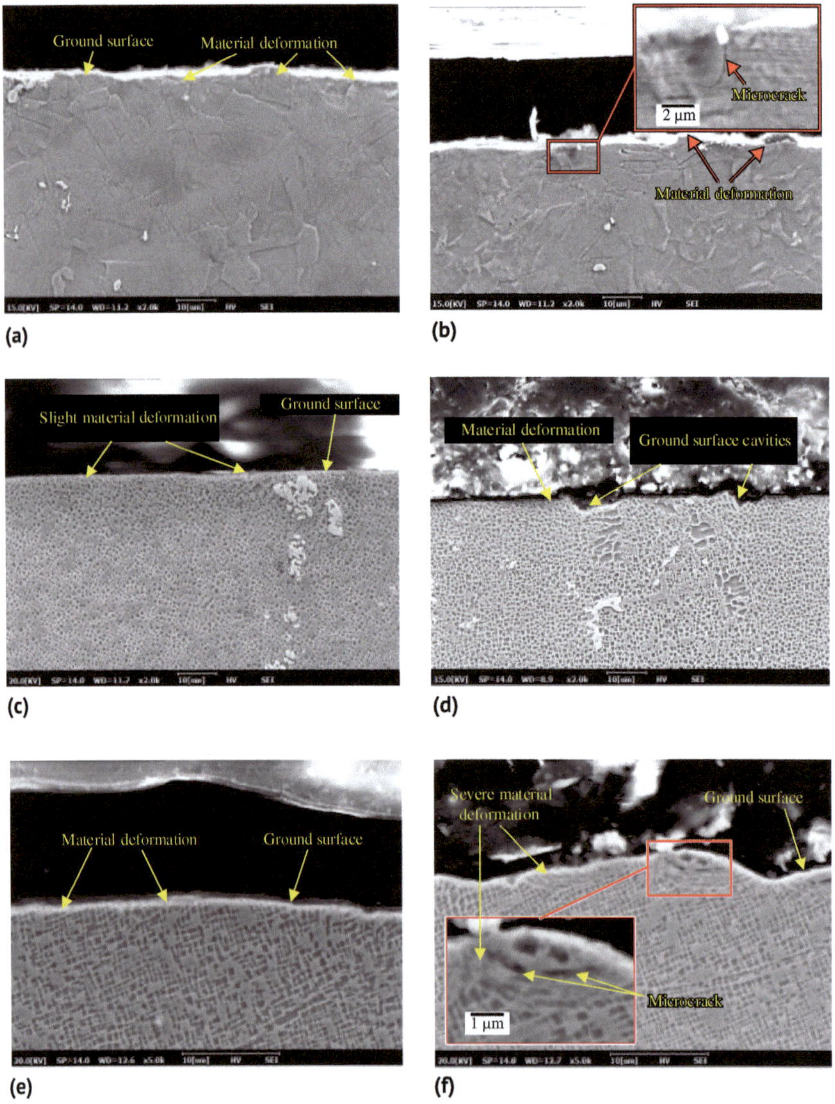

Figure 7.3 Ground subsurface microstructure of different materials: (a) GH4169 when $v_s = 25$ m/s, $v_w = 100$ mm/min, $a_p = 1.2$ mm; (b) GH4169 when $v_s = 25$ m/s, $v_w = 240$ mm/min, $a_p = 1.5$ mm; (c) DZ408 when $v_s = 25$ m/s, $v_w = 100$ mm/min, $a_p = 1.2$ mm; (d) DZ408 when $v_s = 25$ m/s, $v_w = 240$ mm/min, a_p=1.5 mm; (e) DD6 when $v_s = 25$ m/s, $v_w = 100$ mm/min, $a_p = 1.2$ mm; and (f) DD6 when $v_s = 25$ m/s, $v_w = 240$ mm/min, $a_p = 1.5$ mm.

7.3 Surface Microstructure and Surface Roughness in Profile Grinding

In this chapter, nickel-based superalloys served as the primary materials, with workpieces measuring 21 mm in length, 16 mm in width, and 15 mm in height. The grinding setup is illustrated in Figure 7.4. Three distinct sets of grinding parameters were chosen to produce blade root samples, each allowing for a total material removal of 2.63 mm, as specified in Table 7.1. Notably, the surface roughness after grinding remained below 0.8 μm, while the achieved profile accuracy approximated 10 μm.

Figure 7.4 Experimental setup of creep-feed profile grinding of turbine blade root.

Table 7.1 Details of grinding parameters.

Grinding case group No.	Grinding stages	Grinding tests No.	Grinding speed v_s(m/s)	Workpiece speed v_w(mm/min)	Grinding depth a_p(mm)
I (Without burning)	Rough grinding	1	25	240	1.50
		2	35	150	1.00
	Finish grinding	3	35	180	0.13
II (Without burning)	Rough grinding	1	25	240	1.50
		2	35	420	1.00
	Finish grinding	3	35	600	0.13
III (With burning)	Rough grinding	1	25	600	1.50
		2	35	600	1.00
	Finish grinding	3	35	600	0.13

Following the grinding process, the blade root samples were subjected to electrical discharge machining (EDM) cuts, oriented both parallel and perpendicular to the workpiece feed direction. These samples were then carefully polished using mechanical methods and further refined with ion beam finishing. To investigate the detailed cross-sectional microstructures, electron backscatter diffraction (EBSD) and SEM were employed. For nanoscale structural analysis, lamellae thinner than 100 nm were extracted from the surface of the workpiece using a focused ion beam (FIB) system, specifically the FEI Helios G4 CX. The microstructure of these lamellae was then characterized using a transmission electron microscope (TEM) operated at 200 kV (FEI Tecnai FP 5026), equipped with an energy-dispersive X-ray spectrometer (EDS).

Figure 7.5 illustrates the typical surface morphology of the workpiece root after creep-feed profile grinding. Figure 7.5a and b shows distinct grinding marks

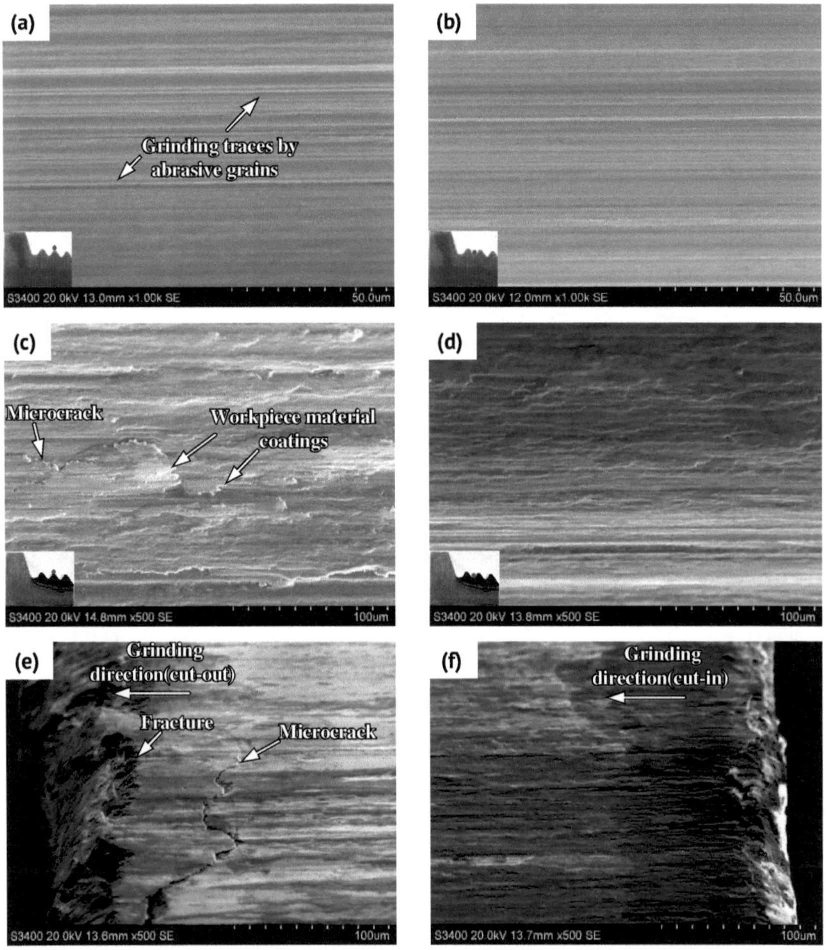

Figure 7.5 Typical topography in different surface areas of blade root: (a) peak region (group I); (b) valley region (group I); (c) peak region (group III); (d) valley region (group III); (e) cut-out (group III); and (f) cut-in (group III).

accompanied by mild plowing of the workpiece material on both peak and valley regions of the root, under the conditions defined by parameter group I. Notably, the absence of surface burnout indicates minimal heat generation in the grinding zone, resulting in negligible thermal effects on the subsurface material, as reported in [8].

Alternatively, surfaces characterized by coarse textures and indistinct patterns were observed under parameter group III, as depicted in Figure 7.5c–f. Dark discoloration on the end face of the root workpiece suggests surface degradation due to high-temperature oxidation. Furthermore, the softened state of the material at elevated temperatures led to significant plastic deformation, exacerbated by abrasive grain friction along the cutting paths in the peak regions (Figure 7.5c). It also displays alluvial-like deposits of workpiece material on the ground surface, contributing to a complex and irregular surface topography. Similar material coatings adhering to abrasive grains due to high grinding temperatures have been observed by Sinha et al. [7]. Indeed, the increased plastic deformation at higher temperatures intensified material accumulation along the cutting path sides, forming coatings under strong abrasive grain pressure. The presence of material coatings in Figure 7.5c can be attributed to the combined effects of debris adherence and material buildup. In the valley regions (Figure 7.5d), discontinuous abrasive grain marks and smeared material were evident, accompanied by micro-cavities likely caused by fragmented abrasive grains, as previously reported in [9]. Additionally, the surface morphology in the entry and exit regions of the workpiece exhibited clear material fractures and microcracks, as shown in Figure 7.5e and f.

In this context, it is plausible to infer that the sliding and plowing actions of individual abrasive grains are the primary mechanisms driving the grinding process under parameter group III. The surfaces generated under these conditions are mainly the result of significant compressive forces and frictional interactions between the abrasive grains and the workpiece material, leading to material fractures. This effect could be exacerbated if the microcracks, oriented parallel to the end face of the workpiece as shown in Figure 7.5, extend further.

As a result, the ground surface exhibiting burnout (produced under parameter group III) displayed a relatively higher average roughness of around 1 μm, compared to the surface without burnout (produced under parameter set I), which measured less than 0.5 μm.

Surface roughness (S_a) was assessed using an optical microscope (OM), with the results depicted in Figure 7.6. In grinding group I, the average surface roughness R_a, measured distinctly in the crest and trough areas of the root, was 0.37 μm and 0.48 μm, respectively. In contrast, grinding group III exhibited a higher S_a value (0.8 μm in the crest area and 1.0 μm in the trough area) than group I, due to increased smearing and lateral material displacement of the workpiece. Notably, in both conditions, the surface roughness in the trough area exceeded that of the crest area by approximately 25%. This difference in surface quality may stem from variations in the operational state of the wheel surface during profile grinding.

During creep-feed profile grinding, initial engagement occurs between the wheel's crest areas and the workpiece, resulting in substantial impact forces and moments due to the expanding contact zone and increased number of active abrasive grains.

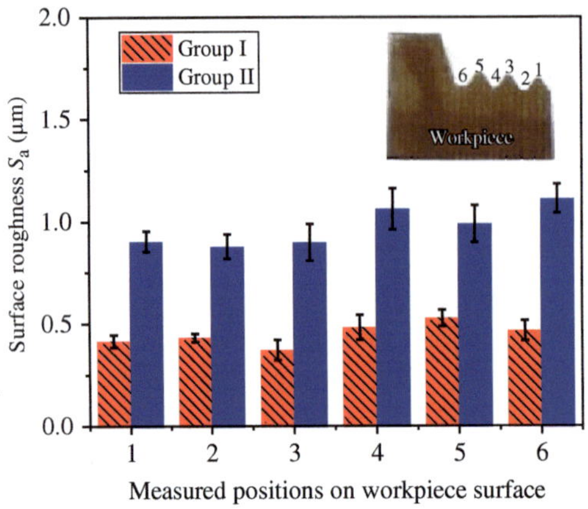

Figure 7.6 Surface roughness in different surface regions of blade root workpiece.

Wear in the wheel's crest areas manifests as grain flattening, workpiece material adherence, grain and bonding bridge fractures. Fracture wear is particularly prevalent when grinding demanding materials like nickel-based superalloys. Consequently, the crest areas on the wheel surface undergo accelerated wear compared to the trough areas, potentially causing dimensional inaccuracies and compromised surface quality of the root workpiece. For example, fractured grains act as abrasive particles, introducing imperfections on the workpiece's trough areas, explaining the smoother surface observed in the crest areas. Additionally, while the sixth grinding pass used smaller parameters, the significant forces imparted by the fourth and fifth passes could induce substantial plastic deformation in the subsurface layer, partially mitigated by the sixth pass. Moreover, in creep-feed profile grinding of blade root, variations in machining quality on the ground surface were observed. The underlying factor can be attributed to the uneven distribution of grinding loads along the workpiece's profile in the normal cross section. To ensure consistent surface quality of the workpiece, periodic dressing of the grinding wheel to rejuvenate the worn tool's profile geometry is essential.

7.4 Subsurface Microstructure in Profile Grinding

7.4.1 Blade Root Workpiece Produced by Parameter Group I

During the grinding process, substantial grinding forces and elevated temperatures can lead to significant changes in the subsurface microstructure and mechanical properties of the workpiece material. Figure 7.7 presents typical cross-sectional images of the workpiece root using parameter set I. Specifically, Figure 7.7a illustrates well-defined dendritic and interdendritic structures, devoid of micro-voids, fractures, or recrystallization. Furthermore, Figure 7.7b–m shows slight curvature in the grids within the superficial layer, resulting from plastic deformation caused by mechanical pressure from abrasive grains. The variation in thickness of the deformed layer is minimal, ranging from 2 to 4 μm, which is attributed to the moderate temperatures and forces characteristic of creep-feed grinding.

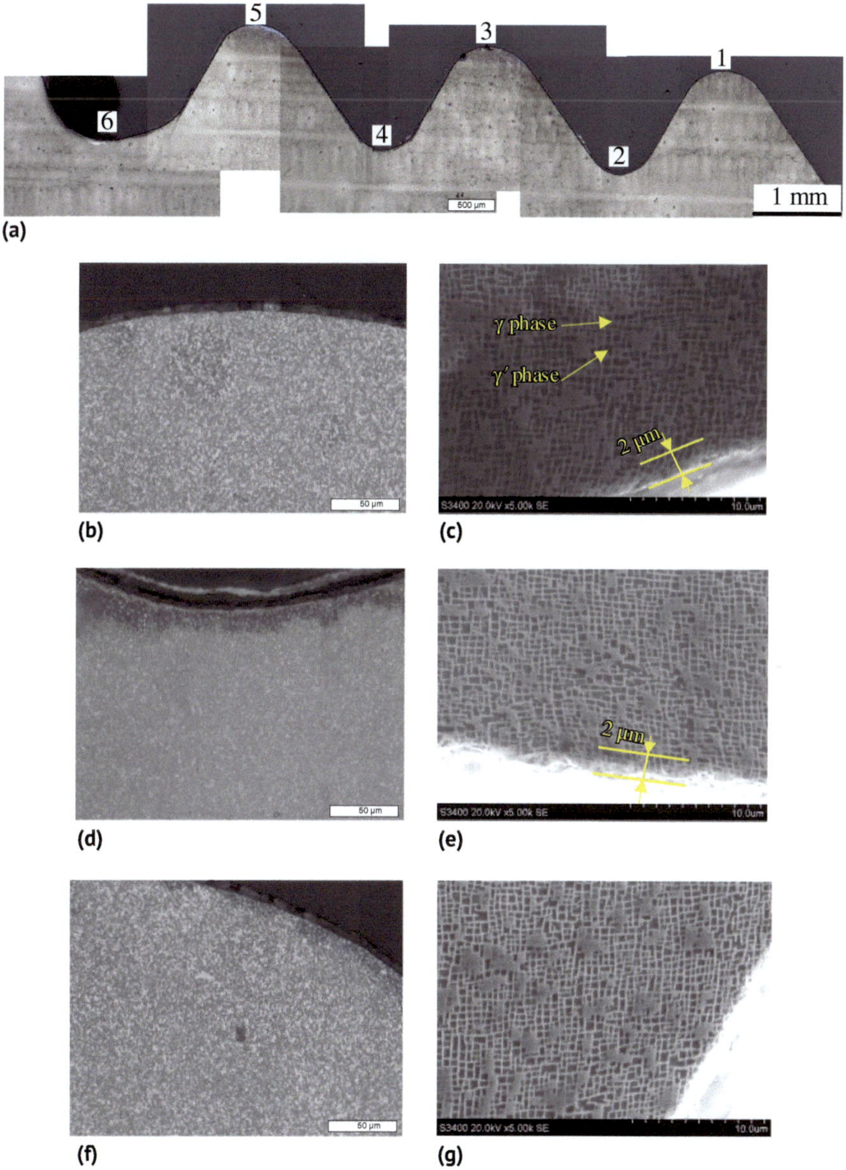

Figure 7.7 Cross-sectional microstructures at different positions of the blade root workpiece produced by parameter group I: (a) overall image; (b) Position No. 1 (OM); (c) Position No. 1 (SEM); (d) Position No. 2 (OM); (e) Position No. 2 (SEM); (f) Position No. 3 (OM); (g) Position No. 3 (SEM).

(Continued)

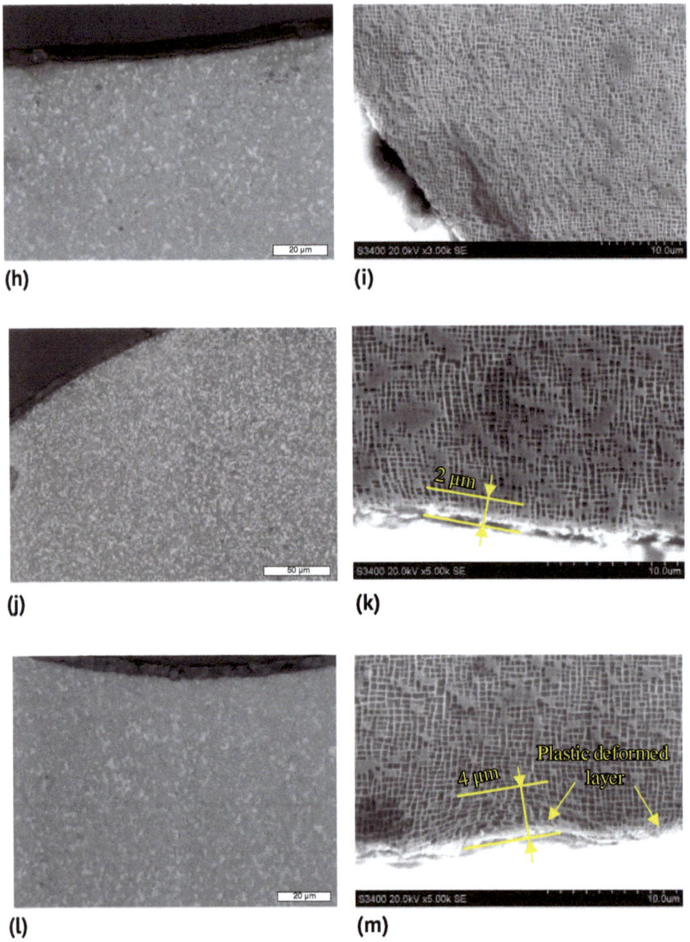

Figure 7.7 (Cont'd) (h) Position No. 4 (OM); (i) Position No. 4 (SEM); (j) Position No. 5 (OM); (k) Position No. 5 (SEM); (l) Position No. 6 (OM); and (m) Position No. 6 (SEM).

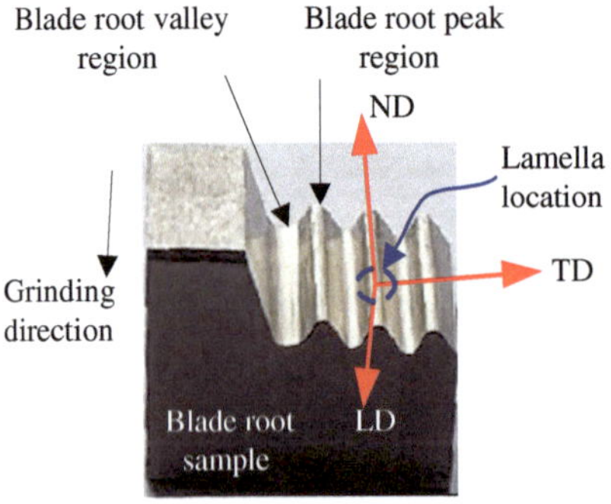

Figure 7.8 Ground blade root sample.

7.4.2 Blade Root Workpiece Produced by Parameter Group II

To aid in the analysis, the sample's orientation is described using three directions: normal direction (ND), transversal direction (TD), and longitudinal direction (LD), as illustrated in Figure 7.8. Previous research has indicated that the grinding loads are greater in the root peak region than in the root valley region during blade root profile grinding [10, 11]. Based on this, the lamellar structure in the root peak area, positioned at the center of the blade root sample, was selected for TEM analysis, as this region experiences more significant plastic deformation.

Upon inspecting the surface layer of the blade root fabricated using parameter group I (Figure 7.9), no defects are apparent on the cross section (TD-ND plane). In contrast, Figure 7.10a, which represents the sample produced with group II, clearly depicts the dendritic structures within the bulk material on the TD-ND plane. Figure 7.10b, a magnified micrograph of Region A in Figure 7.10a, reveals a deformed layer near the ground surface, extending to a depth of approximately 5 μm. Below a depth of 1 μm, the ultrafine structure is indistinct by SEM. Between 1 and 2 μm depth, plastic deformation is evident, with grids composed of γ' precipitates and γ channels exhibiting significant bending, which diminishes with less deformation. Beyond 5 μm depth, the microstructure resembles the bulk material (Region B in Figure 7.10c). This gradient deformation microstructure is also observable on the LD-ND plane in Figure 7.10d.

Additionally, an EBSD analysis was conducted on the TD-ND plane with a scanning step size of 1 μm, as shown

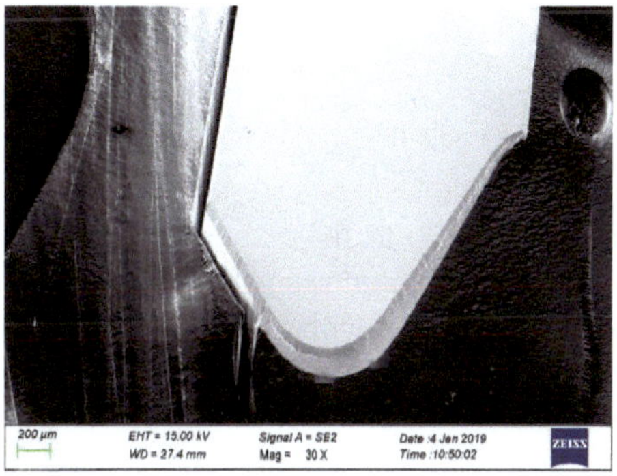

Figure 7.9 Cross-sectional SEM of sample I (TD-ND plane).

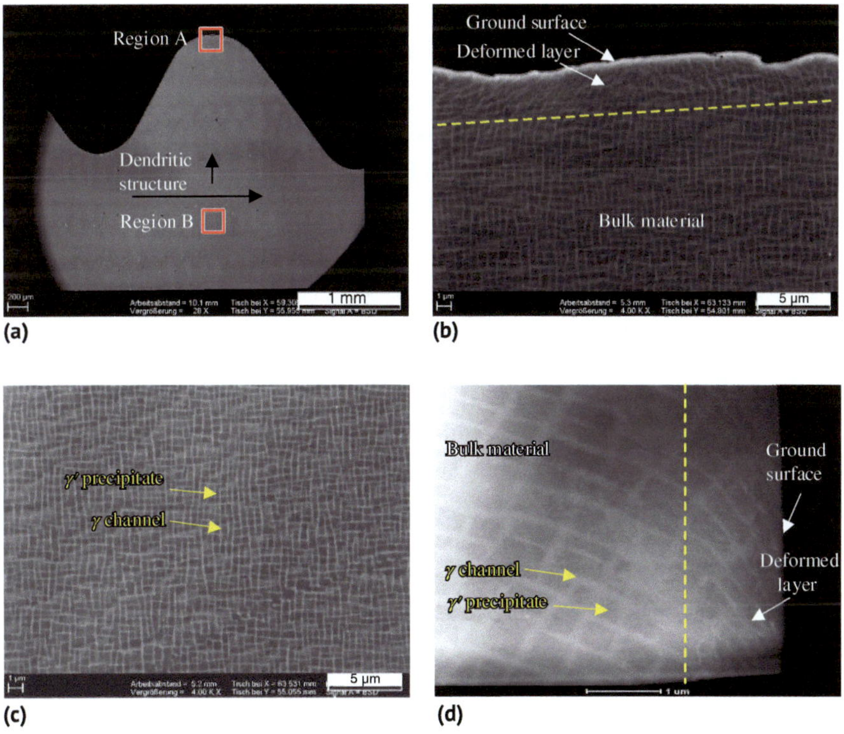

Figure 7.10 Cross-sectional SEM of blade root sample II: (a) whole image on TD-ND plane; (b) Region A on TD-ND plane; (c) Region B TD-ND plane; and (d) microstructure on LD-ND plane.

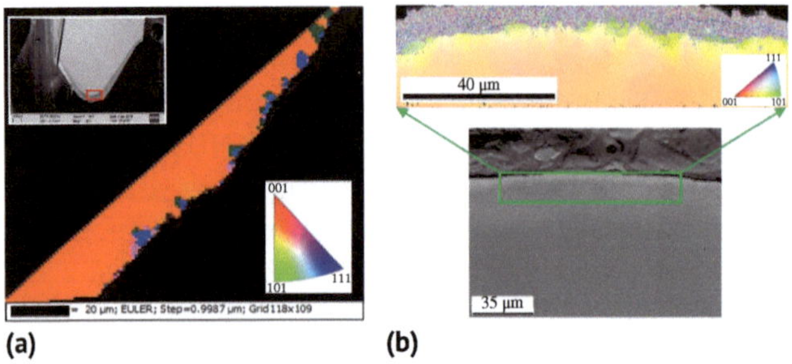

Figure 7.11 Cross-sectional EBSD maps of blade root sample on TD-ND plane: (a) sample I and (b) sample II.

in Figure 7.11. In the case of the blade root fabricated using group I parameters (Figure 7.11a), grains smaller than 2 μm in size are visible near the ground surface, exhibiting orientations that deviate from those of the bulk material. However, this microstructural characteristic is absent in the sample produced with group II parameters (Figure 7.11b), even with a reduced EBSD scanning step size of 0.06 μm. Nevertheless, it is unequivocally established that the grinding process alters the microstructure of the superficial layer. The observed divergence in grain orientation compared to the bulk is likely due to the noncoincidence of the grain growth trajectory with the preferred crystallographic axes.

7.4.3 Blade Root Workpiece Produced by Parameter Group III

When parameter group III is used, grinding burn is observed (Figure 7.12), with OM revealing a white layer in the root peak regions. In Figure 7.12a, the depths of the white layer at positions 1, 3, and 5 are approximately 0.2, 0.3, and 0.2 millimeters, respectively. In contrast, no white layer is present in the root valley regions (positions 2, 4, and 6) (Figure 7.12d, h, and l), where only compacted grids of γ-channels and γ'-particles are visible (Figure 7.12e, i, and m). This indicates that the root peak regions experience higher thermal loads, leading to the formation of a more prominent white layer. Additionally, various subsurface defects, including micro-pits (Figure 7.12b, f, and g), material peeling (Figure 7.12c and k), microcracks (Figure 7.12c), and recrystallization (Figure 7.12j), are induced. Notably, cracks within the workpiece material are visible (Figure 7.12g and k), exhibiting curved shapes that extend toward the ground surface. The direction of these cracks promotes material fracture and peeling, which in turn compromises both surface quality and mechanical properties.

It is worth noting that SX alloy rarely undergoes recrystallization unless high temperatures and significant plastic deformation are present [12]. Under such conditions, a soft subsurface forms, enhancing plastic deformation, which provides the groundwork for recrystallization in the surface layer. Prior literature [12, 13] suggests that recrystallized grain boundaries in SX alloy can readily initiate cracks. From Figure 7.12j and k, it is evident that crack propagation aligns with recrystallized grain boundaries, implying that cracks in Figure 7.12k may originate from surface layer recrystallization. Once recrystallization occurs, the fatigue strength decreases significantly, as emphasized by various researchers [14]. Therefore, in creep-feed profile grinding of turbine blade roots, surface burn of the SX alloy should be minimized to prevent such detrimental effects.

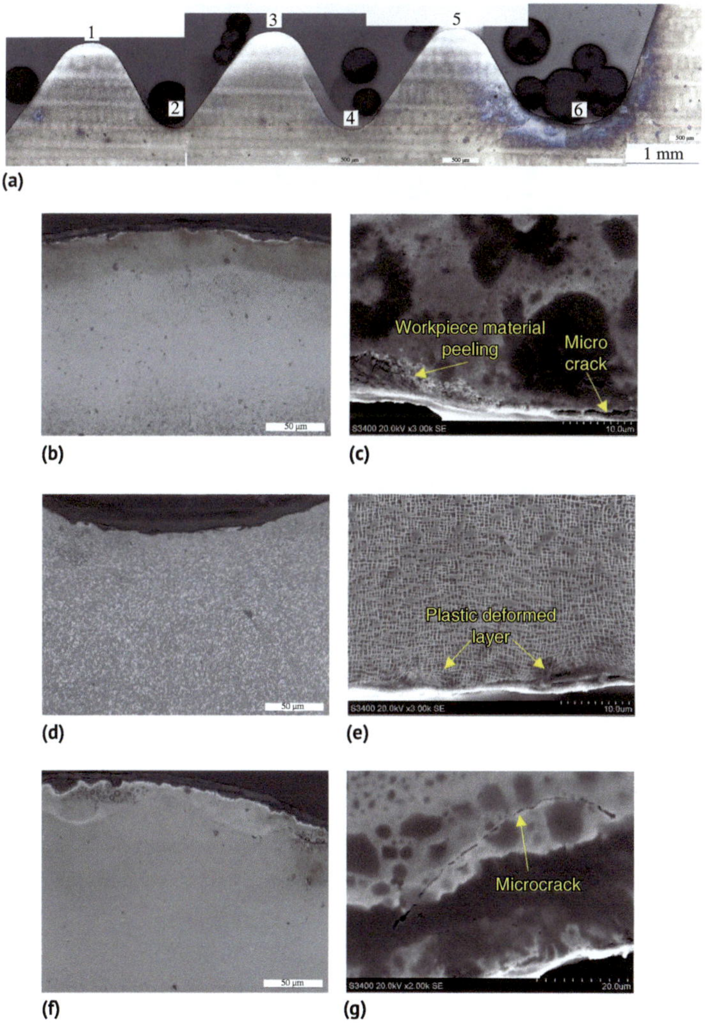

Figure 7.12 Cross-sectional microstructures at different positions of the blade root workpiece produced by parameter group III: (a) overall image; (b) Position No. 1 (OM); (c) Position No. 1 (SEM); (d) Position No. 2 (OM); (e) Position No. 2 (SEM); (f) Position No. 3 (OM); (g) Position No. 3 (SEM);

(Continued)

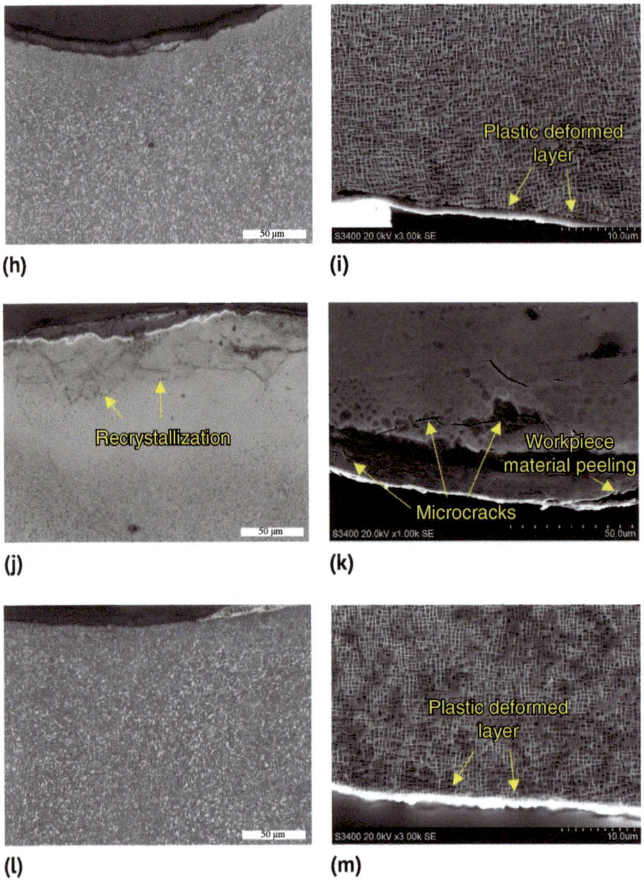

Figure 7.12 (Cont'd) (h) Position No. 4 (OM); (i) Position No. 4 (SEM); (j) Position No. 5 (OM); (k) Position No. 5 (SEM); (l) Position No. 6 (OM); and (m) Position No. 6 (SEM).

References

1. Thakur, A. and Gangopadhyay, S. (2015). State-of-the-art in surface integrity in machining of nickel-based super alloys. *International Journal of Machine Tools and Manufacture* 100: 25–54.
2. Li, Z.Y., Wei, X.T., Guo, Y.B. et al. (2015). State-of-art, challenges, and outlook on manufacturing of cooling holes for turbine blades. *Machining Science and Technology* 19 (3): 361–399.
3. Soo, S.L., Antar, M.T., Aspinwall, D.K. et al. (2013). The effect of wire electrical discharge machining on the fatigue life of Ti–6Al–2Sn–4Zr–6Mo aerospace alloy. *Procedia CIRP* 6: 215–219.
4. Klocke, F., Klink, A., Veselova, D. et al. (2014). Turbomachinery component manufacture by application of electrochemical, electro-physical and photonic processes. *CIRP Annals-Manufacturing Technology* 63: 703–726.

5 Herbert, C.R.J., Kwong, J., Kong, M.C. et al. (2012). An evaluation of the evolution of workpiece surface integrity in hold making operations for a nickel-based superalloy. *Journal of Materials Processing Technology* 212 (8): 1723–1730.

6 Seo, S.M., Kim, I.S., Lee, J.H. et al. (2009). Grain structure and texture evolutions during single crystal casting of the Ni-base superalloy CMSX-4. *Metals and Materials International* 15 (3): 391–398.

7 Sinha, M.K., Setti, D., Ghosh, S. et al. (2016). An investigation on surface burn during grinding of Inconel 718. *Journal of Manufacturing Processes* 21: 124–133.

8 Ye, N.E. and Pearce, T.R.A. (1983). Some observations on profile wear in creep-feed grinding. *Wear* 92: 51–66.

9 Miao, Q., Ding, W.F., Kuang, W.J. et al. (2021). Grinding force and surface quality in creep feed profile grinding of turbine blade root of nickel-based superalloy with microcrystalline alumina abrasive wheels. *Chinese Journal of Aeronautics* 34 (2): 576–585.

10 Miao, Q., Li, H.N., and Ding, W.F. (2020). On the temperature field in the creep feed grinding of turbine blade root: simulation and experiments. *International Journal of Heat and Mass Transfer* 147: 118957.

11 Miao, Q., Ding, W.F., Kuang, W.J. et al. (2020). Tool wear behavior of vitrified microcrystalline alumina wheels in creep feed profile grinding of turbine blade root of single crystal nickel. *Tribology International* 145: 106144–106144.

12 Li, Z.L., Xu, Q.Y., and Liu, B.C. (2016). Experimental investigation on recrystallization mechanism of a Ni-base single crystal superalloy. *Journal of Alloys and Compounds* 672: 457–469.

13 Miao, Q., Ding, W.F., Xu, J.H., et al. (2021). Creep feed grinding induced gradient microstructures in the superficial layer of turbine blade root of single crystal nickel-based superalloy. *International Journal of Extreme Manufacturing*, 3: 045102.

14 Malkin, S. (1976). Selection of operating parameters in surface grinding of steels. *Journal of Engineering for Industry-Transactions of the ASME* 98: 56–62.

8

Subsurface Nanostructures in High-efficiency Grinding

8.1 Subsurface Nanostructures in Profile Grinding

8.1.1 TEM Observations of Sample I

To gain a deeper understanding of the microstructural characteristics within the superficial layer (arising from the parameter set detailed in Chapter 7), a Transmission Electron Microscopy (TEM) analysis was conducted on the aforementioned samples. As illustrated in Figure 8.1, Sample I's superficial layer, which measures approximately 3 μm in thickness, exhibits distinct deformation features. Specifically, the outermost layer, approximately 1 μm thick, undergoes severe plastic deformation, rendering the γ' and γ phase arrangements indistinguishable. In contrast, the layer situated between 1 and 3 μm in depth, as depicted in Figure 8.1, reveals notable changes in the microstructure of this plastically deformed region, including grid sliding, distortion, bending, and partial grid fracture or elongation. Below a depth of 3 μm, the grid shape begins to recover from deformation, albeit accompanied by a significant number of dislocations. Based on these unique microstructural attributes, the superficial layer can be stratified into three zones: the dislocation-accumulated layer (DAL), the deformed layer (DFL), and the severely deformed layer (SDL). Nanoscale inspections focused on two specific zones designated as Region I and Region II in Figure 8.1.

8.1.1.1 Microstructure of DAL (Depth > 3 μm)

Within the DAL, as illustrated in Figure 8.2, grinding-induced deformation manifests as dislocation morphologies. This zone exhibits initial dislocation multiplication and movement adjacent to the unaffected bulk material, followed by a sharp increase in dislocation density as plastic strain intensifies. Figure 8.2a displays numerous dislocation lines and tangles entrapped within the γ' phases, with additional dislocations clustering around γ phases. Further multiplication, motion, clustering, and interaction of dislocations lead to the formation of dense dislocation walls within grains, potentially segmenting γ' phases into multiple fragments. The Selected Area Electron Diffraction (SAED-1) pattern in Figure 8.2b reveals a clear and distinct crystal lattice in the DAL, indicating that despite undergoing plastic strain during grinding, the material retains its single-crystal structure.

Grinding of Single-Crystal Superalloys: Fundamentals and Technologies, First Edition.
Wenfeng Ding et al.
© 2026 Wiley-VCH GmbH. Published 2026 by Wiley-VCH GmbH.

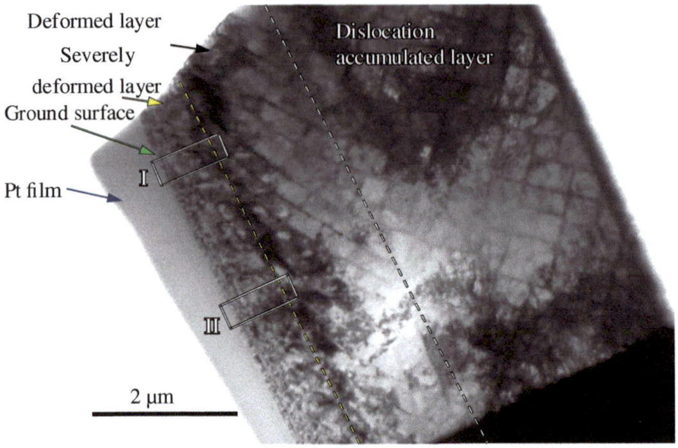

Figure 8.1 Cross-sectional TEM image showing overall microstructure in blade root Sample I on TD-ND plane.

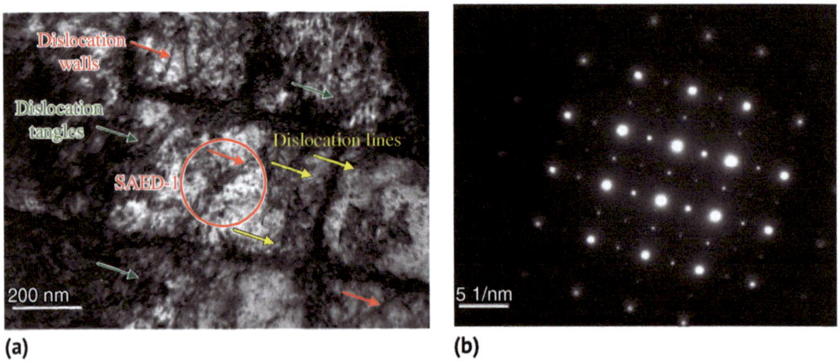

Figure 8.2 Cross-sectional TEM images of the microstructure at about 5 μm deep from ground surface in Sample I on TD-ND plane: (a) dislocation structures and (b) SAED-1.

8.1.1.2 Microstructure of DFL (Depth: 1–3 μm)

In the depth range of 1–3 μm (depicted in Figure 8.3), a high concentration of dislocations is observed, with γ'-precipitates and γ-channels bent along the grinding direction. Consequently, the grids comprising these precipitates and channels appear elongated, resembling bent rectangles (Figure 8.3a). Adjacent to the DFL, near the SDL, γ'-precipitates exhibit a general reduction in thickness (ranging from 200 to 500 nanometers [nm]), ultimately resulting in fracture. This is attributed to the further accumulation of dislocations within γ'-precipitates, exacerbating splitting as deformation intensity increases in the DFL. The SAED-2 pattern in Figure 8.3b, displaying rotating diffraction spots, suggests the possibility that the single-crystal structure may be twisted or that new grains with slight misorientation angles have emerged.

8.1.1.3 Microstructure of SDL (Depth < 1 μm)

Within a 1 μm proximity to the surface (refer to Figure 8.4), the microstructure reveals finely fragmented grains originating from the disintegration of γ'-precipitates

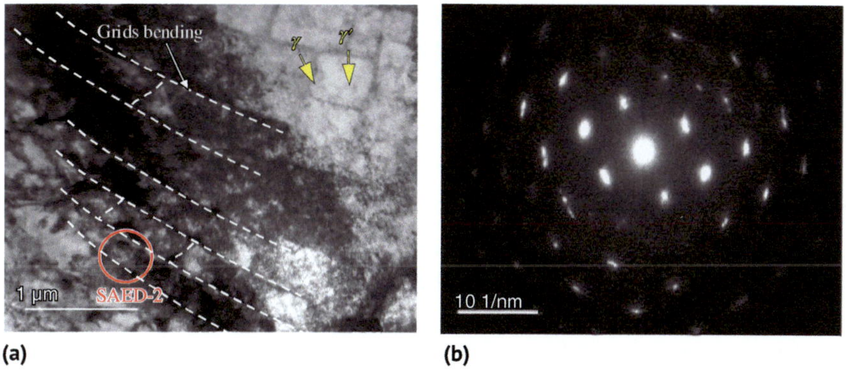

Figure 8.3 Cross-sectional TEM images of the microstructure at about 2 μm deep from ground surface in Sample I on TD-ND plane: (a) deformation of γ' and γ phases and (b) SAED-2.

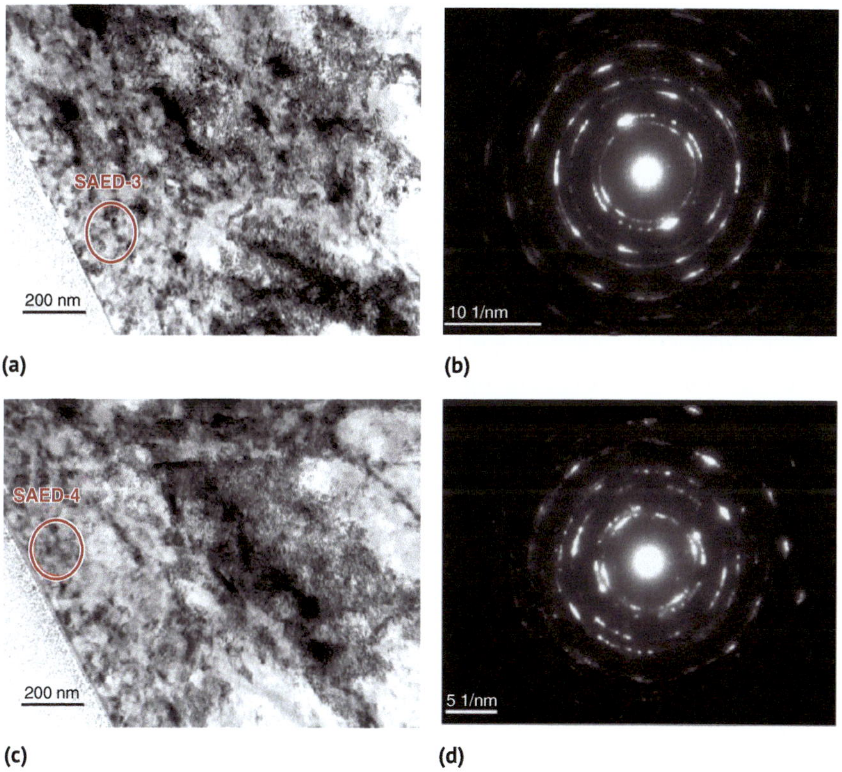

Figure 8.4 Cross-sectional TEM images of the microstructure at about 0.2 μm deep from ground surface in Sample I on TD-ND plane: (a) nanostructures in Region I; (b) SAED-3; (c) nanostructures in Region II; and (d) SAED-4.

and γ-channels. These grains, measuring less than 200 nm, exhibit a uniform distribution across the region, as depicted in Figure 8.4a and 8.4c. Electron diffraction patterns, showcasing nearly continuous rings in Figure 8.4b and 8.4d, indicate the presence of polycrystalline structures in the outermost layer of the nickel-based single-crystal superalloy. This finding suggests a higher level of shear strain in the severely deformed layer compared to the DFL. Moderate variations in the intensity of the diffraction rings imply the existence of low-angle grain boundaries, as reported in [1].

8.1.2 TEM Observations of Sample II

8.1.2.1 Nanostructure in the Superficial Layer

Using a similar TEM analysis approach, the deformation structures on the TD-ND plane of Sample II (processed with the parameter set described in Chapter 7) were examined, as shown in Figure 8.5. Two distinct layers—designated DFL and SDL—are clearly identifiable with well-defined boundaries (Figure 8.5a). The DFL typically contains dislocations and submicron grains that result from these dislocations (Figure 8.5b). Notably, dislocation walls remain within submicron-sized nucleation regions, which can promote further grain refinement and the formation of submicron grain boundaries with continued deformation. On the other hand, the SDL consists of uniformly distributed equiaxed nanograins, each smaller than 100 nm. The electron diffraction patterns in Figure 8.5c show circular rings, characteristic of nanograins with random orientations. Interestingly, nanoscale white particles are observed on the outermost surface in Figure 8.5b. Energy-dispersive Spectroscopy (EDS) analysis of points A (corresponding to the white particle) and B (representing a nanograin), as marked in Figure 8.5b, is presented in Figure 8.6a and b, respectively. The higher concentrations of O and Al at point A compared to point B suggest that the white particles may be oxides. Previous research has reported the formation of oxygen-related cavities or pores near the γ' phase on machined surfaces [1]. However, in this study, the O elements are likely sourced from air or alumina grains, because the particles' location is at the ground surface edge where has the significant strain and heat generated during grinding. This facilitates oxidation between the alumina grains and the workpiece materials [2].

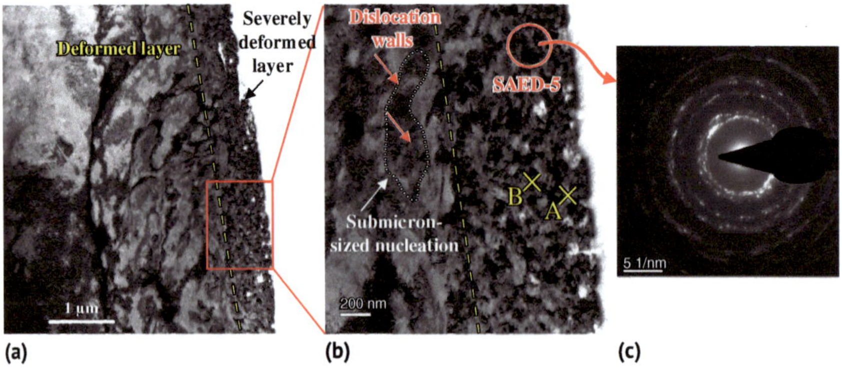

Figure 8.5 TEM images of the cross-sectional deformation microstructure on the TD-ND plane of Sample II: (a) whole; (b) regional; and (c) SAED-5.

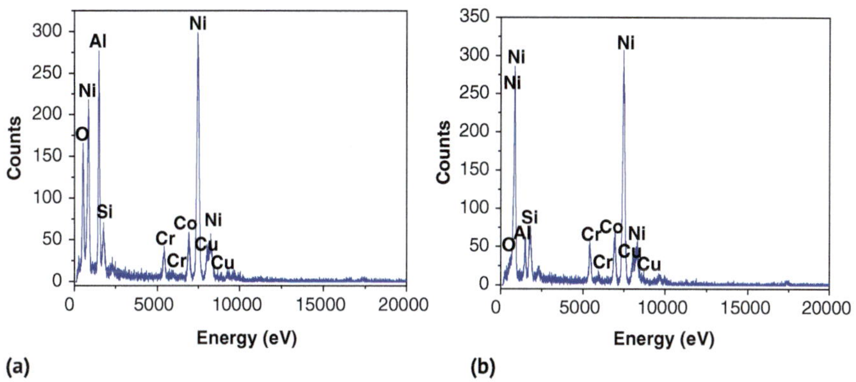

Figure 8.6 EDS analysis on (a) points A and (b) B marked in Figure 8.5b.

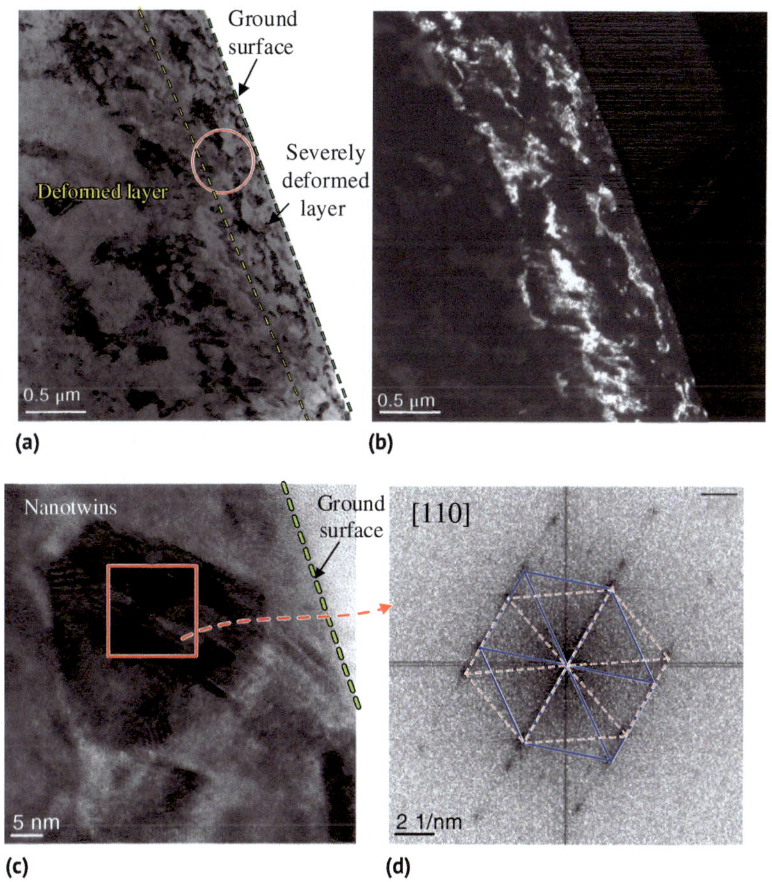

Figure 8.7 TEM observation of nanotwins in the superficial layer of blade root Sample II on LD-ND plane: (a) bright-field image; (b) dark-field image; (c) HRTEM image; and (d) Fast Fourier Transform (FFT) image.

8.1.2.2 Nanotwins in the Superficial Layer

The nano-twinned structure, shown in Figure 8.7, is observed within the surface layer of Sample II's blade root, particularly on the LD-ND plane. The morphology of both nanograins and submicron grains in the ground layer is clearly evident in the bright/dark-field images presented in Figure 8.7a and b. Nanotwins, situated about 10 nm beneath the surface, are identified by high-resolution transmission electron microscopy (HRTEM) in Figure 8.7c, exhibiting equiaxed crystals with distinct twin boundaries. These nanotwins are approximately 30 nm in size. A symmetrical crystal plane, derived from Fast Fourier Transform (FFT) analysis, is shown in Figure 8.7d.

Previous investigations have demonstrated that the nano-twinned structure exhibits exceptional physical characteristics, notably superior thermal stability and enhanced strength retention, when compared to fine-grained metals [3–5]. However, despite rigorous efforts in this research, this distinctive structure is rarely observed in other parts of the sample, indicating limited nanotwin formation within the blade root workpiece under the specified grinding conditions. Lu [4] proposed that minimizing twin boundary energies facilitate nanoscale twin formation. Therefore, twinning is more easily achieved in face-centered cubic (FCC) metals with low stacking fault energies (SFEs), whereas metals with high SFEs face significant challenges in producing twins [6]. For single-crystal nickel-based alloys, the underlying reason of limited twinning induction in the superficial layer remains unknown and deserves further exploration.

8.2 Analysis on Formation Mechanism of Nanostructures

Microstructures resulting from grain refinement due to plastic deformation typically include dislocations, submicron grains, and nanograins. These structures can be deliberately created using various methods such as cold rolling [7], mechanical attrition [8], and grinding [9]. Research indicates that the strain, strain rate, and strain gradient generated by these mechanical processes play a crucial role in the refinement of grains [10]. To investigate the grain refinement mechanism in this study, the strain and strain rate during grinding were estimated based on Merchant's theory of metal cutting mechanics. In this approach, it is assumed that material removal occurs through interaction with a single abrasive grain per unit volume, and the shear strain ε in the cutting zone is generated by a single grain per unit time. The shear strain ε and strain rate were then computed using equations outlined in Ref. [11].

$$\varepsilon = \frac{\cos \alpha_n}{\sin \varphi_n \cos(\varphi_n - \alpha_n)} \tag{8.1}$$

$$\varepsilon = \frac{\lambda v_s \cos \alpha_n \sin \varphi_n}{a_e \cos(\varphi_n - \alpha_n)} \tag{8.2}$$

$$a_e \approx 0.5a_{gmax} = \left[\frac{4}{C \cdot N_d} \cdot \frac{v_w}{v_s} \cdot \left(\frac{a_p}{d_s}\right)^{1/2}\right]^{1/2} \tag{8.3}$$

where α_n represents the abrasive grain rank angle; φ_n is the shear angle, as per references [12, 13]; λ represents the aspect ratio of the shear zone, which typically ranges from 6 to 12. The parameter a_e denotes the thickness of the shear zone, which is half of the maximum undeformed chip thickness (a_{gmax}). C is a constant associated with the abrasive rake angle, N_d refers to the number of active abrasive grains, and d_s represents the diameter of the grinding wheel.

The parameters utilized in this study are presented in Table 8.1. By integrating these parameters into Eqs. (8-1)–(8-3), one can approximate the shear strain and strain rate during the final grinding phase. Notably, the ground surface undergoes substantial shear deformation, reaching a maximum strain of 6.67 and a high strain rate of approximately 8.17×10^{-1}, which surpasses those induced by cutting processes by approximately three orders of magnitude, as reported in [14]. Additionally, due to repeated grinding of the blade root, a significant strain gradient accumulates within the superficial layer, promoting dislocation formation and increasing dislocation density in the workpiece.

8.2.1 Formation of Submicron-sized Grains

The results outlined above highlight the significant role of dislocation activity in modifying the microstructure within the ground layer. Specifically, to accommodate a higher density of dislocations, numerous lamellar structures form from cell blocks and dislocation walls, as shown in Figure 8.8. With increasing plastic strain, these lamellar structures become thinner and more numerous as additional

Table 8.1 Calculation of strain and strain rate in the shear zone during grinding.

Group No.	I	II
Grinding stages	Finish grinding	Finish grinding
Grain rank angle α_n	$-35°$	$-35°$
Shear angle φ_n	12°	10°
Constant C	1.428	1.428
Active grain number N_d	$6.1\,mm^{-2}$	$6.1\,mm^{-2}$
Grinding wheel diameter	387 mm	387 mm
Aspect ratio of shear zone λ	7	9
Maximum undeformed chip thickness a_{gmax}	0.85 μm	1.55 μm
Shear strain	5.78	6.67
Strain rate	$7.89 \times 10^7 \cdot s^{-1}$	$8.17 \times 10^7 \cdot s^{-1}$

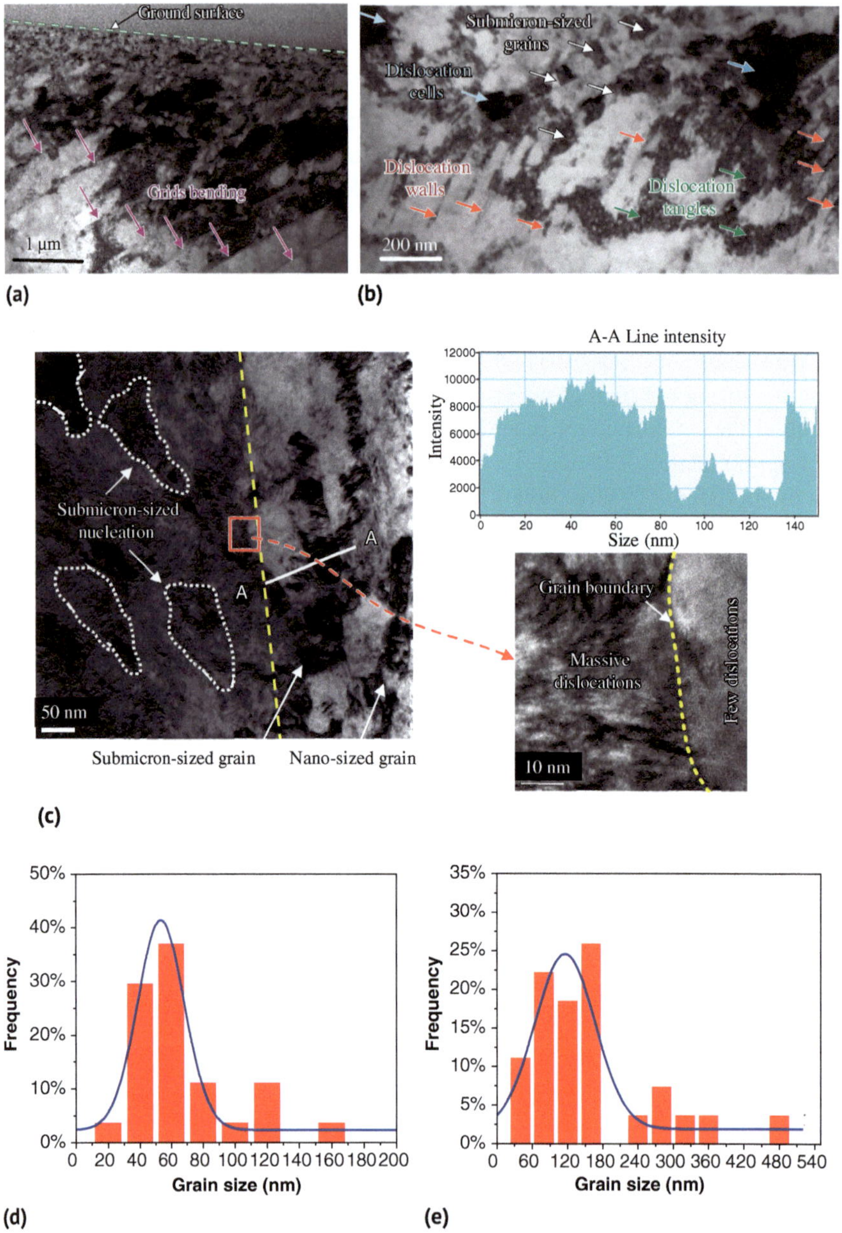

Figure 8.8 Formation of submicron-sized grains in the ground layer of blade root Sample I on LD-ND plane: (a) microstructure at a depth of 5 μm; (b) microstructure at a depth of 2 μm; (c) microstructure at a depth of 1 μm; (d) width of submicron-sized grain; and (e) length of submicron-sized grain.

boundaries develop (Figure 8.8a). These lamellar structures can further subdivide into several parallel layers through the expansion of dislocation walls that intersect the lamellar boundaries (Figure 8.8b). As a result, the spacing between the boundaries decreases, which promotes the nucleation and growth of submicron grains

(Figure 8.8c), potentially originating from the subdivided lamellar regions. This process is facilitated by the ongoing accumulation of dislocations at the boundaries. Furthermore, due to the uneven distribution of these extended boundaries, the submicron grains exhibit varied widths and lengths, with average dimensions of 66 nm and 158 nm, respectively, yielding an aspect ratio of approximately 2.5 (Figure 8.8c–e). In comparison to the initial nucleation stage (Figure 8.8c), these relatively coarse submicron grains display similar aspect ratios but have more defined boundaries and lower dislocation densities. Nevertheless, further grain refinement can be promoted by the dispersed dislocations present within the submicron grains.

This process is considered as the key mechanism, which is responsible for the refinement of grain structure and the emergence of submicron grain boundaries in the ground region of single-crystal nickel-based superalloys. As noted, grain formation is closely linked to the misorientation of grain boundaries. As strain increases, adjacent boundaries begin to slip along different crystal slip systems once the grain boundary misorientation surpasses a critical threshold. The continuous incorporation of dislocations into the grain boundaries under higher strains results in changes to the grain morphology and texture, alongside increased misorientation of the boundaries [15]. This leads to the subdivision of grains. It can thus be inferred that grain boundary misorientation is a consequence of the micron-scale splitting of single-crystal nickel-based superalloys.

8.2.2 Formation of Nano-sized Grains

The surface of the workpiece experiences significant plastic deformation, leading to substantial dislocation and grain boundary formation. This process provides favorable conditions for the development of nanoscale grains and the evolution of texture. The grinding process induces high strain, resulting in the formation of numerous dislocations. Due to the high strain rate and the limited time available for dislocation annihilation, the dislocation density on the surface rises sharply, as highlighted by Zhang and Shim [16]. Moreover, the large strain gradient necessitates the accumulation of geometrically necessary dislocations to maintain crystal continuity and deformation compatibility [17]. As a consequence, creep-feed grinding in the present study proves effective in inducing and accumulating dislocations, which in turn drives the formation of densely packed low-angle grain boundaries. To accommodate the accumulating dislocations, various dislocation structures, such as lines, tangles, walls, and cells, form sequentially and nucleate extensively, fostering grain boundary growth.

To minimize stored energy, these dislocation configurations gradually evolve into low-angle grain boundaries. The formation of nanograins is a direct consequence of the ongoing and intense interaction between dislocations and submicron grain boundaries. The fragmentation of submicron structures is illustrated in Figure 8.9a, which displays distinct nanograin morphologies and triple junctions, particularly in Figure 8.9b. The presence of nanograins with an aspect ratio near unity (48 nm in width and 67 nm in length) suggests the development of equiaxed crystalline structures (Figure 8.9c and d). Ding et al. [18] noted that

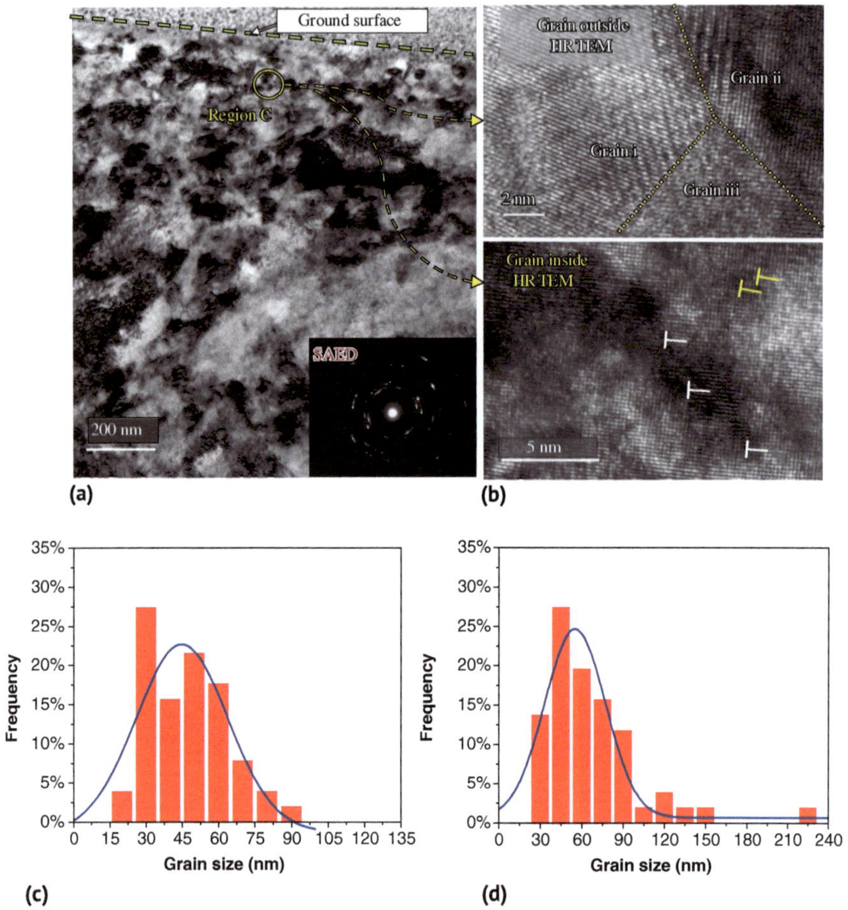

Figure 8.9 Formation of nano-sized grains in the ground layer of blade root Sample I on TD-ND plane: (a) microstructure at the topmost surface; (b) TEM image with high magnification of Region C; (c) width of nano-sized grain; and (d) length of nano-sized grain.

nanograins often exhibit an equiaxed shape and few dislocations. However, in this study, dislocations with varying Burgers vectors (denoted as "T" symbols in Figure 8.9b) are observed near the grain boundaries. The limited dislocation activity indicates that the nanograins are not entirely free of defects, leaving room for further grain segmentation and refinement.

Similar mechanisms for surface nanograin formation have been observed in the deformation of Cu [19] and Al alloys [1], where dislocation-grain boundary interactions are key factors. However, the formation of nanograins in the ground layer of the single-crystal nickel-based superalloy appears to follow a distinct process. This unique behavior may result from the gradual fragmentation of the γ and γ' phase lattices by grain boundaries originating from intense dislocation structures, along with the generation and redistribution of grain orientations.

8.3 Plastic Deformation and Microstructure Evolution of Single-crystal Nickel Superalloy

8.3.1 Transition of Single Crystal to Polycrystalline Grain

Broadly, the primary mechanisms accommodating plastic deformation encompass dislocation glide and deformation twinning. Analysis reveals that dislocation activity is crucial in the evolution of surface microstructure as a result of creep-feed grinding of single-crystal nickel-based superalloys. Notably, the temperatures encountered during creep-feed grinding, typically under 120°C [20], exert minimal impact on dislocations and are thus omitted from this study. For metals exhibiting low SFEs, deformation twinning is favored due to the hindrance of dislocation glide [21]. Conversely, metals with high SFEs primarily induce dislocations on specific low-index crystal planes initially [22]. In single-crystal nickel-based superalloys, dislocations are more prone to glide on the {111} crystal plane due to its heightened self-energy levels, facilitating dislocation glide in multiple directions within the same plane. This presents a significant opportunity for subdividing an initial coarse single crystal into numerous polycrystalline nanograins with random orientations.

The evolution traits of surface gradient microstructures in single-crystal nickel-based superalloys, induced by creep feed grinding, can be delineated into three stages, as illustrated in Figure 8.10:

- Stage I: Surface plastic deformation initiates the accumulation of dislocations at the microscopic level, followed by crystal twisting and rotation at the macroscopic level, accompanied by adjustments in the crystal orientation of the ground layer.
- Stage II: Through dislocation accumulation in both γ-channels and γ'-precipitates, the single crystal is subdivided into submicron grains characterized by low-angle boundaries.

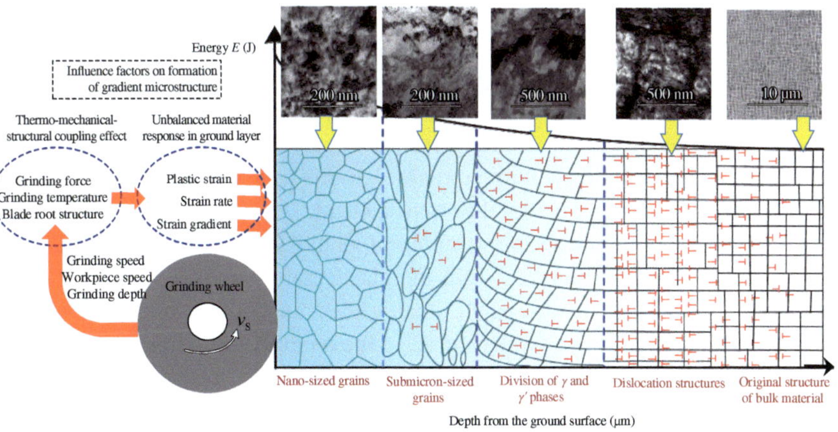

Figure 8.10 Transition process of single crystal to polycrystalline grain in the ground layer of single-crystal nickel-based superalloy.

- Stage III: As strains intensify, the submicron grains undergo further refinement, resulting in decreased grain size, increased grain misorientation, and a transition from irregular to equiaxed shapes.

The evolution of gradient microstructures in the superficial layer entails a transition from a coarse single crystal to nanoscale grains and from a uniform crystal orientation to a random polycrystalline orientation. This structural metamorphosis is primarily fueled by dislocation activities, contrasting with prior research on polycrystalline Inconel 718 alloy, where mechanical twinning subdivided initial coarse grains [23, 24].

Furthermore, numerous studies have shown that gradient structured materials, characterized by a nano-grained surface layer and a coarse-grained interior, exhibit unique mechanical properties. For instance, interstitial-free steel with gradient microstructures demonstrates enhanced yield strength [25], while copper with gradient microstructures exhibits a tensile strength tenfold higher than its non-gradient counterpart [26]. These observations suggest that incorporating gradient microstructures into the superficial layers of alloys could be a promising strategy to substantially enhance the performance of turbine blade roots. However, profile grinding complicates this process due to the intricate interaction of thermal, mechanical, and structural factors (as depicted in Figure 8.10). During profile grinding, significant variations arise in strain, strain rate, and strain gradient, rendering them challenging to predict accurately. Hence, to effectively control the development of gradient microstructures, it is crucial to establish a qualitative link between plastic deformation and the thermo-mechanical-structural coupling effect by tuning grinding parameters such as grinding speed, workpiece velocity, and grinding depth. This understanding is foundational for achieving optimal results in the manufacturing of turbine blade roots.

8.3.2 Nanostructure in High-generation Single-crystal Nickel Alloys

8.3.2.1 Conventional Grinding

For the blade root of the third-generation single-crystal nickel alloy DD90N (Figure 8.11), the thickness of the plastic deformation layer exceeds 3 μm. Within a depth of 1 μm from the grinding surface, due to severe plastic deformation, the small grids composed of the γ' strengthening phase and the γ matrix phase are broken and recombined, resulting in a new microstructure. SAED patterns (Figure 8.11 SADE-1) indicate that nano-sized grains have formed in these recombined regions, with grain sizes smaller than 100 nm.

For the blade root of the second-generation single-crystal nickel alloy DD98M (Figure 8.12), the microstructure within the plastic deformation layer is similar to that in Figure 8.11, with no observable small grids composed of the γ' strengthening phase and the γ matrix phase. However, in the region 1–2 μm from the grinding surface, a large number of dislocation structures are still detectable in the small grid areas. SAED patterns of the near-surface region show that nano-grained areas (approximately below 500 nm from the grinding surface) have formed, and dislocations are still present at the newly formed grain boundaries (Figure 8.12 HRTEM).

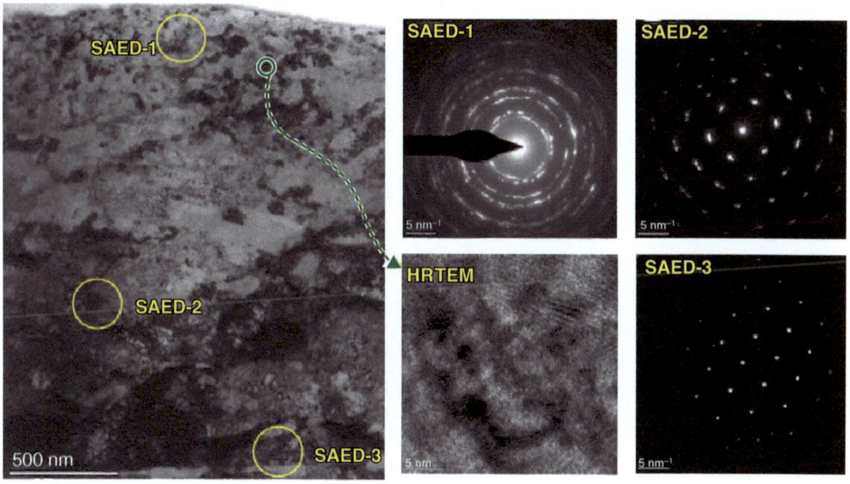

Figure 8.11 Micro-nano structure of the DD90N blade root surface layer under conventional grinding (grinding group I with mild parameters).

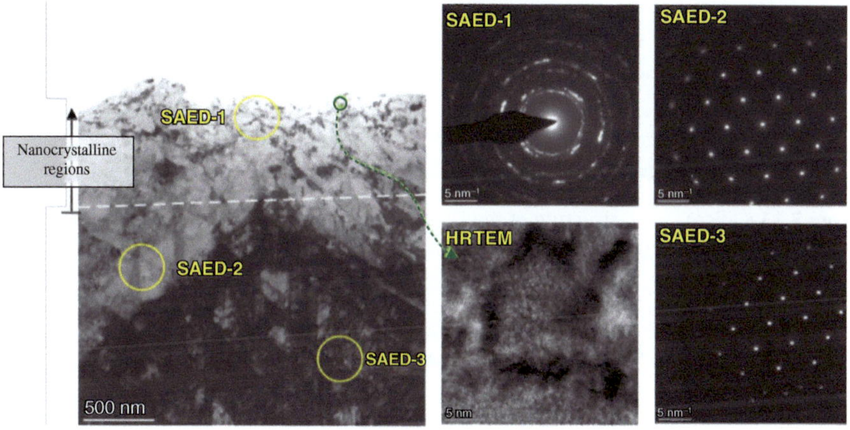

Figure 8.12 Micro-nano structure of the DD98M blade root surface layer under conventional grinding (grinding group I with mild parameters).

For the DD90N and DD98M blade root machining surface layers (Figures 8.13 and 8.14), in addition to the large dislocation structures in the dislocation and plastic deformation zones (Figures 8.11 and 8.12), SAED patterns can be detected in the near-grinding surface region (within approximately 2 μm). These diffraction spots show a relatively uniform ring pattern, indicating that the grains have been refined, with orientations close to random, and nano-crystalline particles have formed, with sizes ranging from 50 to 200 nm. Moreover, EDS surface scanning results reveal the formation of a large amount of oxidation products in the near-grinding surface region, as shown in Figures 8.13 and 8.14. The formation of these oxides generally has an adverse impact on the high-temperature service performance of the blade root, such as reducing material strength and fatigue life. This phenomenon can also be observed near the grinding surface of DD6 blade roots (Figure 8.5b), where some

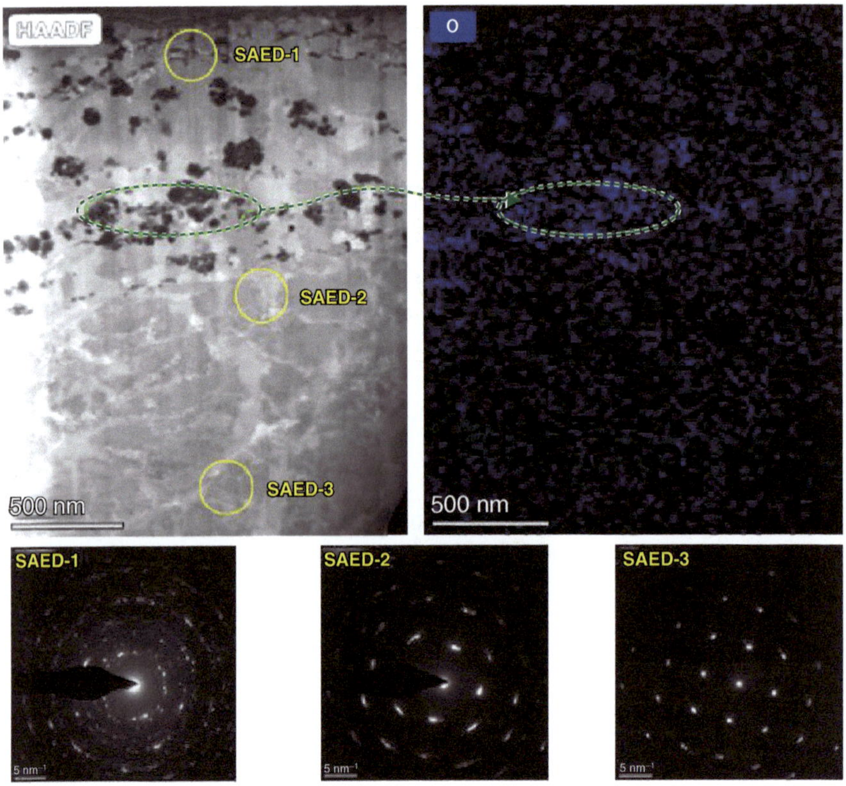

Figure 8.13 Micro–nano structure of the DD90N blade root surface layer under conventional grinding (grinding group II with aggressive parameters).

"white" nanoscale particles are present. EDS point scanning analysis was conducted on one of the "white" particles (particle A) (Figure 8.6a), and for comparison, the same analysis was performed on the nano-crystal B in the same area (Figure 8.6b). It can be observed that the oxygen content at point A is significantly higher than at point B, suggesting that particle A is an oxide. Related literature has reported the presence of oxidation pits or hole defects in the micro–nano structure of the ground surface of nickel-based alloys, and these defects were found in the γ' strengthening phase regions. This indicates that the oxide content in the γ' strengthening phase is lower than in the γ matrix phase. However, the source of the oxygen element has not been explained. Considering the complex force-thermal-structural coupling during the grinding process of the blade root, and the distribution of oxide particles at the very edge of the near-grinding surface, it is possible that the oxide formation is due to an oxidation reaction between the grinding wheel abrasives and the workpiece material during intense friction, which allows oxygen from the air to enter the grinding surface layer of the workpiece. In other words, the formed oxides are caused by external oxygen rather than oxygen elements originating from the workpiece material.

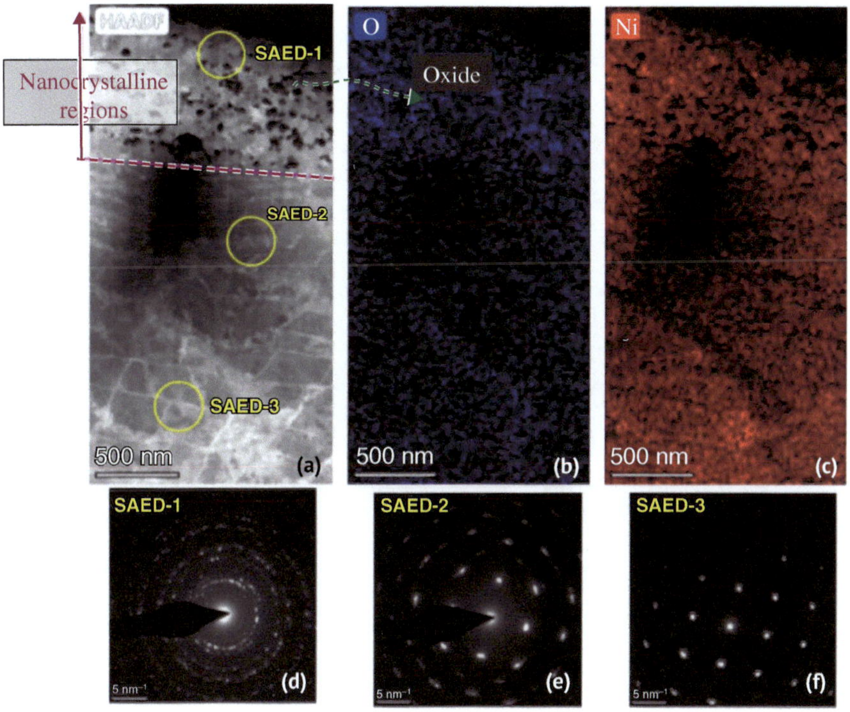

Figure 8.14 Micro–nano structure of the DD98M blade root surface layer under conventional grinding (grinding group II with aggressive parameters).

8.3.2.2 Vibration Grinding

The single-crystal blade root samples under vibration-grinding conditions were examined. When grinding conditions were mild (Figure 8.15), the plastic deformation zone at the surface of the DD90N blade root was approximately 1.5 μm in depth, and this zone was not uniform. This indicates that the microstructural deformation in the blade root is inconsistent due to the uneven external load distribution across different regions of the blade root. Additionally, the rotating electron diffraction pattern shown in SADE-2 suggests that sub-microcrystals may have formed within the plastic deformation zone, along with small-angle misorientations. Combining the bright-field TEM, dark-field TEM, and SAED patterns from the near-grinding surface region, it can be inferred that nano-grains with sizes less than 100 nm have formed in this area. A similar phenomenon was also observed in the surface layer of the DD98M blade root, as shown in Figure 8.16. Near the grinding surface, the small grids composed of γ phase and γ' phase gradually bend with depth, eventually returning to the matrix structure. The plastic deformation zone was also about 1.5 μm in depth.

When vibration was applied under aggressive grinding parameters (Figures 8.17 and 8.18), the plastic deformation zone in both the DD90N and DD98M blade roots increased to more than 3 μm, and the range of nano-crystal formation expanded

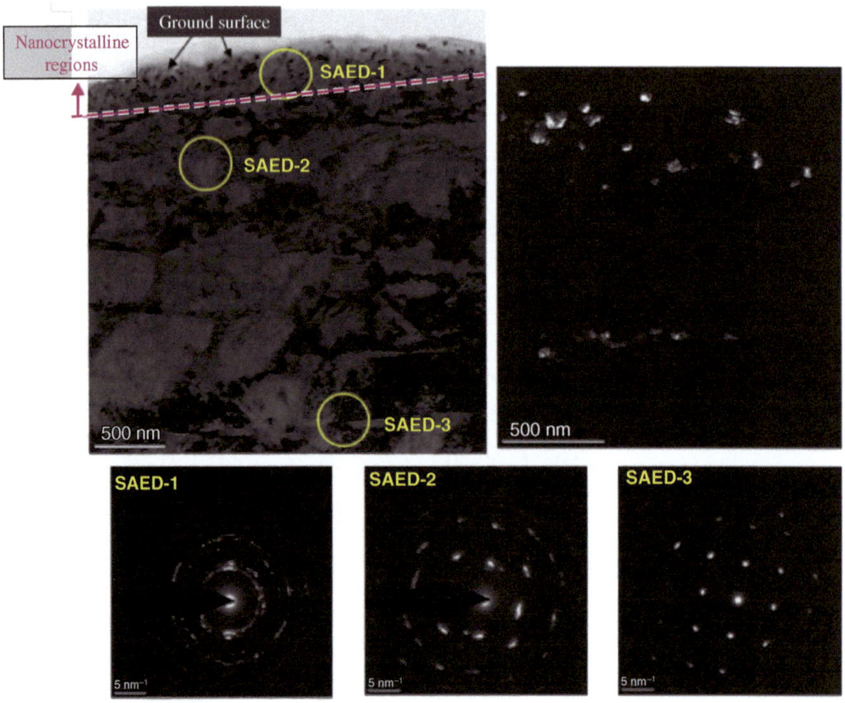

Figure 8.15 Micro–nano structure of the DD90N blade root surface layer under vibration grinding (grinding group I with mild parameters).

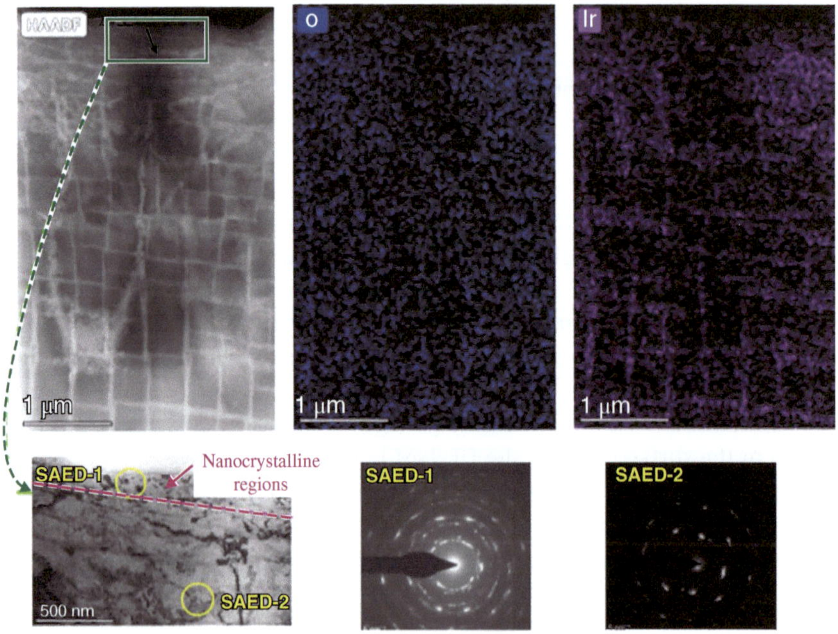

Figure 8.16 Micro–nano structure of the DD98M blade root surface layer under vibration grinding (grinding group I with mild parameters).

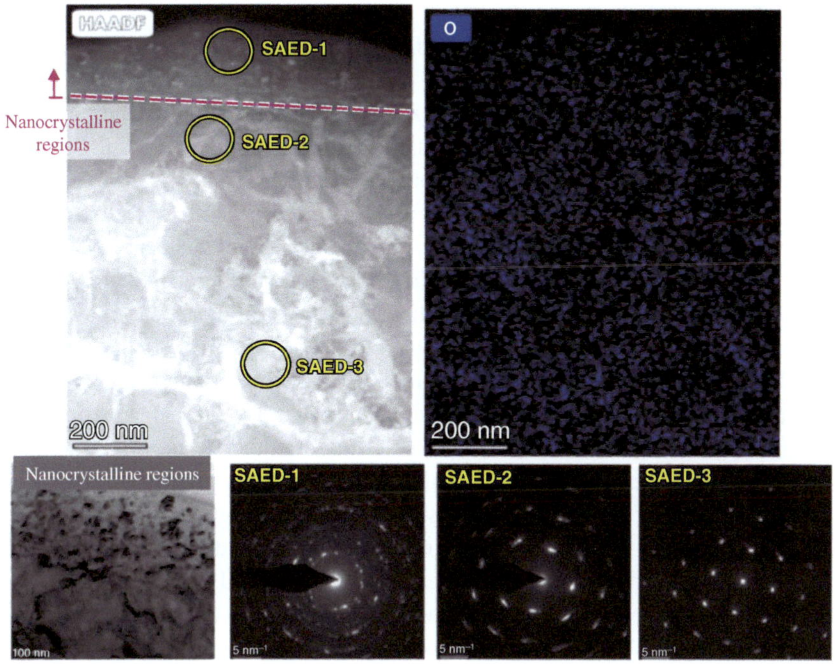

Figure 8.17 Micro–nano structure of the DD90N blade root surface layer under vibration grinding (grinding group II with aggressive parameters).

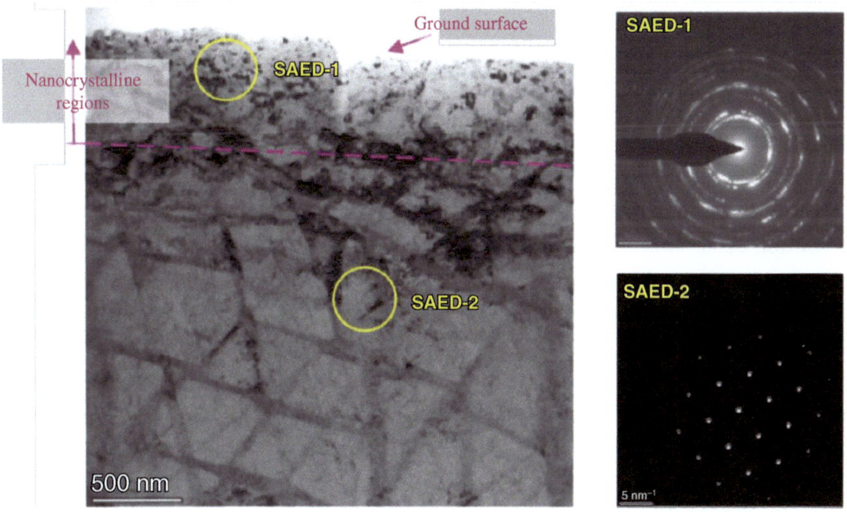

Figure 8.18 Micro–nano structure of the DD98M blade root surface layer under vibration grinding (grinding group II with aggressive parameters).

from 200 nm to approximately 500 nm. The nano-grain sizes remained similar to those under grinding group I, all being smaller than 100 nm. Despite the increased grinding load, EDS surface scanning results (Figures 8.17 and 8.18) revealed that no significant amount of oxidation products formed at the blade root surface layer.

8.3.2.3 Comparative Analysis

Based on the above detection results, both conventional grinding and vibration-grinding methods can introduce plastic deformation into the surface layer of the single-crystal blade root, and the structural distribution along the depth can be divided into the following zones: matrix → dislocation zone → plastic deformation zone → severe plastic deformation zone → grinding surface.

In the dislocation zone (Figures 8.15–8.17), small strains near the strain-free matrix initially lead to the proliferation and movement of dislocations. As the region gradually approaches the plastic deformation zone, the dislocation density increases further, and a large number of dislocation lines and dislocation entanglements become visible. These structures are mostly present in the γ' cubic phase, with dislocation accumulation also observed around the γ matrix phase. As the deformation process continues, with sustained energy input, these dislocations proliferate, move, accumulate, and interact, even undergoing dynamic recovery and annihilation, leading to the formation of thicker dislocation walls. These dislocations, in the form of single dislocations, dislocation entanglements, and dislocation walls, are stored within the crystal grains, serving to hinder dislocation motion and thus improving the macroscopic strength of the material. SAED patterns show sharp and clear lattice points, which are typical of γ' phase diffraction, indicating that although the dislocation zone is affected by plastic deformation, it still retains a single-crystal structure.

In the plastic deformation zone (Figures 8.15–8.17), as the material undergoes further deformation, the γ matrix phase and γ' cubic phase, which resemble channels in single-crystal alloys, undergo significant deformation due to intense extrusion and cutting action by the abrasives. The small grids transform into elongated rectangles. Near the end of the severe plastic deformation zone, the width of the γ' phase gradually decreases or even disappears, while near the dislocation zone, the γ' phase width gradually recovers to normal. The strong shear forces force a large number of dislocations to accumulate and entangle in the γ' cubic phase, forming high-density dislocation walls, which divide the original grains into submicron scale dislocation cells. This cutting action of the γ' cubic phase is one of the mechanisms for forming new grain boundaries and grains. The rotating diffraction patterns in the SAED maps confirm that the single crystal has undergone torsion or that new grains may have formed, with small-angle misorientations between them.

In the near-grinding surface region, regardless of the processing method, a comparison of the micro–nano structure of the obtained grinding surface layers reveals common characteristics. In the grinding surface layer, a gradient structure is formed along the depth direction. In the plastic deformation zone, significant sub-microcrystalline structures are observed, and in the severe plastic deformation zone, nano-crystals form with almost random orientations. The differences are that, in conventional grinding, due to the higher material removal rate under grinding parameters ii compared to condition i, the degree of plastic deformation increases, leading to a larger plastic deformation zone (over 3 μm). The nano-crystal formation region is around 1.5 μm, and oxidation products appear near the grinding surface. In contrast, when vibration grinding is used, the depth of plastic deformation induced (about 1.5 μm with grinding parameters i) is smaller than that induced by conventional grinding (about 2 μm with grinding parameters i), and the corresponding nano-crystal region decreases from 500 nm to 200 nm. This indicates that during the machining of single-crystal blade roots, the vibration-grinding method results in a thinner plastic deformation layer than conventional grinding, which helps reduce the adverse effects on the single-crystal workpiece.

References

1 Chen, Y.X., Yang, Y.Q., Feng, Z.Q. et al. (2017). Surface gradient nanostructures in high speed machined 7055 aluminum alloy. *Journal of Alloys and Compounds* 726: 367–377.

2 Wu, C.J., Guo, W.C., Li, R. et al. (2020). Thermal effect on oxidation layer evolution and phase transformation in grinding of Fe-Ni super alloy. *Materials Letters* 275: 128072.

3 Cheng, Z., Zhou, H.F., Lu, Q.H. et al. (2018). Extra strengthening and work hardening in gradient nanotwinned metals. *Science* 362 (6414): eaau1925.

4 Lu K. Stabilizing nanostructures in metals using grain and twin boundary architectures. Nature Reviews Materials, 2016, 1(5): 1–29.

5 Sun, L.G., Wu, G., Wang, Q. et al. (2020). Nanostructural metallic materials: Structures and mechanical properties. *Materials Today* 38: 114–135.

6 Liu, X.C., Zhang, H.W., and Lu, K. (2015). Formation of nano-laminated structure in nickel by means of surface mechanical grinding treatment. *Acta Materialia* 96: 24–36.

7 Wang, M., Jiang, Y.B., Li, Z. et al. (2020). Microstructure evolution and deformation behaviour of Cu-10 wt% Fe alloy during cold rolling. *Materials Science and Engineering: A* 801: 140379.

8 Liu, X.M., Nakatani, M., Gao, H.L. et al. (2021). Effect of staking fault energy on deformation mechanisms in Cu and Cu-30% Zn alloy with gradient structure obtained by SMAT. *Journal of Alloys and Compounds* 865: 158863.

9 Li, C., Cui, W.F., and Zhang, Y.S. (2017). Surface self-nanocrystallization of $\alpha + \beta$ titanium alloy by surface mechanical grinding treatment. *Metals and Materials International* 23 (3): 512–518.

10 Lu, K. (2014). Making strong nanomaterials ductile with gradients. *Science* 345: 1455–1456.

11 Malkin, S. and Guo, C.S. (2008). Grinding technology: theory and applications of machining with abrasives. *Mechanical Engineering*, 2e 130 (7): 43.

12 Yan, L., Li, X.K., Jiang, F. et al. (2009). Research on microscopic grain-workpiece interaction in grinding through micro-cutting simulation, part 2: Factorial study. *Advanced Materials Research* 849 (76–78): 15–20.

13 Ohbuchi, Y. and Obikawa, T. (2003). Finite element modeling of chip formation in the domain of negative rake angle cutting. *Journal of Engineering Materials and Technology* 125: 324–332.

14 Liao, Z.R., Polyakov, M., Diaz, O.G. et al. (2019). Grain refinement mechanism of nickel-based superalloy by severe plastic deformation—mechanical machining case. *Acta Materialia* 180: 2–14.

15 Hughes, D.A. and Hansen, N. (1997). High angle boundaries formed by grain subdivision mechanisms. *Acta Materialia* 45: 3871–3886.

16 Zhang, B. and Shim, V.P.W. (2010). Effect of strain rate on microstructure of polycrystalline oxygen-free high conductivity copper severely deformed at liquid nitrogen temperature. *Acta Materialia* 58: 6810–6827.

17 Yang, M.X., Pan, Y., Yuan, F.P. et al. (2016). Back stress strengthening and strain hardening in gradient structure. *Materials Research Letters* 4: 145–151.

18 Ding, J., Li, Q., Li, J. et al. (2018). Mechanical behavior of structurally gradient nickel alloy. *Acta Materialia* 149: 57–67.

19 Li, W.L., Tao, N.R., and Lu, K. (2008). Fabrication of a gradient nano-micro-structured surface layer on bulk copper by means of a surface mechanical grinding treatment. *Scripta Materialia* 59: 546–549.

20 Rowe, W. (2017). Temperatures in grinding—a review. *Journal of Manufacturing Science and Engineering-Transactions of the Asme* 139: 121001.

21 Lu, K., Lu, L., and Suresh, S. (2009). Strengthening materials by engineering coherent internal boundaries at the nanoscale. *Science* 324: 349–352.

22 Lu, L. (2004). Ultrahigh strength and high electrical conductivity in copper. *Science* 304: 422–426.

23 Laplanche, G., Wieczorek, N., Fox, F. et al. (2018). On the influence of crystallography and dendritic microstructure on micro shear behavior of single crystal Ni-based superalloys. *Acta Materialia* 160: 173–184.

24 Luo, J.T., Yu, W.L., Xi, C.Y. et al. (2019). Preparation of ultrafine-grained GH4169 superalloy by high-pressure torsion and analysis of grain refinement mechanism. *Journal of Alloys and Compounds* 777: 157–164.

25 Wu, X.L., Jiang, P., Chen, L., et al. (2014). Synergetic strengthening by gradient structure. *Materials Research Letters* 2: 185–191.

26 Fang, T.H., Li, W.L., Tao, N.R. et al. (2011). Revealing extraordinary intrinsic tensile plasticity in gradient nano-grained copper. *Science* 331: 1587–1590.

9

Microhardness and Residual Stresses in High-efficiency Grinding

9.1 Microhardness in Surface Grinding

The microhardness profiles, obtained perpendicular to the feed direction under diverse grinding conditions, are illustrated in Figure 9.1. Specifically, at a grinding speed of 25 m/s, a feed rate of 100 mm/min, and an axial grinding depth of 1.2 mm, DZ408 specimens reached a maximum microhardness of 440 $HV_{0.05}$, surpassing the bulk hardness by 3.3% and spanning approximately 20 μm in depth (depicted in Figure 9.1a). Conversely, GH4169 specimens demonstrated a slight softening of roughly 6.5%, while DD6 specimens exhibited minimal hardening with negligible microhardness variations. However, when the grinding parameters were adjusted to 25 m/s, 240 mm/min, and 1.5 mm, respectively, more notable alterations in microhardness were observed across all four workpiece types, as shown in Figure 9.1b. Under these altered conditions, the workpiece materials underwent thermal softening due to the predominant effect of thermal energy compared to mechanical deformation [1], with softening extents reaching 10%, 8.8%, and 6.4% for GH4169, DZ408, and DD6 specimens, respectively, exceeding a depth of 40 μm.

9.2 Microhardness in Profile Grinding

It is a well-documented fact that high grinding forces typically induce subsurface work hardening via plastic deformation [2, 3], whereas the extensive heat produced during grinding frequently results in subsurface thermal softening [4–6]. The interplay between these two mechanisms dictates the ultimate state of the subsurface. In blade root grinding, the final pass aims to eliminate grinding burrs and enhance subsurface mechanical properties using conservative parameters (such as decreased grinding depth in the sixth pass). Consequently, subsurface work hardening or thermal softening from previous passes may be partially alleviated. Ideally, a moderate degree of subsurface work hardening is favored, as it leads to compressive residual stress and improved wear resistance.

The fluctuations in microhardness beneath the ground surface for grinding cases I and II are presented in Figure 9.2a and b, respectively. In grinding case I, the

Grinding of Single-Crystal Superalloys: Fundamentals and Technologies, First Edition.
Wenfeng Ding et al.
© 2026 Wiley-VCH GmbH. Published 2026 by Wiley-VCH GmbH.

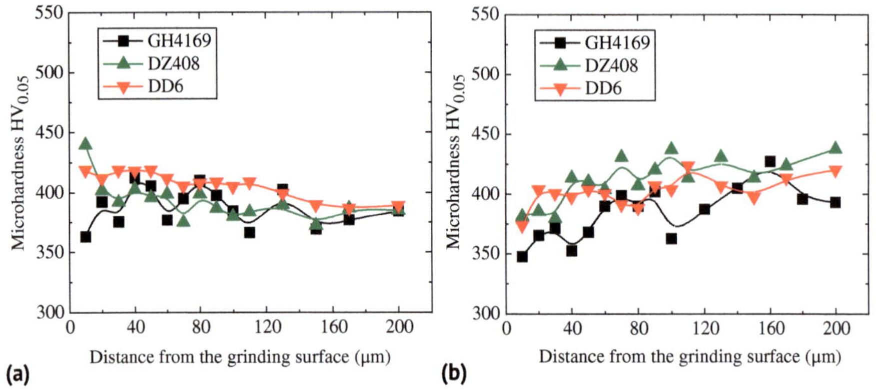

Figure 9.1 Microhardness variation beneath the ground surface: (a) $v_s = 25$ m/s, $v_w = 100$ mm/min, $a_p = 1.2$ mm and (b) $v_s = 25$ m/s, $v_w = 240$ mm/min, $a_p = 1.5$ mm.

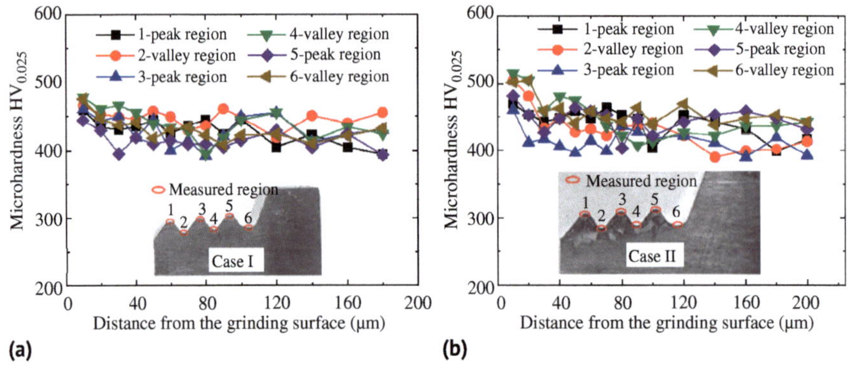

Figure 9.2 Microhardness variation on the cross section of the root workpiece: (a) grinding case I and (b) grinding case II.

near-surface microhardness of the root workpiece ranged from 444 to 478 $HV_{0.025}$, indicating work hardening with a confined surface hardening level of approximately 3.2–11.1% to a depth of less than 10 μm. In contrast, Figure 9.2b exhibits more substantial changes in microhardness (ranging from 469 to 516 $HV_{0.025}$) in grinding case II, with surface hardening levels of 9.1–20% to a depth of approximately 30 μm. These findings suggest a more prominent subsurface work hardening compared to thermal softening.

Additionally, microhardness measurements were conducted across different regions of the blade root. Under grinding condition I, the root peak region exhibited an average microhardness of 455 $HV_{0.025}$, reflecting a work hardening level of 5.8%. In contrast, the root valley region showed a higher degree of work hardening, reaching up to 10%. For grinding condition II, the average microhardness values increased to 466 $HV_{0.025}$ at the root peak and 507 $HV_{0.025}$ at the root valley, corresponding to work hardening levels of 8.4% and 17.9%, respectively. These findings suggest that, regardless of the grinding parameters, the root valley consistently experienced approximately twice the work hardening level of the root peak. This was attributed to the intensified plastic deformation from grinding forces concentrated in the valley areas.

9.3 Residual Stresses in Profile Grinding

Compressive residual stresses are known to improve the fatigue resistance of stress-sensitive components, such as turbine blade roots. However, due to the small size of the root workpiece and the specific testing requirements for measuring residual stress in single-crystal nickel-based superalloys, obtaining stress data from the root valley region using conventional stress analyzers is difficult. Therefore, the residual stress measurements presented in Figure 9.3 focus solely on the root peak regions of DD6 alloys under two different grinding conditions. As shown in Figure 9.3, the principal stresses σ_{11} and σ_{22} were −86 and −138 MPa for grinding case I, while for case II, they reached −185 and −440 MPa, respectively. In both cases, compressive residual stresses were observed.

The development of residual stresses on the ground surface is primarily driven by factors such as thermal expansion, phase changes induced by grinding temperatures, and plastic deformation caused by the grinding forces [7]. Upon examining subsurface microstructural changes and work hardening, it appears that grinding forces played a more significant role than grinding temperatures in generating these stresses in both grinding cases. Therefore, the residual stresses were mainly produced by the mechanical interactions between the grinding wheel and the workpiece. The observed variations in subsurface mechanical properties are attributed to plastic deformation resulting from these forces. Furthermore, as seen in Figure 9.3, more aggressive grinding parameters (such as those in grinding case II) lead to the removal of a larger volume of material, which increases the grinding forces and alters the workpiece's surface characteristics. Considering the surface topography, subsurface plastic deformation, and work hardening, the grinding parameters in case I may be more suitable for achieving a higher-quality machined surface in the creep-feed profile grinding of blade root components, even though case II generates higher compressive stresses on the ground surface.

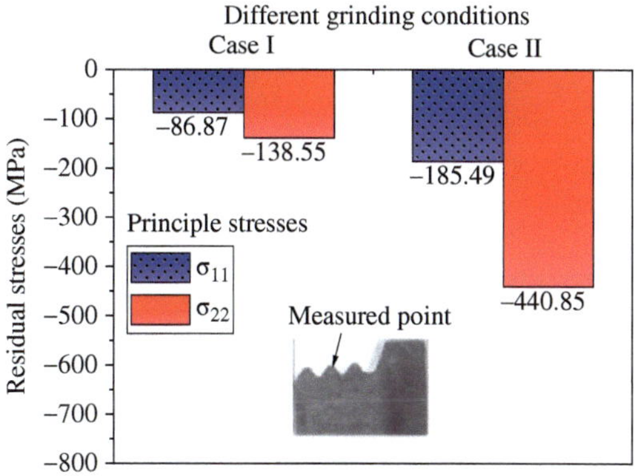

Figure 9.3 Residual stresses of the ground surface of the DD6 root workpiece.

References

1 Karpuschewski, B., Knoche, H., and Hipke, M. (2008). Gear finishing by abrasive process. *CIRP Annals - Manufacturing Technology* 57 (2): 621–640.
2 Zhang, Z.Y., Cui, J.F., Chang K.K. et al. (2019). Deformation induced new pathways in silicon. *Nanoscale* 11 (20): 9862–9868.
3 Xu, W.X. and Zhang, L.C. (2018). Tool wear and its effect on the surface integrity in the machining of fiber-reinforced polymer composites. *Composite Structures* 188: 257–265.
4 An, Q.L., Chen, J., Cai, X.J. et al. (2018). Thermal characteristics of unidirectional carbon fiber reinforced polymer laminates during orthogonal cutting. *Journal of Reinforced Plastics and Composites* 37: 905–916.
5 Liu, W.D, Liu, M., and Zhang, L.C. (2019). Thermally induced surface deterioration of glassy carbons. *Materials Research Express* 6 (7): 075607.
6 Ding, W.F., Xu, J.H., Chen, Z.Z. et al. (2010). Grindability and surface integrity of cast nickel-based superalloy in creep feed grinding with brazed CBN abrasive wheels. *Chinese Journal of Aeronautics* 23 (4): 501–510.
7 Chen, X., Rowe, W.B., and McCormack, D. (2000). Analysis of the transitional temperature for tensile residual stress in grinding. *Journal of Materials Processing Technology* 107 (1–3): 216–221.

10

Fretting Wear Behavior of the Machined Surface in High-efficiency Grinding

10.1 Friction Coefficient, Wear Volume, and Wear Rate During Fretting

The schematic representation of the fretting wear test setup is depicted in Figure 10.1. The tests were conducted using an oscillating friction tester, specifically the SRV-IV model from Optimal in Germany, under controlled conditions of 20 °C temperature and 50% relative humidity. The friction pair consisted of a YG6 ball made of hard alloy steel and rectangular specimens of nickel-based superalloy DD6. During the experiment, the YG6 ball, positioned as the upper specimen, underwent tangential reciprocation under a normal force designated as F_n. The parameters associated with the tests are listed in Table 10.1. Microhardness measurements were carried out using a Vickers indenter with an HXS-1000AK hardness tester, applying a load of 50 g for a duration of 15 seconds, as cited in [1].

Figure 10.2 illustrates the friction coefficient variation across different cycle numbers during the fretting tests. Notably, across all tests, an initial sharp increase in the friction coefficient was observed, followed by a stabilization phase. For polished samples, as the normal force increased from 50 to 180 N, the number of cycles required to reach a steady state decreased from 52,000 to 8000. This indicates that a higher normal force facilitates a quicker attainment of a steady state in the friction process. Additionally, the steady-state friction coefficients decreased as the normal force increased, specifically 1.00, 0.50, and 0.40 under normal forces of 50, 100, and 180 N, respectively. In contrast, for ground surfaces, the number of cycles to achieve a steady state was 1800, with a steady-state friction coefficient of 0.23. Both values were lower compared to those in polished samples. These findings align with the mildest fretting wear observed on ground surfaces, attributed to the roughness imparted by grinding.

To better understand the evolution of the friction coefficient, Figure 10.3 illustrates its fluctuation on a polished surface subjected to a 180 N load over 15,000 cycles. The fretting process can be broken down into three distinct stages: In Phase I, the friction coefficient increases rapidly within the first 1500 cycles, marking the initial contact and interaction between the two surfaces, which leads

Grinding of Single-Crystal Superalloys: Fundamentals and Technologies, First Edition.
Wenfeng Ding et al.
© 2026 Wiley-VCH GmbH. Published 2026 by Wiley-VCH GmbH.

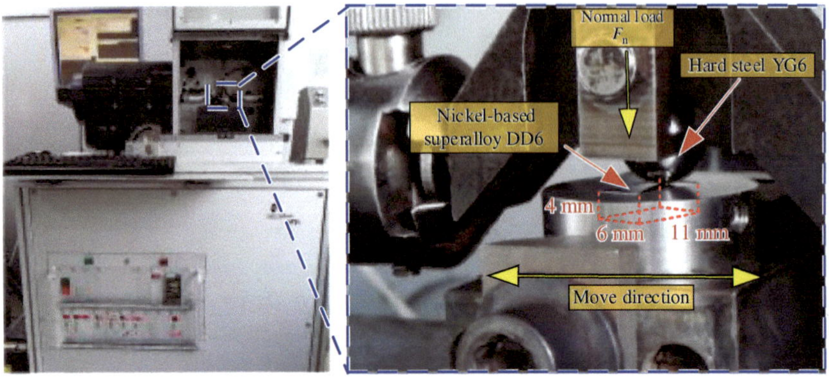

Figure 10.1 Setup of fretting wear test.

Table 10.1 Fretting wear tests parameters.

No.	Surface type	Normal load F_n (N)	Frequency (Hz)	Cycle number N	Displacement amplitude A (μm)
1	Polished	50	50	1×10^5	60
2	Polished	100			
3	Polished	180			
4	Ground	100			

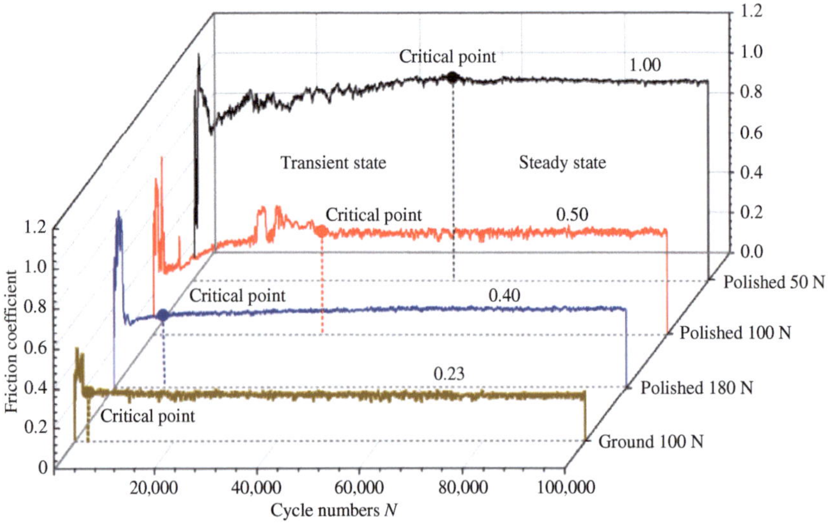

Figure 10.2 Variation of friction coefficient with cycle numbers.

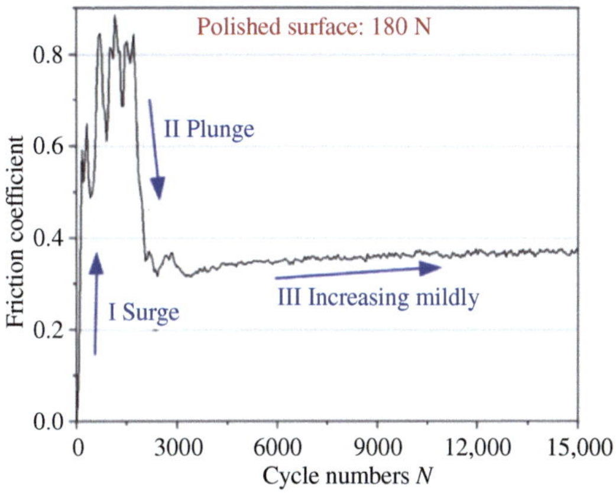

Figure 10.3 Variation of friction coefficient with cycle numbers in the early stage on polished surface.

to significant wear debris generation. During Phase II, the coefficient begins to decrease as some of the wear debris is expelled from the contact region. This expulsion reduces the friction, as the shear stress between the remaining debris and the surfaces is lower than that between the workpiece materials [2]. During these first two phases, the friction coefficient rises from 0 to 0.8 before dropping to 0.3, representing a transient period. In Phase III, the curve gradually ascends as the friction coefficient increases from 0.3 to 0.4, signaling the establishment of a steady-state friction condition. This phase is also associated with the development of tribolayers, whose structure influences the friction coefficient. For example, tribolayers with a high density of granular asperities lead to a higher friction coefficient than those with smoother surfaces. Previous studies have also shown that the friction coefficient is affected by the number of asperities in contact [3, 4].

Figures 10.4 and 10.5 depict the relationship between wear volume and wear rate, respectively, as a function of normal force. Both exhibit a similar trend: as the normal force increases, both wear volume and wear rate decrease. On polished surfaces, from 50 to 100 N, there is a significant reduction in wear volume by 92% (from $1.12 \times 10^6\ \mu m^3$) and wear rate by 96% (from $9.33 \times 10^6\ Pa^{-1}$). As mentioned earlier, a compacted layer forms under 100 N, which is absent at 50 N. Experimental results indicate that friction and wear decrease due to this protective layer. Consequently, from 50 to 100 N, the wear volume decreases by 92% due to the protective effect of the compacted layer formed at 100 N. When the normal force rises from 100 to 180 N, the wear volume increases slightly (to $0.13 \times 10^6\ \mu m^3$), suggesting a decrease in wear intensity despite the volume increase. For ground surfaces, both wear volume and wear rate attain their minimum values ($0.08 \times 10^6\ \mu m^3$ and $0.08 \times 10^6\ Pa^{-1}$, respectively) across all tests. For polished surfaces, the most severe wear occurs at 50 N, indicated by the significant decrease in volume and rate with increasing load up to 180 N. In contrast, the ground surface exhibits the lowest wear intensity under 100 N in the fretting wear test.

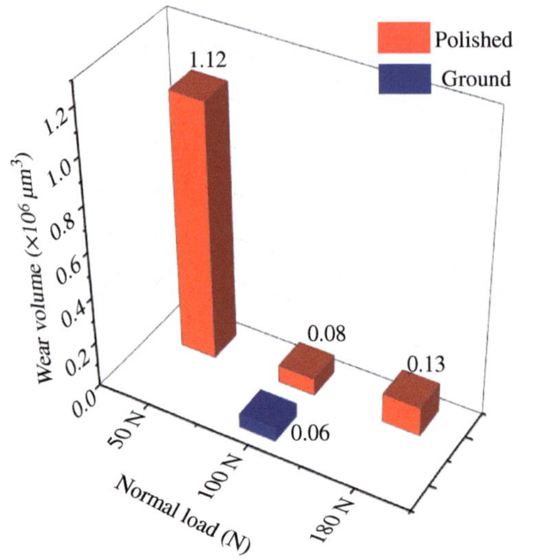

Figure 10.4 Variation of wear volume with normal load.

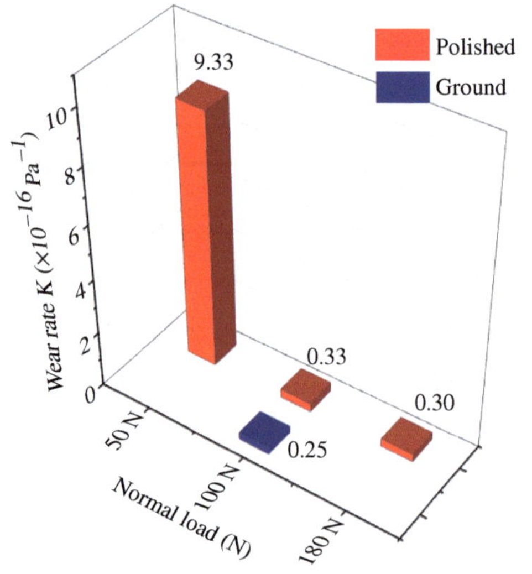

Figure 10.5 Variation of wear rate with normal load.

10.2 Surface and Subsurface Microstructure During Fretting

Figure 10.6 presents scanning electron microscopy (SEM) images and energy-dispersive spectroscopy (EDS) mapping patterns of wear scars on the ground surface after 100 N normal force, 50 Hz frequency, 60 μm displacement, and 1×10^5 cycles. The primary wear mechanism observed is abrasive wear, indicated

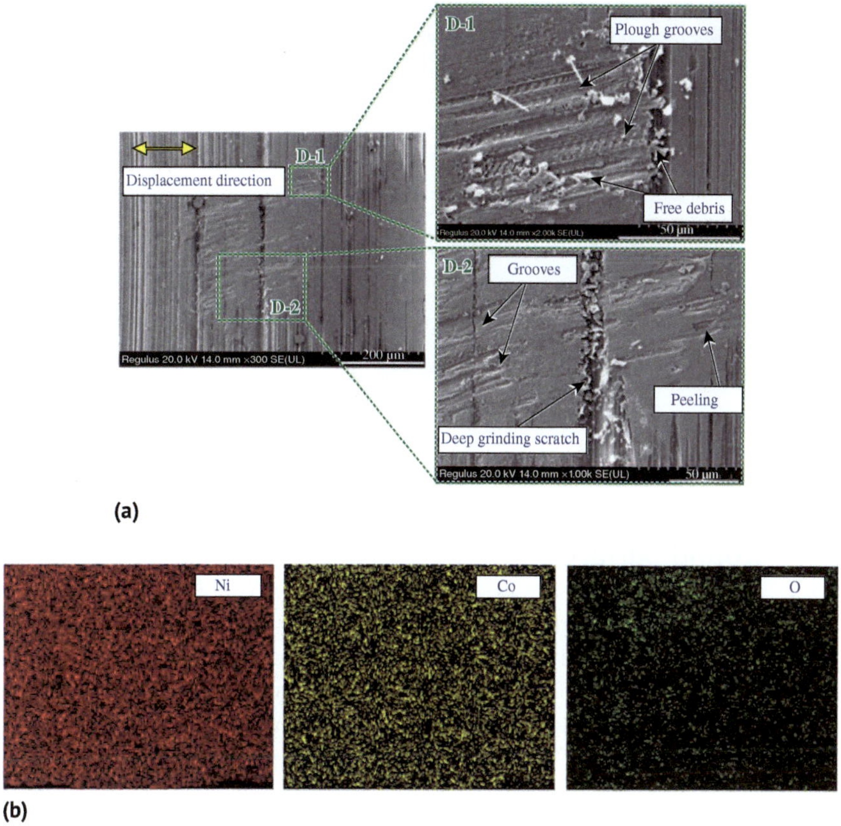

Figure 10.6 (a) SEM and (b) EDS mapping images of the wear scar on the ground surface.

by plough grooves aligned in the direction of sliding (Figure 10.6a). In area D-2, a deep grinding scratch is visible, filled with wear debris. The generated debris can be classified into three categories: debris that moves freely on the surface, debris that forms a compacted tribolayer under friction, and debris expelled into adjacent hollows and valleys.

Under the same testing conditions, the ground surface shows less wear compared to the polished surface, suggesting that the ground surface enhances resistance to fretting wear by better accommodating the debris formed during friction. This observation aligns with findings in [4, 5]. A decrease in wear debris during fretting wear reduces third-body abrasion between the contacting surfaces [6, 7]. The EDS mapping images (Figure 10.6b) show no signs of oxidative wear, as the concentrations of Ni, Co, and O remain consistent, and the wear scar maintains a shiny appearance (Figure 10.6a). SEM analysis of the wear scar on the ground surface (Figure 10.7) shows only slight plastic deformation near the flat surface, with no significant wear features. This is consistent with the topographical characteristics of the wear scar on the polished surface (Figure 10.7).

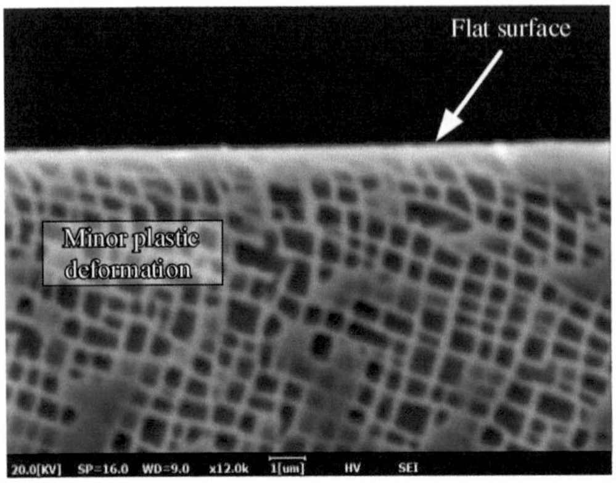

Figure 10.7 Cross-sectional SEM images of the wear scar on the ground surface.

10.3 Analysis of Fretting Wear Evolution on the Ground Surface

In conclusion, the ground surface demonstrates superior wear resistance in all fretting tests. Understanding the mechanism of wear debris evolution on the ground surface (Figure 10.8) during fretting wear tests could further enhance wear resistance. Therefore, this section delineates the evolution of wear debris into three stages as shown in Figure 10.9:

- Stage I: Initial contact and severe friction generate a large amount of wear debris. Some debris is expelled from the wear area without further interaction.
- Stage II: Remaining debris is compressed between the contact areas, causing plough grooves (Figure 10.8). Other debris is embedded in grinding scratches formed during grinding.
- Stage III: Deep scratches keep debris away from the fretting process. Accumulated debris in deep scratches remains unaffected during friction. Overflowing debris compresses to form a compacted tribolayer, enhancing surface wear resistance. Debris in shallow scratches continues to participate in the fretting process.

Figure 10.10 illustrates the hardness of both polished and ground surfaces, as measured along the depth direction. The results indicate that the ground surface exhibits a higher hardness, specifically 452 $HV_{0.05}$, which surpasses the polished surface's hardness by 13%. This disparity is attributed to the more pronounced plastic deformation observed near the surface of the ground samples, resulting from the grinding process. Additionally, such plastic deformation enhances the hardness, thereby contributing positively to the improvement of the wear-resistance capacity, as referenced in [8].

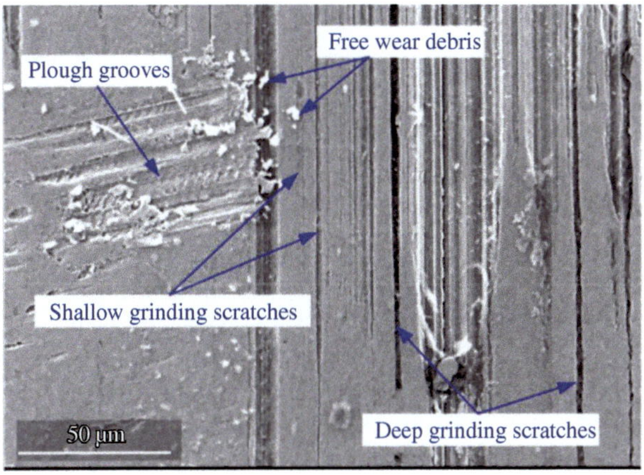

Figure 10.8 SEM image of localized wear scar on the ground surface.

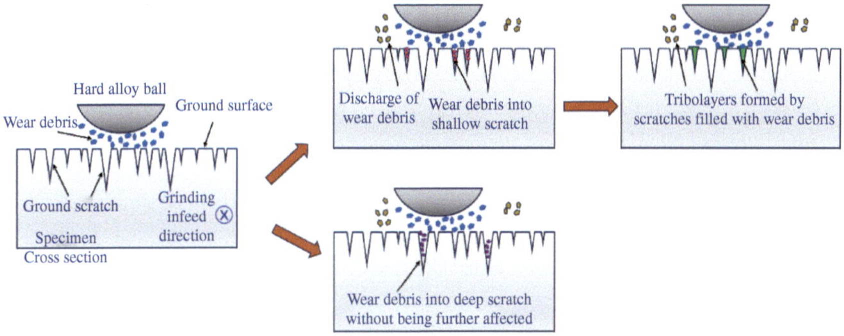

Figure 10.9 Schematic illustration of the evolution of fretting wear debris.

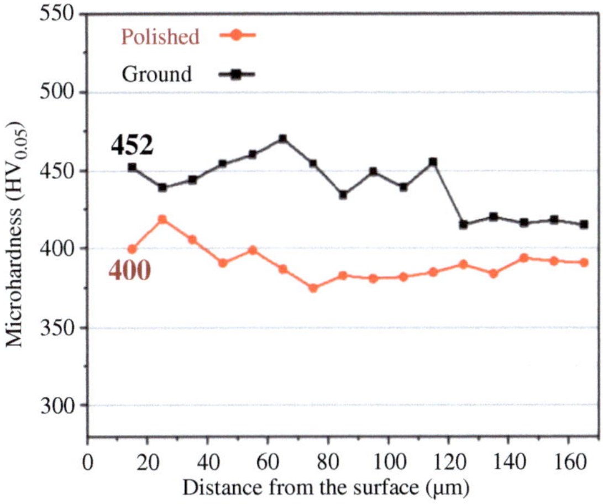

Figure 10.10 Hardness at different surfaces along depth.

References

1 Zhao, Y.J., Xu, W.H., and Xi, C.Z. (2021). Automatic and accurate measurement of microhardness profile based on image processing. *IEEE Transactions on Instrumentation and Measurement* 70: 1–9.

2 Varenberg, M., Halperin, G., and Etsion, I. (2002). Different aspects of the role of wear debris in fretting wear. *Wear* 252 (11–12): 902–910.

3 Schmitz, T.L., Action, J.E., and John, C. (2005). The difficulty of measuring low friction: uncertainty analysis for friction coefficient measurements. *Journal of Tribology* 127 (3): 673–678.

4 Amanov, A., Umarov, R., and Amanov, T. (2018). Increase in strength and fretting resistance of alloy 718 using the surface modification process. *Materials* 11 (8): 1366.

5 Pereira, K., Yue, T., and Wahab, M.A. (2017). Multiscale analysis of the effect of roughness on fretting wear. *Tribology International* 110: 222–231.

6 Amanov A. and Pyun Y-S. (2017). Local heat treatment with and without ultrasonic nanocrystal surface modification of Ti-6Al-4V alloy: mechanical and tribological properties. *Surface and Coatings Technology* 326: 343–354.

7 Amanov, A., Tsuboi, R., and Oe, H. (2013). The influence of bulges produced by laser surface texturing on the sliding friction and wear behavior. *Tribology International* 60: 216–223.

8 Bai, H.Q., Zhong, L.S., Kang, L. et al. (2021). A review on wear-resistant coating with high hardness and high toughness on the surface of titanium. *Journal of Alloys and Compounds* 882: 160645.

Part IV

Surface Integrity in Micro-grinding

The surface integrity of single-crystal nickel alloys in micro-grinding encompasses texture roughness, surface topography, subsurface deformation layer, and surface layer hardening during grinding. Grinding surface quality, a crucial indicator for evaluating the quality of grinding operations, significantly impacts the strength and mechanical properties of parts.

In contrast to traditional grinding, micro-grinding operations exhibit a size effect phenomenon. The rationale behind this stems from the fact that the disparity between the grinding wheel's diameter and the grinding depth remains comparatively minimal, thereby rendering precise control over texture roughness problematic during micro-grinding operations. Moreover, texture roughness in micro-grinding is more sensitive to the wear of the grinding wheel and the material properties of the workpiece. Additionally, the material removal mechanism during the grinding of single-crystal nickel alloys differs from that of polycrystalline materials. Single-crystal nickel alloys are typically considered hard-to-grind materials.

Numerous researchers have attached significant importance to exploring the surface integrity in the cutting processes of nickel-based superalloys. Nevertheless, systematic investigation into the micro-grinding surface quality of these materials remains relatively scarce. Consequently, it is imperative to conduct in-depth studies on the texture roughness of nickel-based single-crystal superalloys at the micro-grinding scale.

Within this section, the mathematical predictive model for grinding texture roughness, the regression simulation framework, and the effects of grinding process parameters and material characteristics on micro-grinding texture roughness are systematically explored. This investigation aims to establish a basis for predicting and regulating the micro-grinding texture roughness of nickel-based single-crystal superalloys. This study further investigates how grinding process parameters influence the surface and subsurface integrity during micro-grinding of single-crystal superalloys. It analyzes the contributing factors to microhardness alterations on the micro-ground surfaces of nickel-based single-crystal superalloys, while proposing strategies to suppress recrystallization in workpieces subjected to such micro-grinding operations. This research holds substantial importance for deepening the understanding of the micro-grinding mechanism in nickel-based single-crystal superalloys and for prolonging the service life of mechanical components.

11

Surface Roughness in Micro-grinding

11.1 Theoretical Model of Surface Roughness

Malkin et al. [1] proposed a grinding texture roughness model based on the interference motion mechanism between abrasive cutting edges and the workpiece. Two key assumptions were adopted during model formulation [2–4]: (i) material along the trajectory of each abrasive cutting edge was assumed to be completely removed, with the resulting surface texture exclusively formed through the cutting action; (ii) abrasive grains were considered to be uniformly distributed on the grinding wheel surface with a fixed spacing L, and all grains exhibited consistent protrusion heights from the wheel matrix.

Malkin established a peak-valley roughness theoretical model based on the geometric relationship in Figure 11.1:

$$R_t = \frac{1}{4}\left(\frac{v_w L}{v_s d_g^{1/2}}\right)^2 \tag{11.1}$$

The arithmetic mean roughness of the profile can be expressed as Eq. (11.2) [5]:

$$R_a = \frac{1}{9\sqrt{3}}\left(\frac{v_w L}{v_s d_g^{1/2}}\right)^2 \tag{11.2}$$

In the theoretical derivation of texture roughness in micro-grinding, Cheng et al. [6] considered the micro-scale size effect and introduced a size effect coefficient. The texture roughness model can be calculated as Eq. (11.3).

$$R_t = \frac{1}{2}M_r\left\{d_s - \left[d_s^2 - \left(\frac{v_w L}{v_s}\right)^2\right]^{1/2}\right\} \tag{11.3}$$

Malkin pointed out that the calculated value of Eq. (11.2) differs significantly from the experimental value by nearly 1000 times. The model proposed by Cheng et al. can only roughly describe the surface in the micro-grinding of hard and brittle materials. These models are independent of the grinding depth (a_p), yet the grinding depth has a significant impact on texture roughness in micro-grinding, which

Grinding of Single-Crystal Superalloys: Fundamentals and Technologies, First Edition.
Wenfeng Ding et al.
© 2026 Wiley-VCH GmbH. Published 2026 by Wiley-VCH GmbH.

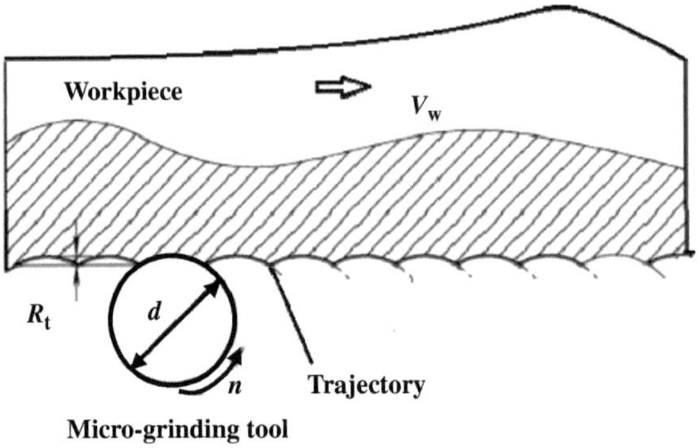

Figure 11.1 Surface texture and texture roughness of micro-grinding.

cannot be ignored. Moreover, the protrusion height of abrasive grains on the surface of the micro-grinding rod is nonuniform.

Thus, this study develops a theoretical framework for the micro-grinding of nickel-based single-crystal superalloys, in which the protrusion heights of abrasive grains on the micro-grinding tool surface are characterized by a Rayleigh distribution. Furthermore, the real geometric characteristics of abrasive grains are explicitly incorporated into the model formulation.

During the electroplating fabrication of CBN micro-grinding rods, the surface abrasive grain distribution exhibits stochastic characteristics, while the abrasive grain dimensions display inherent heterogeneity. This leads to an uneven protrusion height of the abrasive particles on the surface of the micro-grinding rod. It is assumed that the protrusion height of abrasive grains follows a normal distribution, and Hecker assumed it obeys a Rayleigh distribution. The normal distribution requires two parameters, μ and σ, to be determined, while the Rayleigh distribution only needs one parameter. Thus, relatively speaking, the Rayleigh distribution is easier to determine [7]. This study assumes that the abrasive grain protrusion heights on the micro-grinding tool surface obey a Rayleigh distribution. Consequently, the undeformed chip thickness (h) of individual abrasive grains also adheres to this probabilistic distribution pattern. The mean value of the Rayleigh distribution is defined by Eq. (11-4), presented as follows:

$$E(h^2) = \int_0^{\infty} h^2 f(h) \mathrm{d}h = \int_0^{\infty} (h^3/\beta^2) e^{-h^2/2\beta^2} \mathrm{d}h = 2\beta^2 \tag{11.4}$$

Considering the geometric characteristics of abrasive grains on the micro-grinding tool surface, four key assumptions are put forward as follows:

1. The abrasive grains are frustum-shaped, with a taper angle of γ.
2. All abrasive grains in the grinding area on the micro-grinding rod participate in grinding and material removal.
3. The abrasive grains are arranged parallel to the surface of the micro-grinding rod.
4. There is no spring back on the grinding surface.

The arithmetic average of grinding texture roughness is:

$$R_a = \frac{1}{l}\int_{y_{min}}^{y_{max}} |z - z_{CL}| P(z)dz \tag{11.5}$$

In this context, the symbol l denotes the sampling length, while z signifies the contour height generated by abrasive grains on the workpiece surface. Let the probability of an abrasive grain exhibiting a specific protrusion height z be expressed as $P(z)$. The parameters y_{max} and y_{min} correspond to the maximum and minimum values of the groove depth within the grinding surface profile, respectively.

Figure 11.2 illustrates the theoretical surface topography generated as an abrasive grain slides across the workpiece surface, where a reference central line L exists at $Z = Z_{CL}$, a position that ensures the shaded regions above and below the line possess equal area. The surface roughness value can be computed based on the areas of these upper and lower shaded regions relative to the Z_{CL} reference line in Figure 11.2.

According to the definition of R_z, the area of the shadow above line L is equal to the area of the shadow below line L. The p' and p'', respectively, represent the probabilities of grinding groove depths being lower and higher than Z_{CL}, which can be obtained as Eqs. (11-6) and (11-7).

$$p'E(A') + 2p''E(A''_{top}) = p''E(A''_{bottom}) \tag{11.6}$$

$$p' = \int_0^{z_{CL}} f(h)\mathrm{d}h \tag{11.7}$$

$$p'' = \int_{z_{CL}}^{\infty} f(h)\mathrm{d}h \tag{11.8}$$

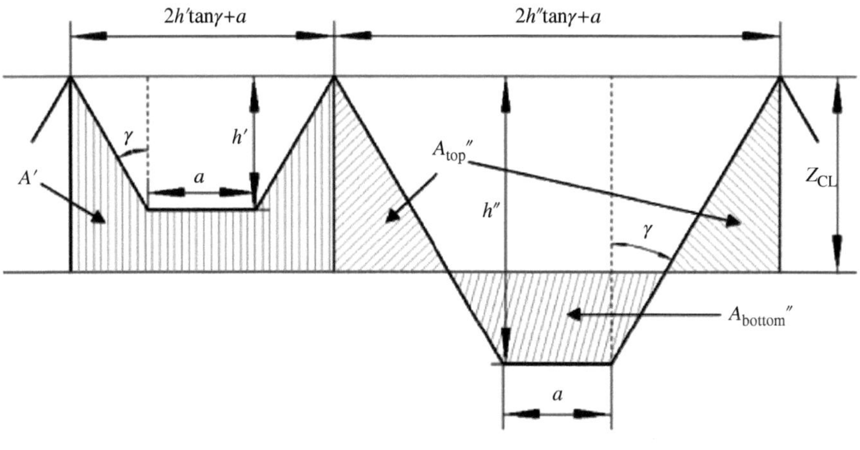

Figure 11.2 Theoretical morphology of the abrasive slipping over the workpiece surface.

According to the definition of the texture roughness of the section profile, the expected value of R_a for a groove depth less than Z_{CL} in Figure 11.2 is obtained as Eqs. (11.9) and (11.10).

$$E(R_a'') = \frac{\frac{a}{2}E(h'') + 2z_{CL}{}^2\tan\gamma}{2E(h'')\tan\gamma + a} + 0.5E(h') - z_{CL} \tag{11.9}$$

$$\int_{x_1}^{x_2} hf(h)\mathrm{d}h = \left[-he^{-\frac{h^2}{2\beta^2}} + \frac{\sqrt{2\pi}}{2}\sigma Erf\left(\frac{h}{\sqrt{2\sigma^2}}\right)\right]_{x_1}^{x_2} \tag{11.10}$$

11.2 Influence of Grinding Parameters

Figure 11.3 shows the impact of different spindle speeds on the experimental texture roughness R_a and the predicted texture roughness R_a for nickel-based single-crystal superalloys (DD98 and DD5) micro-grinding. Figure 11.4 presents the three-dimensional (3D) topography, cross-sectional profile, and the probability distribution of 3D profile heights of the DD98 grinding surface at various spindle speeds, with a clear surface texture.

As illustrated in Figure 11.4, at lower spindle speeds, the contour grooves on the ground surface exhibit broader and deeper profiles, accompanied by a darker topography appearance and a more pronounced height variation in the cross-sectional contour. As spindle speed increases, the 3D surface topography transitions to a progressively smoother morphology, reflecting reduced groove depth and improved surface uniformity. The height difference of the cross-sectional contour decreases, and the

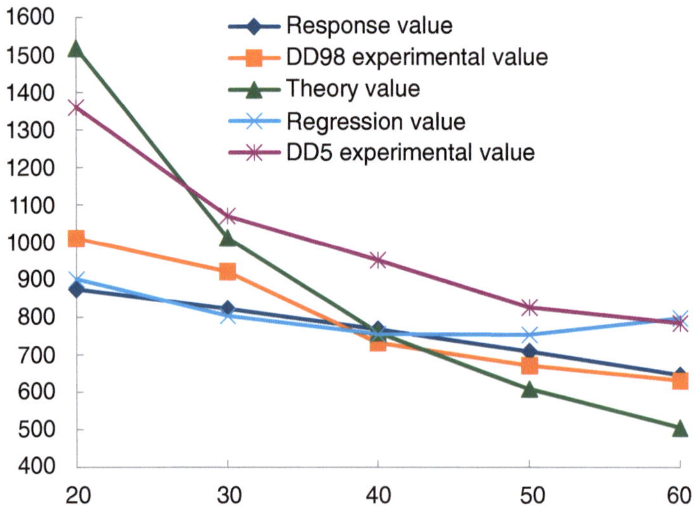

Figure 11.3 Influence of spindle speed on texture roughness.

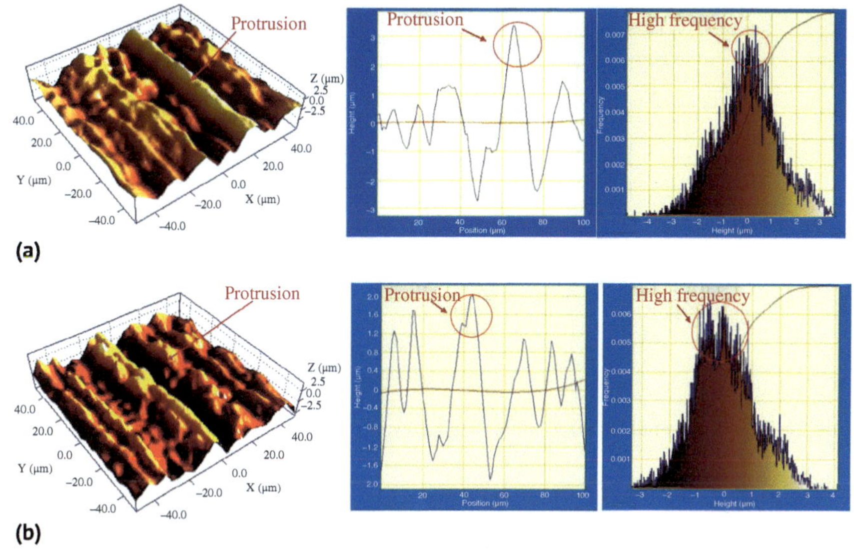

Figure 11.4 3D grinding surface topography with different spindle speeds: (a) $v_s = 20$ kr/min, $v_w = 120$ μm/s, $a_p = 9$ μm and (b) $v_s = 30$ kr/min, $v_w = 120$ μm/s, $a_p = 9$ μm.

height distribution of the 3D topography becomes more concentrated. Although some parts of the contour may be prominent and high due to slight vibrations during grinding, the overall surface is smooth, with shallow grooves and small roughness values.

From Figures 11.3 and 11.4, it can be seen that as v_s increases, both the test and predicted values of the texture roughness in the micro-grinding of nickel-based single-crystal superalloys decrease, and the decreasing trend becomes more gradual. The reasons are as follows: a higher spindle speed means a larger linear velocity of abrasive particles. At lower spindle speeds, the increased number of abrasive grains engaging with the workpiece per unit time diminishes the undeformed cutting thickness of individual grains, thereby reducing grinding forces. Conversely, at higher speeds, the propagation rate of material plastic deformation lags behind the grinding linear velocity, enabling material removal without significant plastic flow. This facilitates timely chip evacuation, as most grinding heat is carried away by the discharged chips, minimizing thermal accumulation. Consequently, both grinding forces and the likelihood of chip redeposition are substantially reduced, leading to finer texture roughness. However, once the spindle speed exceeds an optimal threshold, its effect on roughness diminishes progressively. Excessively high speeds may induce machine tool vibrations, which in turn exacerbate surface roughness. In contrast, low-speed operations result in larger single-grit cutting thicknesses, causing inadequate chip evacuation and increased plowing effects, ultimately producing a coarser ground surface.

Figure 11.5 demonstrates the influence of varying workpiece feed rates on both the measured and predicted R_a values of the micro-ground surfaces for nickel-based single-crystal superalloys (DD98 and DD5). Conversely, Figure 11.6 displays the 3D surface morphology, cross-sectional profiles, and probability distribution of 3D profile heights for the DD98 ground surface under different workpiece feed rates.

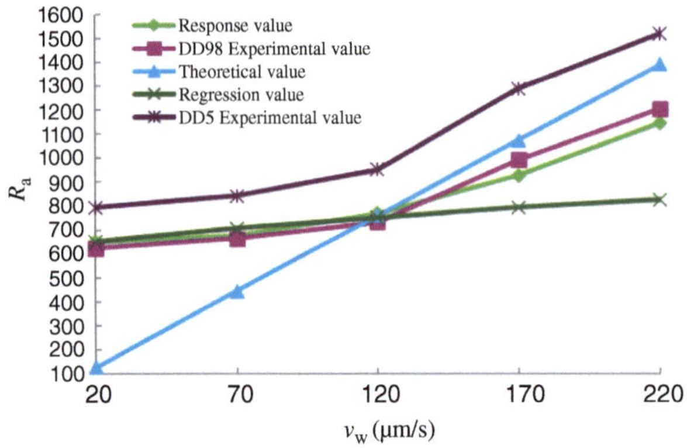

Figure 11.5 Influence of feeding rate on texture roughness.

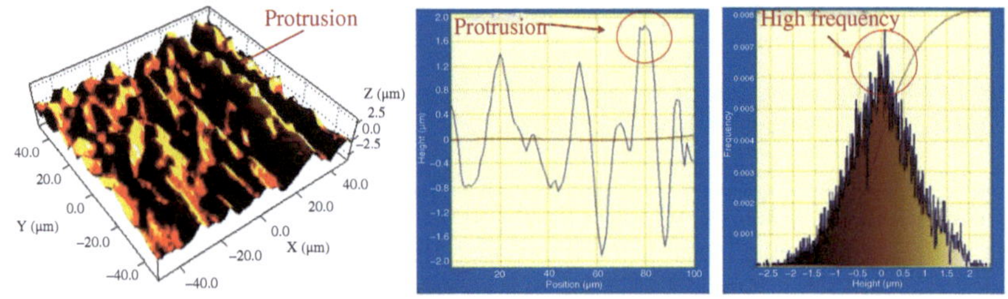

(a)

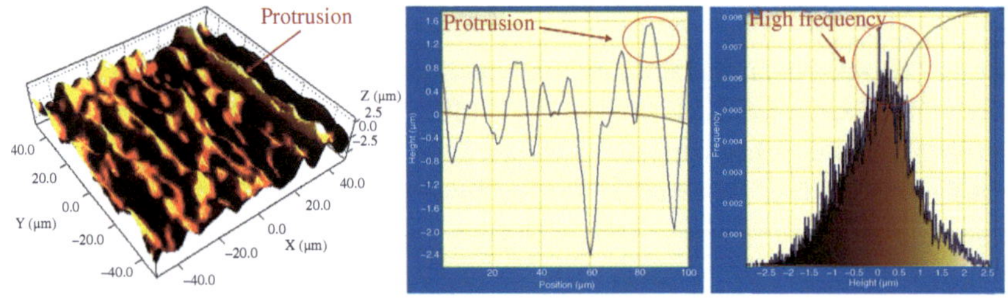

(b)

Figure 11.6 3D grinding surface topography with different feeding rates: (a) $v_s = 40$ kr/min, $v_w = 20$ μm/s, $a_p = 9$ μm and (b) $v_s = 40$ kr/min, $v_w = 70$ μm/s, $a_p = 9$ μm.

As evident from Figure 11.6, at low feed speeds, the grinding surface profile exhibits narrow and shallow grooves with minimal depth, presenting a relatively smooth surface topography. The cross-sectional profile features a small height variation, and the distribution of profile heights centered around the zero reference exhibits a narrow range. In contrast, as the feed speed increases, the 3D surface topography becomes progressively rougher. The grooves in the 3D morphology deepen and widen, the height differential in the cross-sectional profile increases significantly, and the distribution range of profile heights broadens, markedly distinct from the narrow, low-amplitude distribution observed at lower feed rates.

As observed in Figure 11.6, lower feed speeds result in grinding surface profiles with narrow, shallow grooves and a relatively smooth topography, characterized by minimal cross-sectional height variation and a narrow distribution of profile heights centered around the 0 reference plane. In contrast, increasing the feed speed leads to a progressive deterioration in surface finish, where the three-dimensional topography exhibits broader, deeper grooves and a marked expansion in the range of profile height distributions—signaling a transition toward rougher surface textures. The gullies in the 3D topography become wider and deeper, the height difference of the profile section increases, and the width of the 3D profile height distribution becomes wider, unlike the narrower and lower-range distribution when the feed speed is low.

Figure 11.7 shows the effect of different grinding depths on the measured and predicted R_a values of the micro-grinding surface of nickel-based single-crystal superalloys (DD98 and DD5).

As evident from Figure 11.8, smaller grinding depths yield grinding surface profiles with narrow, shallow grooves and a relatively smooth topography, characterized by minimal cross-sectional height variation and a confined distribution of contour heights centered around the 0 reference plane. Conversely, as the grinding depth increases, the 3D surface morphology undergoes a notable transformation: the topography becomes progressively rougher, with deepened and broadened grooves evident in the 3D profile. This is accompanied by a significant increase in cross-sectional height differential and a widening of the contour height distribution range—markedly distinct from the narrow, low-amplitude distribution observed at lower depths. The intensification of surface roughness with increasing depth is further reflected in the darker visual appearance of the 3D topography, directly correlating with the enhanced groove dimensions and height fluctuations.

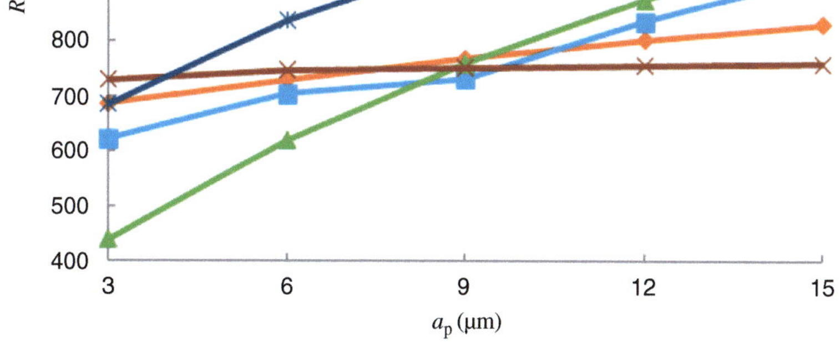

Figure 11.7 Influence of grinding depth on surface roughness.

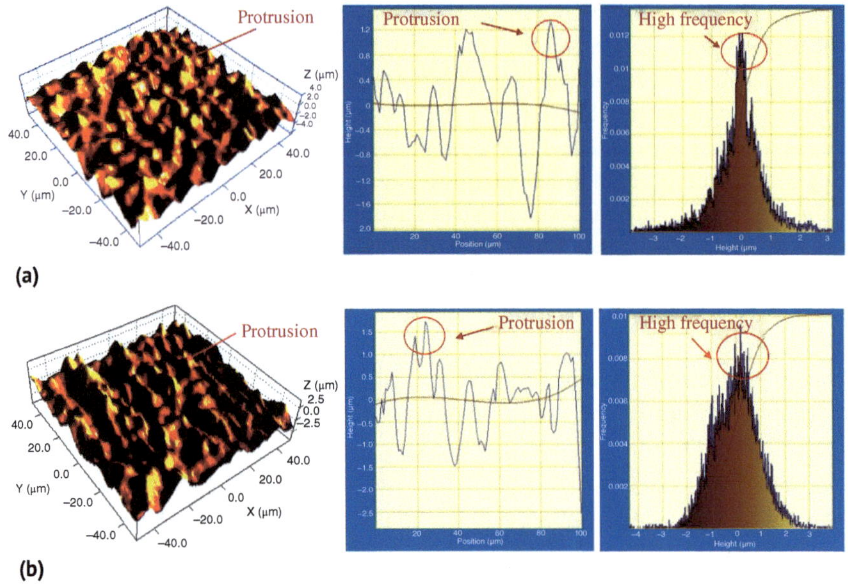

Figure 11.8 3D grinding surface topography with different grinding depths: (a) v_s = 40 kr/min, v_w = 120 μm/s, a_p = 3 μm and (b) v_s = 40 kr/min, v_w = 120 μm/s, a_p = 6 μm.

As illustrated in Figures 11.7 and 11.8, both experimental and predicted values of micro-grinding texture roughness for nickel-based single-crystal superalloys exhibit a consistent upward trend with increasing grinding depth (a_p). This phenomenon can be attributed to two primary factors: first, as a_p increases, the contact arc length within the grinding zone expands, impeding efficient heat dissipation and leading to elevated thermal accumulation. Second, the undeformed cutting thickness of individual abrasive grains rises proportionally with deeper grinding depths, enabling each grain to remove a larger volume of material. Consequently, this results in deeper and wider grooves on the ground surface, thereby elevating the overall texture roughness. The combined effect of intensified thermal-mechanical interactions and increased material removal per grit directly drives the observed roughness growth, aligning well with both experimental measurements and theoretical predictions.

11.3 Influence of Material Anisotropy of Nickel-based Single-crystal Superalloy

The fundamental characteristic of crystals is the periodicity of their structural arrangement. Atoms in a crystal are arranged periodically in a certain pattern. The arrangement and density of atoms vary in different crystal planes and orientations, and so does the atomic binding force. The spacing between different crystal planes also differs, resulting in different physical and chemical properties in various directions, which constitutes the anisotropy of the crystal. Generally, the elastic modulus and shear modulus of different crystal planes and orientations are not the same. Crystal anisotropy has a significant impact on processing. Selecting an appropriate processing direction may greatly improve the quality of the processed surface.

11.3.1 Influence of Different Crystal Faces

This study selects three representative crystallographic plane materials of nickel-based single-crystal superalloys: (001), (110), and (111) orientations. Their atomic lattice structures and the corresponding as-cast dendritic cross-sectional morphologies of the alloys are illustrated in Figure 11.9a and b, featuring a regular "cross-shaped" dendritic microstructure typical of single-crystal superalloys. Grinding experiments are conducted under the following parameters: spindle speed $v_s = 3000$ r/min, workpiece feed rate $v_w = 100$ μm/s, and grinding depth $a_p = 10$ μm. The specified grinding directions align with three crystallographic orientations:

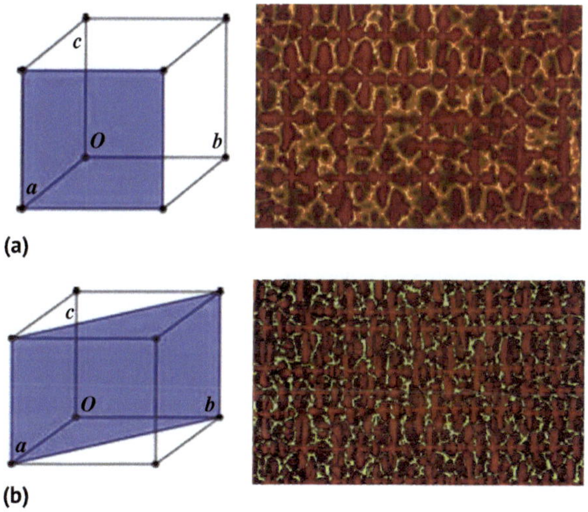

Figure 11.9 (a) (001) and (b) (110) crystal plane structure and dendrite section metallographic structure. [8]/Springer Nature.

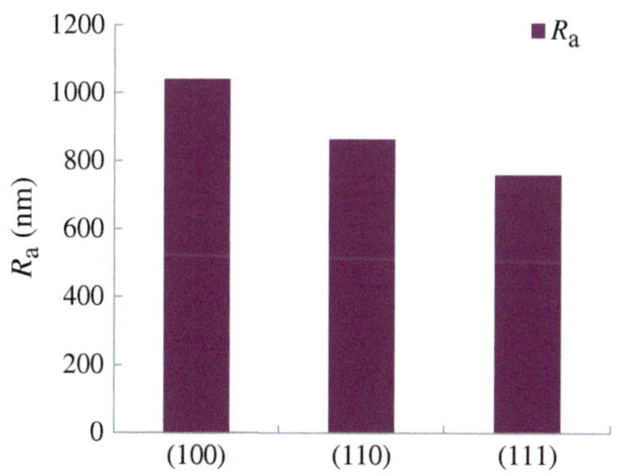

Figure 11.10 Grinding surface roughness along different crystal planes.

along the (001) plane, (011) plane, and (111) plane. A smooth grinding method is employed to investigate surface integrity under controlled kinematic conditions.

As shown in Figure 11.10, surface roughness values during grinding along different crystallographic planes exhibit significant differences: the micro-grinding roughness along the (001) plane is the highest, while that along the (111) plane is the lowest. This discrepancy arises primarily because the maximum undeformed chip thickness per abrasive grain during micro-grinding reaches the nanometer scale—even smaller than the unit cell lattice parameter—causing material removal through slip fracture within the crystal lattice. Crystallographic slip does not occur randomly but is preferentially activated along specific slip planes determined by crystal geometry and atomic bonding characteristics.

The surface topographies generated during grinding along different crystallographic planes are presented in Figures 11.11 and 11.12. For face-centered cubic

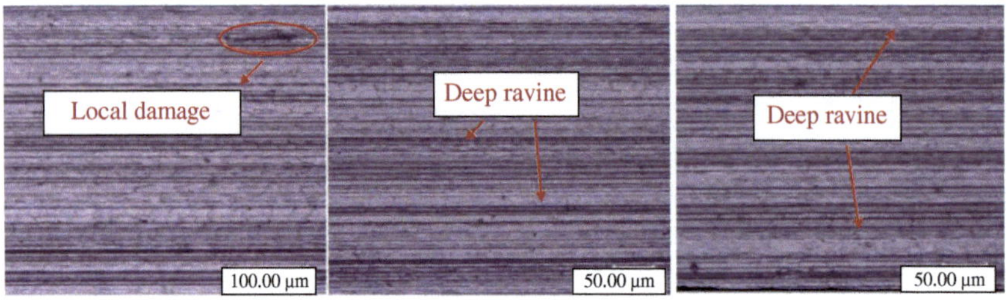

Figure 11.11 Surface morphology grinding along different crystal planes.

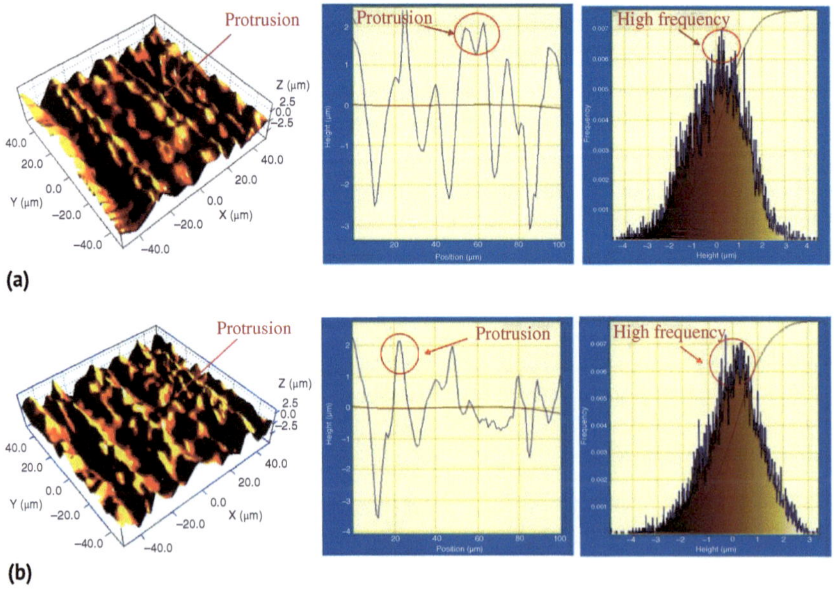

Figure 11.12 Grinding surface and their surface profile of different crystal planes: (a) Micro-grinding surface and profile along (001) surface and (b) Micro-grinding surface and profile along (110) surface.

(FCC) crystals, the primary slip planes belong to the {111} family, with the slip direction aligned along the atomic close-packed [110] orientation. These slip planes exhibit the highest atomic packing density, leading to strong interatomic bonding within the plane. Concurrently, however, the interplanar spacing of {111} planes is relatively large, resulting in weaker interlaminar bonding forces between adjacent crystal planes. This structural feature inherently weakens the material's mechanical resistance across these planes, making them preferential sites for plastic deformation and damage initiation during grinding.

During the micro-grinding of nickel-based single-crystal superalloys, shear slip initiates when the resolved shear force along the slip plane exceeds a critical threshold. Based on crystal structure geometry, the component of unit grinding force transmitted onto the (111) slip plane from the (001) crystallographic direction is smaller than that from the (110) direction. In turn, the (110)-directional force component on the (111) plane is smaller than the unit grinding force applied along the (111) plane itself. Notably, grinding along the (001) orientation generates the highest normal force component perpendicular to the (111) shear-slip plane. This mechanical asymmetry leads to a hierarchical force requirement for slip-induced fracture: the (001)-oriented material demands more force to initiate slip fracture than the (110) orientation, which in turn requires more force than the (111)-oriented material. The disparity arises from the crystallographic misorientation between the grinding direction and slip system, affecting how applied forces resolve into shear and normal components on critical slip planes. Specifically, orientations with higher misalignment (such as (001)) result in greater normal force contributions that impede slip, whereas aligned orientations (such as (111)) facilitate shear deformation by maximizing resolved shear stress on the {111} slip planes.

Therefore, under the same grinding conditions, the grinding force along the (111) crystal plane is relatively small, the material is prone to shear fracture, and the texture roughness during grinding is small. When grinding along the (001) crystal plane, the maximum grinding force is applied, the material is not easily removed, and the texture roughness during grinding is relatively high.

11.4 Influence of Different Crystal Orientations of Nickel-based Single-crystal Superalloy

Atoms are arranged differently in various crystal directions on the same crystal plane, which also results in different mechanical properties of the material. Generally, single-crystal high-temperature alloys oriented along the (001) plane have excellent comprehensive mechanical properties. This article takes the (001) crystal plane as an example. Grinding is carried out along directions that form different angles θ with the [100] crystal direction. By observing the surface quality, we explore the cutting direction to achieve better surface quality.

Extensive experimental characterization revealed pronounced anisotropy in machined surface roughness within the (001) plane. The roughness values exhibited periodic fluctuations with a fundamental period of $\pi/2$ radians, as illustrated in

Figure 11.13. This periodicity aligns with the fourfold symmetry of the (001) lattice, where equivalent crystallographic directions repeat every 90 ° ($\pi/2$). The observed angular dependence of roughness arises from the orientation-dependent interaction between the grinding tool and crystallographic slip systems, highlighting how atomic-level structural symmetry directly influences microscale material removal mechanisms during machining.

The roughness value undergoes periodic changes, with a fluctuation period of $\pi/2$ as shown in Figure 11.13.

As depicted in Figure 11.13, within the range of 0 °–45 °, the micro-grinding texture roughness decreases as θ increases. In the range of 45 °–90 °, the micro-grinding texture roughness shows an increasing trend with the increase of θ. The texture roughness is minimized when grinding along the $\theta = 45°$ direction in the (001) crystal plane.

As shown in Figure 11.14, the probability maps of the surface morphology height distribution in Figure 11.14a are more dispersed than those in Figure 11.14b. The surface morphology height in Figure 11.14a is mainly distributed between −1.5 and 2 μm, which is relatively concentrated. The surface morphology in Figure 11.14b is better than that in Figure 11.14a. The differences in the surface 3D morphology and the probability distribution of height between Figure 11.14a and b are similar. The following will explain the reasons for this phenomenon in detail.

Based on the analysis of the shear modulus of nickel-based single-crystal high-temperature alloy materials, the shear modulus within the (001) crystal plane is shown in Eq. (11.11).

$$G_{(001)} = \frac{1}{S_{44} + \frac{1}{2}\left[(S_{11} - S_{12}) - \frac{1}{2}S_{44}\right] - \left[(S_{11} - S_{12}) - \frac{1}{2}S_{44}\right]\frac{\cos 4\theta}{2}} \tag{11.11}$$

In this formulation, θ denotes the angular difference between arbitrary crystallographic orientations within the (001) plane and the [100] reference orientation. The elastic compliance coefficients are defined as follows: $S_{11} = 1/E$, $S_{44} = 1/G$, $S_{12} = -\mu/E$, where E is Young's modulus, G is the shear modulus, and μ represents Poisson's ratio. For the nickel-based single-crystal superalloy DD98 grown along the

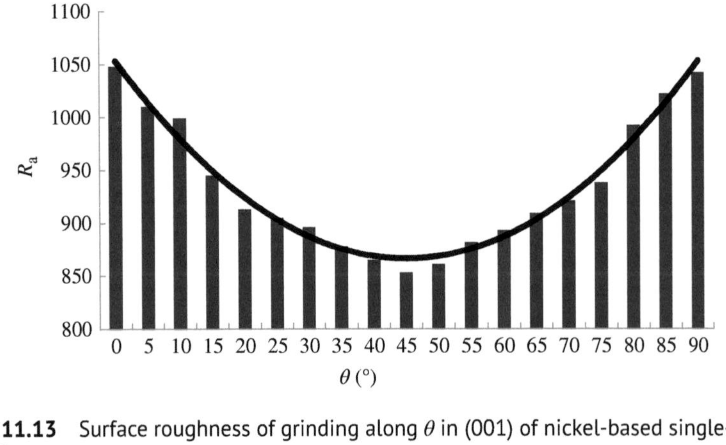

Figure 11.13 Surface roughness of grinding along θ in (001) of nickel-based single-crystal superalloy.

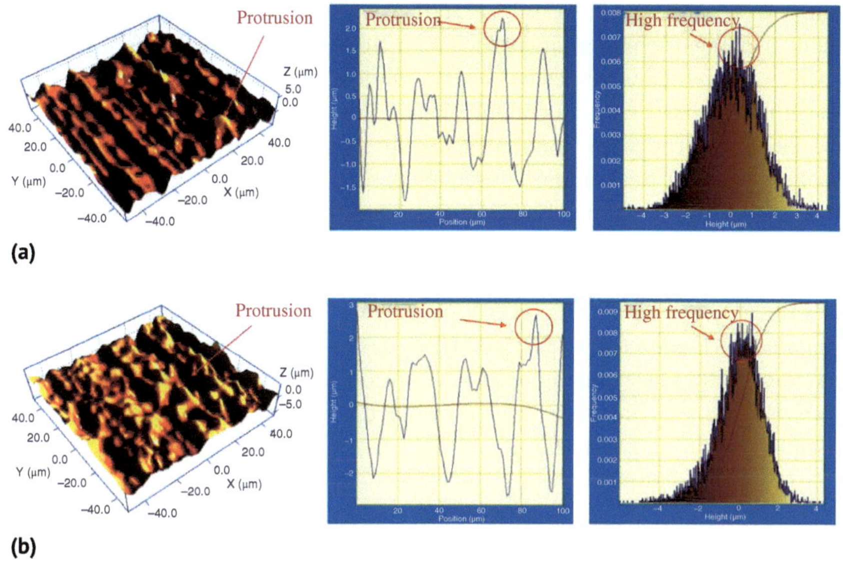

Figure 11.14 3D surface profile and their section diagram when grinding along θ in (001) crystal plane: (a) $\theta = 0°$ and (b) $\theta = 15°$.

(001) crystallographic orientation at room temperature, the material properties are specified as: Young's modulus $E = 131.5$ Gpa and shear modulus $G = 137$ GPa.

$$S_{11} - S_{12} - \frac{1}{2}S_{44} > 0 \tag{11.12}$$

Based on Eq. (11.12), the shear modulus of DD98 displays cyclic fluctuations with a $\pi/2$ periodicity within the (001) crystal plane orientation. Consequently, the texture roughness during DD98 micro-grinding process experiences periodic variations corresponding to the cruciform dendritic architecture of nickel-based single-crystal superalloys.

From Eq. (11.12), during micro-grinding experiments on the (001) crystal plane of nickel-based single-crystal superalloys, the shear modulus of the alloy reaches its minimum when the grinding direction forms angles of 45°, 135°, 225°, and 315° with the [100] crystallographic direction. This material demonstrates a tendency toward shear fracture, which contributes to improved micro-grinding surface quality. Conversely, when the grinding direction aligns with 0°, 90°, 180°, and 270° relative to the [100] crystallographic direction, the shear modulus of nickel-based single-crystal superalloys attains its peak value. During the chip-forming process in grinding, the mechanical load applied to individual abrasive grains becomes significantly higher under these conditions, imposing greater resistance to material removal and causing degradation in micro-scale grinding surface integrity.

The machined surface quality of nickel-based single-crystal superalloys exhibits an anisotropic distribution pattern. In practical manufacturing operations, selecting an appropriate machining direction is essential to achieve a high-quality processed surface, which provides valuable practical guidance for industrial production. The proposed approach can be utilized to forecast the variation trend of surface roughness on machined surfaces. In practical manufacturing, it is advisable to steer clear of grinding orientations associated with greater surface roughness and instead opt for directions characterized by lower roughness levels. During the grinding process within the (001) crystallographic plane of nickel-based single-crystal superalloys, the surface roughness parameter measured along the [110] crystal direction exhibits the lowest magnitude. Consequently, within the micro-grinding of these alloys, machining along the [110] orientation within the (001) plane yields a superior surface quality compared to other directions.

In the directional solidification process of single-crystal components, precise control over the longitudinal growth orientation of the workpiece can be complemented by implementing targeted strategies to regulate the lateral crystalline orientation within designated processing zones. This dual-axis control mechanism enhances the surface integrity of the material, thereby optimizing its suitability for subsequent machining operations.

11.5 Influence of Grinding Methods

In the micro-grinding operation, the grinding factors of $v_s = 40000\,\text{r/min}$, $v_w = 120\,\mu\text{m/s}$, and $a_p = 9\,\mu\text{m}$ were used to compare the texture roughness values of nickel-based single-crystal high-temperature alloy during microscale down and up grinding. Through detection, it was found that R_a of down grinding is 794 nm and R_a of back grinding is 726 nm, as shown in Figure 11.15. The roughness under down grinding is greater than that under up grinding.

This phenomenon primarily arises because during the up-grinding process, the single-abrasive-grain cutting thickness increases from zero to a peak value before rapidly decreasing back to zero. The entire grinding action progresses through three sequential stages: sliding, plowing, and cutting. Material at the peak cutting

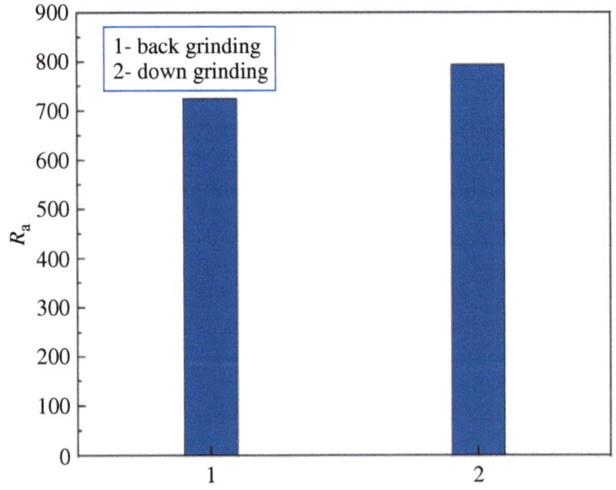

Figure 11.15 The contrast in R_a of nickel-based single-crystal high-temperature alloy machined by back and down grinding process.

thickness undergoes initial heating during the sliding phase, which leads to a reduction in its mechanical strength. Additionally, the abrupt decrease of the single-grit cutting thickness from its maximum to zero causes the resulting chips to fracture more readily. This combination results in lower grinding forces and temperatures, thereby enabling the achievement of superior surface quality during the machining operation.

However, during the down grinding operation, the cutting thickness of the abrasive particles is relatively large when they first enter the workpiece. At this time, the abrasive particles enter the plowing and cutting stages almost without sliding friction during the grinding operation and may even directly enter the cutting stage. The abrasive particles are subjected to high grinding resistance, and the debris needs to pass through the entire grinding area before being discharged. In contrast, the material at the fracture zone during cutting does not undergo preheating in the sliding phase, thereby maintaining higher mechanical strength. As a result, the grinding forces generated during the process are significantly greater, leading to substantial grinding-induced heat and surface deformation that cause the ground surface quality to be inferior to that of up-grinding. Thus, it is evident that during the micro-scale grinding of DD98, the texture roughness parameter in up-grinding conditions exceeds that observed in down-grinding operations.

References

1 Malkin, S. (1989). *Grinding Technology: Theory and Applications of Machining with Abrasives*. Ellis Horwood Limited.
2 Zhou, Y., Ma, L., Gong, Y. et al. (2019). Study on the mechanism of chip forming and the microhardness of micro-grinding nickel-based single-crystal superalloy. *International Journal of Machine Tools and Manufacture* 103: 281–295.
3 Li, C., Li, X., Wu, Y. et al. (2019). Deformation mechanism and force modelling of the grinding of YAG single crystals. *International Journal of Machine Tools and Manufacture* 143: 23–37.
4 Aurich, J.C., Carrella, M., and Walk, M. (2015). Micro grinding with ultra-small micro pencil grinding tools using an integrated machine tool. *Cirp Annals-Manufacturing Technology* 64: 325–328.
5 Zhou, Y.G. (2017). *Study on the Technological Theory and Key Technology in Micro Grinding Nickel-based Single Crystal Superalloy*. Doctoral Dissertation, Northeastern University China.
6 Cheng, J., Gong, Y.D., Wu, Z.Z. et al. (2012). Experimental study on the formation mechanism of micro-cutting surface for hard and brittle materials. *Journal of Mechanical Engineering* 48 (21): 190–198.
7 Hecker, R.L. and Liang, S.Y. (2003). Predictive modeling of surface roughness in grinding. *International Journal of Machine Tools and Manufacture* 43: 755–761.
8 Zhou, Y., Gong, Y., Cai, M. et al. (2017). Study on surface quality and subsurface recrystallization of nickel-based single-crystal superalloy in micro-grinding. *International Journal of Advanced Manufacturing Technology* 55 (2): 1749–1768.

12

Ground Surface and Subsurface Damage in Micro-grinding

12.1 Influence of Grinding Parameters

An optical microscope can be used to directly observe the grinding surface. When observing the subsurface quality of micro-grinding, small samples are cut from the micro-ground workpiece. These samples are then mounted, ground, polished, etched, and observed. The subsurface image, obtained using a confocal microscope and scanning electron microscope (SEM), is shown in Figure 12.1. It can be seen that the material near the grinding surface of the single-crystal alloy micro-grinding differs in color from the matrix. This is due to plastic deformation caused by the extrusion of abrasive particles on the workpiece [1, 2].

After the abrasive particles unload, the part of the workpiece that undergoes elastic deformation tends to return to its original shape, while plastic deformation cannot fully recover [3, 4]. It is compressed by the surrounding elastic matrix, forming a plastic deformation zone. In single-crystal alloys, plastic deformation occurs through the proliferation and movement of dislocations under the action of grinding forces [5, 6]. The dislocation density increases rapidly with increasing plastic deformation. Under SEM, the plastic deformation zone shows structural deformation along the grinding direction, and flow lines along this direction are clearly visible, indicating plastic flow of the material. This is because the lattice inside the material is distorted and deformed by grinding forces, resulting in plastic deformation [7]. The plastic deformation during grinding significantly affects the physical and mechanical properties of the workpiece.

At lower spindle speeds, the number of abrasive grains engaged in grinding per unit time is relatively limited. Protrusions created by prior abrasive grains are not rapidly eliminated by subsequent ones, leading to increased protrusion heights, elevated grinding forces, and degraded surface quality [8, 9]. In contrast, as spindle speed increases, a greater number of abrasive grains participate in the grinding process per unit time—enhancing the frequency of material removal events and promoting more uniform surface interactions. This higher engagement density reduces the likelihood of residual protrusions persisting, thereby mitigating force fluctuations and improving surface finish consistency through more balanced abrasive grain interactions. The protrusions are quickly compressed into grooves or even

Grinding of Single-Crystal Superalloys: Fundamentals and Technologies, First Edition.
Wenfeng Ding et al.
© 2026 Wiley-VCH GmbH. Published 2026 by Wiley-VCH GmbH.

Figure 12.1 Subsurface plastic deformation layer of single-crystal superalloy.

removed by other abrasive particles, resulting in lower protrusion height and narrower groove width.

Figure 12.2a–c shows the grinding surface images at different spindle speeds ($v_s = 60{,}000, 50{,}000, 40{,}000, 30{,}000$, and $20{,}000$ r/min; $v_w = 120$ μm/s; and $a_p = 9$ μm). The grinding surface at low spindle speeds shows material accumulation, while the high-speed grinding surface appears smoother with better surface quality.

Figure 12.3a–c shows the surface topography of the workpiece at different feed rates ($v_s = 40{,}000$ r/min; $v_w = 20, 70, 20, 170$, and 220 μm/s; $a_p = 9$ μm). Figure 12.3a shows the variation of the thickness of the subsurface deformation layer with the workpiece feed rate, while Figure 12.3b and c shows the microstructure of the subsurface in different feed rates.

As the feed rate increases, the surface quality deteriorates. Scratches on the surface become deeper, and material flows laterally, causing material to accumulate on both sides of the groove. At higher feed rates, chip removal becomes difficult, leading to slight vibration and surface burn during machining, further worsening the surface quality [10]. Additionally, an increase in the feed rate raises the interaction force between the grinding rod and the workpiece, intensifying plastic deformation on the workpiece surface and increasing the thickness of the subsurface deformation layer. Moreover, as the feed rate increases, the subsurface damage depth becomes more significant, as shown in Figure 12.4a. Figure 12.4b–c depicts the subsurface topography at different feed rates, indicating that at lower feed rates, the grinding surface quality is better and the subsurface damage depth is lower.

Figure 12.5a–c shows the surface topography at different grinding depths ($v_s = 40{,}000$ r/min, $v_w = 120$ μm/s, and $a_p = 3$, 12, and 15 μm). As grinding depth increases, the grinding traces on the surface deepen along the thickness direction and widen perpendicular to the grinding direction, which results in a decline in surface quality. This is accompanied by the appearance of slight thermal discoloration (burning) and particle-like adherents on the ground surface, likely due to intensified frictional heating and chip redeposition. Conversely, at lower grinding depths, the machining process produces a dense, smooth surface characterized by shallow and fine-textured grooves—features that arise from reduced material removal per pass and minimized thermal-mechanical interactions. These morphological differences are clearly illustrated in Figure 12.5a–c, where the transition from coarse, deep traces at high depths to fine, uniform textures at low depths is visually evident.

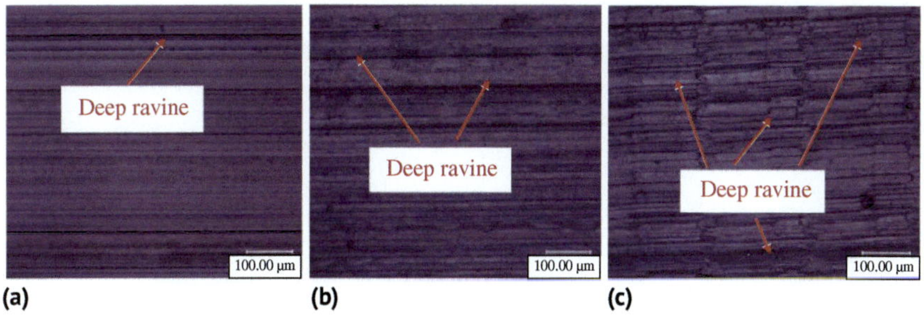

Figure 12.2 Grinding surface with different spindle speed: (a) $v_s = 60{,}000$ r/min; (b) $v_s = 50{,}000$ r/min; and (c) $v_s = 40{,}000$ r/min.

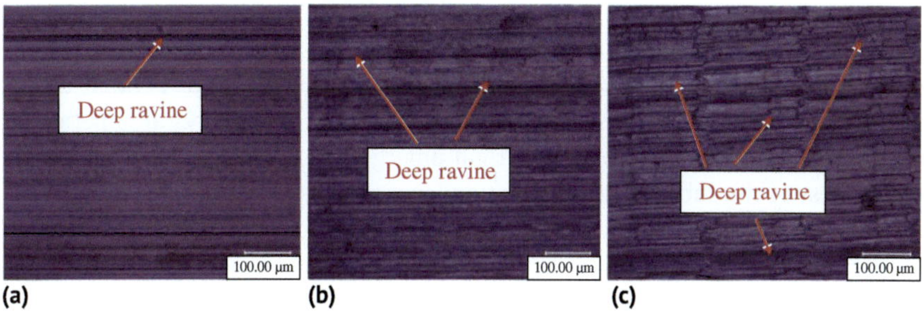

Figure 12.3 Grinding surface with different feeding rate: (a) $v_w = 20$ µm/s; (b) $v_w = 170$ µm/s; and (c) $v_w = 220$ µm/s.

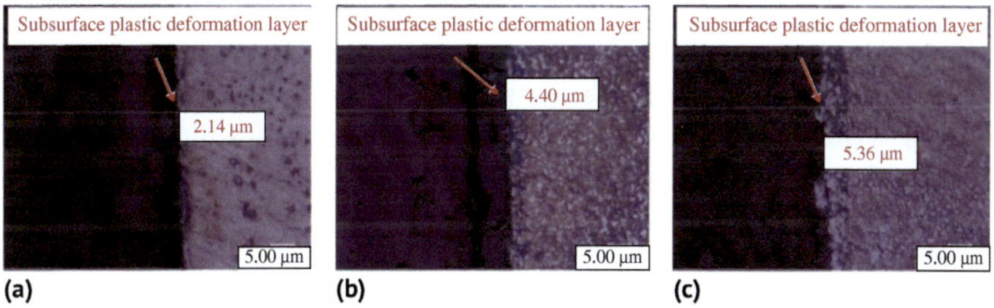

Figure 12.4 Grinding subsurface with different feeding rate: (a) $v_w = 70$ µm/s; (b) $v_w = 170$ µm/s; and (c) $v_w = 220$ µm/s.

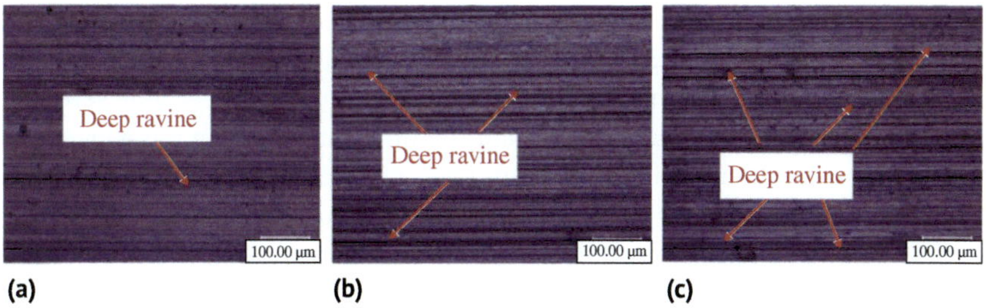

Figure 12.5 Grinding surface with different grinding depths: (a) $a_p = 3$ µm; (b) $a_p = 12$ µm; and (c) $a_p = 15$ µm.

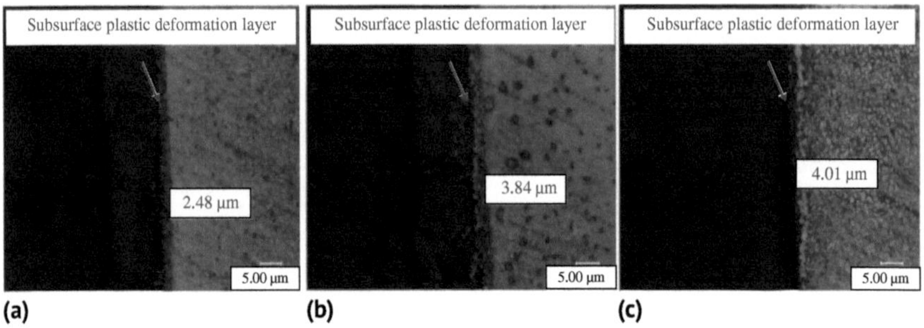

Figure 12.6 Grinding subsurface with different grinding depths: (a) $a_p = 6$ μm; (b) $a_p = 12$ μm; and (c) $a_p = 15$ μm.

Figure 12.6a illustrates the relationship between subsurface deformation layer thickness and grinding depth during micro-grinding, while Figure 12.6b and c depicts the subsurface microstructures of nickel-based single-crystal superalloys at different grinding depths. A clear trend emerges: as grinding depth increases, the thickness of the subsurface deformation layer produced by the grinding process systematically augments. This phenomenon stems from two coupled mechanisms: first, greater material removal at deeper depths generates elevated frictional heat within the grinding zone, and second, the resultant thermomechanical stress—combined with increased grinding forces—drives more pronounced plastic deformation. The intensified heat–force interaction facilitates deeper penetration of deformation-induced dislocations and lattice distortions into the workpiece matrix, thereby extending the subsurface damage depth. Specifically, the cumulative effect of enhanced plastic flow and thermal softening at higher depths allows the deformation front to propagate further below the ground surface, as evident in the progressively thicker deformation layers observed in the microstructural images. This correlation between grinding depth and subsurface damage highlights the critical role of process parameters in controlling both surface and subsurface integrity during micro-grinding of single-crystal superalloys.

12.2 Influence of Working Fluid

Figure 12.7a and b illustrates the morphology of the grinding rod and the corresponding grinding surface under dry grinding conditions. The grinding rod experiences chip clogging and adhesion, while chips accumulate on the grinding surface. As the volume of material removal increases, the grinding rod wears more severely. The surface abrasives gradually fall off, and the tip of the grinding rod is partially ground into a cone shape, rendering it unusable, as depicted in Figure 12.7c.

The reason is that nickel-based single-crystal superalloys are hard-to-machine materials with high viscosity and poor thermal conductivity. Under dry grinding, chips stick to the abrasive peaks, making chip removal difficult [2]. They clog the gaps between abrasives and enclose the cutting edges. When these chips re-enter the grinding zone, the adhered material on the abrasive surface prevents some abrasives from contacting the material, causing further adhesion and worsening the

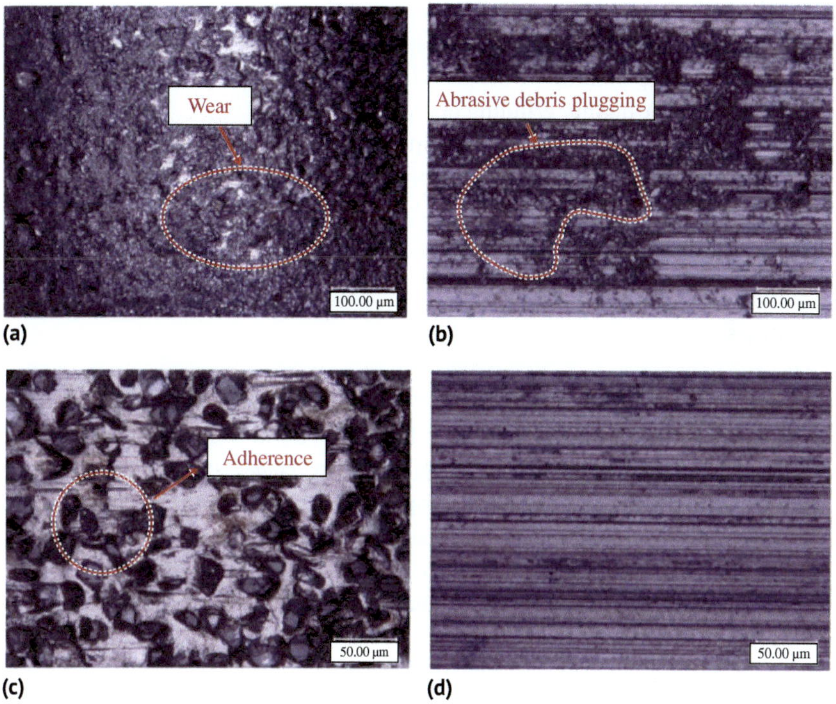

Figure 12.7 Topography of grinding tool and grinding surface under different cooling conditions. [1] / Springer Nature: (a) Micro tool of dry grinding $V = 20\ \text{mm}^3$; (b) corresponding grinding surface; (c) micro tool of wet grinding $V = 20\ \text{mm}^3$; and (d) corresponding grinding surface.

clogging-induced wear of the grinding rod. This vicious cycle leads to the detachment of the electroplated layer on the grinding rod surface and the abrasives. It also makes material removal challenging and causes chip accumulation on the grinding surface, deteriorating the surface quality, as shown in Figure 12.7b.

The increased friction between the clogged grinding rod and the workpiece raises the grinding force, generating more heat. This weakens the electroplated layer's grip on the abrasives, accelerating the detachment of both. As this continues, the grinding rod wears more severely, ultimately resulting in the situation in Figure 12.7c.

Conversely, Figure 12.7d shows the grinding rod and surface morphology under grinding fluid conditions. Clearly, the grinding rod wear and surface quality are much better with grinding fluid than with dry grinding.

The grinding fluid plays a critical role in regulating workpiece surface quality and mitigating grinding rod wear through three synergistic mechanisms. Its cooling action efficiently dissipates frictional heat generated in the grinding arc zone, preventing thermal damage to the workpiece surface. Concurrently, the lubricating function reduces interfacial friction between the micro-grinding rod and workpiece, minimizing abrasive wear and energy dissipation during contact. Additionally, the flushing effect ensures prompt removal of grinding debris, preventing chip redeposition and blocking that could degrade surface finish or induce tool defects. These combined effects create a thermally stable, low-friction machining environment, thereby enhancing both surface integrity and grinding tool lifespan through optimized heat management, reduced mechanical load, and improved debris evacuation. This prevents chip buildup on the grinding surface and clogging of the grinding rod, leading to a better-quality grinding operation.

Based on extensive micro-grinding experiments on DD98, the following practices can be summarized: during grinding, high-pressure grinding fluid should be used to cool the grinding area, and paraffin wax should be periodically applied to the grinding rod surface to prevent clogging.

References

1 Zhou, Y.G., Gong, Y.D., Cai, M. et al. (2017). Study on surface quality and subsurface recrystallization of nickel-based single-crystal superalloy in micro-grinding. *The International Journal of Advanced Manufacturing Technology* 90: 1749–1768.

2 Zhou, Y.G. (2017). Study on the Technological Theory and Key Technology in Micro Grinding Nickel-based Single Crystal Superalloy. Doctoral Dissertation, Northeastern University, China.

3 Sun, Y., Su, Z.P., and Gong, Y.D. (2021). Analytical and experimental study on micro grinding surface generated mechanism of single crystal superalloy using micro diamond pencil grinding tool. *Archives of Civil and Mechanical Engineering* 21: 1–22.

4 Cheng, J., and Gong, Y.D. (2014). Experimental study of surface generation and force modeling in micro-grinding of single crystal silicon considering crystallographic effects. *International Journal of Machine Tools and Manufacture* 77: 1–15.

5 Cheng, J., Wang, C., Wen, X.L. et al. (2014). Modeling and experimental study on micro-fracture behavior and restraining technology in micro-grinding of glass. *International Journal of Machine Tools and Manufacture* 85: 36–48.

6 Sun, Yao., Su, Z., Jin, L. et al. (2021). Modelling and analysis of micro-grinding surface generation of hard brittle material machined by micro abrasive tools with helical chip pocket. *Journal of Materials Processing Technology* 297(1), 117242.

7 Zhou, Y.G., Gong, Y.D., Cai, M. et al. (2017). Study on surface quality and subsurface recrystallization of nickel-based single-crystal superalloy in micro-grinding. *International Journal of Advanced Manufacturing Technology* 90: 1749–1768.

8 Zhang, M.L., Cai, M., Gong, Y.D. et al. (2024). Experimental study on the effect of minimum quantity lubrication on the grinding surface/subsurface of nickel-based single-crystal superalloys. *International Journal of Advanced Manufacturing Technology* 134:5375–5390.

9 Zhou, Y., Ma, L., Gong, Y. et al. (2019). Study on the mechanism of chip forming and the microhardness of micro-grinding nickel-based single-crystal superalloy. *International Journal of Advanced Manufacturing Technology* 103 (1–4):281–295.

10 Zhou, Y., Gong, Y., Zhu, Z. et al. (2016). Modelling and optimization of surface roughness from micro-grinding of nickel-based single crystal superalloy using the response surface methodology and genetic algorithm. *International Journal of Advanced Manufacturing Technology* 85 (9–12), 2607–2622.

13

Subsurface Microstructure and Recrystallization in Micro-grinding

13.1 Subsurface Microstructure in the Micro-grinding Process

In order to further investigate the microstructure of the plastic-deformation layer in the subsurface of nickel-based single-crystal superalloy during micro-grinding, this study conducted the transmission electron microscope (TEM) analysis on the subsurface after grinding. This analysis is of great help in understanding the plastic-deformation mechanism and surface hardening during grinding. The grinding factors used in the experiment were: $v_s = 30000$ r/min, $v_w = 100$ μm/s, and $a_p = 10$ μm, and the material used was nickel-based single-crystal superalloy DD98 grown along the (001) crystal direction.

During the grinding process, the deformed material can be categorized into two distinct components: the metal removed by abrasive grits and the subsurface plastic-deformation layer beneath the ground surface. As the grinding rod traverses the workpiece surface at a specified velocity, the workpiece material is subjected to compressive forces from both the abrasive grits' cutting edges and their rake faces. This mechanical interaction induces plastic flow in the material, with a portion being sheared off as chips by the cutting edges while another portion undergoes intensive plastic deformation immediately below the machined surface. The synergy between cutting and plowing actions of the abrasive grits thus creates a dual-layer deformation mechanism, where the upper layer experiences material removal and the subsurface layer accumulates residual stresses and crystallographic defects due to localized plastic strain. The material is mainly subjected to the compressive stress of the abrasive particle, causing a large number of atomic rearrangements and redistributions. This results in a wide range of dislocations and their movements.

Furthermore, during micro-grinding, dislocations generated by surface plastic deformation in the workpiece exhibit nonuniform distribution. The curved trajectory of abrasive grains subjects these dislocations to rotational forces induced by contact compressive stress, causing progressive reorientation of dislocation lines. Consequently, in the subsurface region of single-crystal superalloys subjected to micro-grinding, dislocations intersect and agglomerate due to this orientation

Grinding of Single-Crystal Superalloys: Fundamentals and Technologies, First Edition.
Wenfeng Ding et al.
© 2026 Wiley-VCH GmbH. Published 2026 by Wiley-VCH GmbH.

change—a phenomenon arising from the combined effects of grain trajectory curvature and localized stress fields. This dislocation entanglement forms defect clusters that influence subsurface mechanical properties, highlighting the role of abrasive kinematics in shaping microstructural evolution during microscale material removal. As the micro-grinding operation progresses, a large number of dislocations entangle on the grinding subsurface, forming “dislocation entanglement,” as shown in Figure 13.1a–d. Due to the dislocations generated during micro-grinding of nickel-based single-crystal superalloys, the shear slip between the two parts of the crystal is not a rigid movement of the upper and lower parts but occurs through dislocation movement.

During the micro-grinding of crystalline materials, the plastic-deformation zone exhibits nonuniform metal distribution, where the subsurface of single-crystal alloy workpieces evolves into alternating regions of low and high dislocation density, marking the incipient stage of cell-structure formation. As abrasive grains induce progressive deformation in the subsurface layer, dislocations undergo continuous proliferation and migration under the mechanical action of grinding forces. This leads to the accumulation of dense dislocation networks, particularly at the boundaries between low- and high-density regions, where stress concentrations promote dislocation entanglement. The resulting microstructural heterogeneity—characterized by emerging cellular sub-grains—arises from the dynamic balance

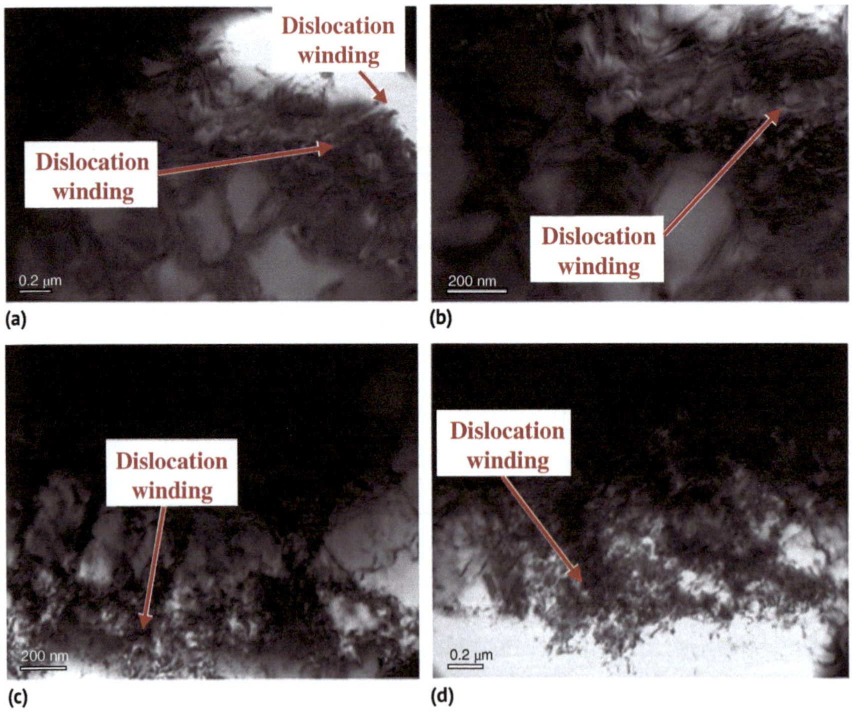

Figure 13.1 Transmission electron microscope (TEM) dark-field image of subsurface of micro-grinding single-crystal superalloy: (a) dislocation tangle area 1; (b) dislocation tangle area 2; (c) dislocation tangle area 3; and (d) dislocation tangle area 4.

between dislocation generation, movement, and interaction during microscale material removal, forming a critical precursor to subsurface damage evolution in single-crystal superalloys. The high-density entangled dislocations at the grinding subsurface concentrate around the cells. The dislocations evolve from entanglement to cell walls, surrounding the regions with lower dislocation densities, thus forming a distinct cellular substructure. There is a slight orientation difference between each cell.

When a dislocation-cell substructure appears in a single-crystal superalloy, the deformed material is divided into many small sub-crystalline structures by dislocation cells with small orientation differences, as shown in Figure 13.2. It shows the subsurface bright-field image of nickel-based single-crystal superalloy micro-grinding and the selected-area electron diffraction (SAED) of the corresponding region. Figure 13.2b, c are TEM bright-field images and SAED at different positions during middle-distance grinding in Figure 13.2a. It can be seen that the size of the sub-crystals near the grinding surface is small and the number is large, whereas with the increase in distance from the grinding surface, the size of the sub-crystals increases and the number decreases. In Figure 13.2b, the diffraction spots display distinct ring-shaped patterns composed of arc segments, indicating the formation of small crystallographic orientation differences within the material's deformation zone. This structural feature signifies the emergence of a sub-grain structure containing numerous small-angle grain boundaries, which arise from localized dislocation rearrangement and tangling during plastic deformation. The curvilinear diffraction arcs reflect gradual misorientations between adjacent sub-grains, a direct consequence of dislocation wall formation as the material accommodates shear strain.

Conversely, Figure 13.2c shows a reduction in the intensity and continuity of these diffraction spot rings, indicating a gradual decrease in plastic deformation within this region. The diminished ring structure corresponds to a lower dislocation density in the sub-grain boundaries, as reduced mechanical loading during grinding leads to less pronounced dislocation proliferation and accumulation. This transition from well-defined ring patterns in (b) to attenuated features in (c) visually demonstrates the gradient of deformation severity across the subsurface, with areas farther from the grinding surface experiencing less dislocation activity and milder microstructural evolution.

The change in the internal structure of the metal in the plastic-deformation area of the grinding subsurface inevitably leads to a change in material properties. The most pronounced alteration occurs in mechanical properties, where elevated dislocation density generally correlates with increased hardness. Microhardness testing confirms that cold-working deformation significantly enhances the strength and hardness of metallic materials—a phenomenon rooted in work-hardening mechanisms. At the microstructural level, work-hardening arises from increased resistance to dislocation motion as machining-induced plastic deformation progresses. As dislocation density in the matrix rises, a critical threshold is reached where dislocations are organized into a hierarchical substructure: the dislocation cellular network. Within this substructure, dislocation cells exhibit a distinct two-phase morphology: cell interiors contain few mobile dislocations, while cell walls consist of dense dislocation tangles forming small-angle boundaries. This architectural

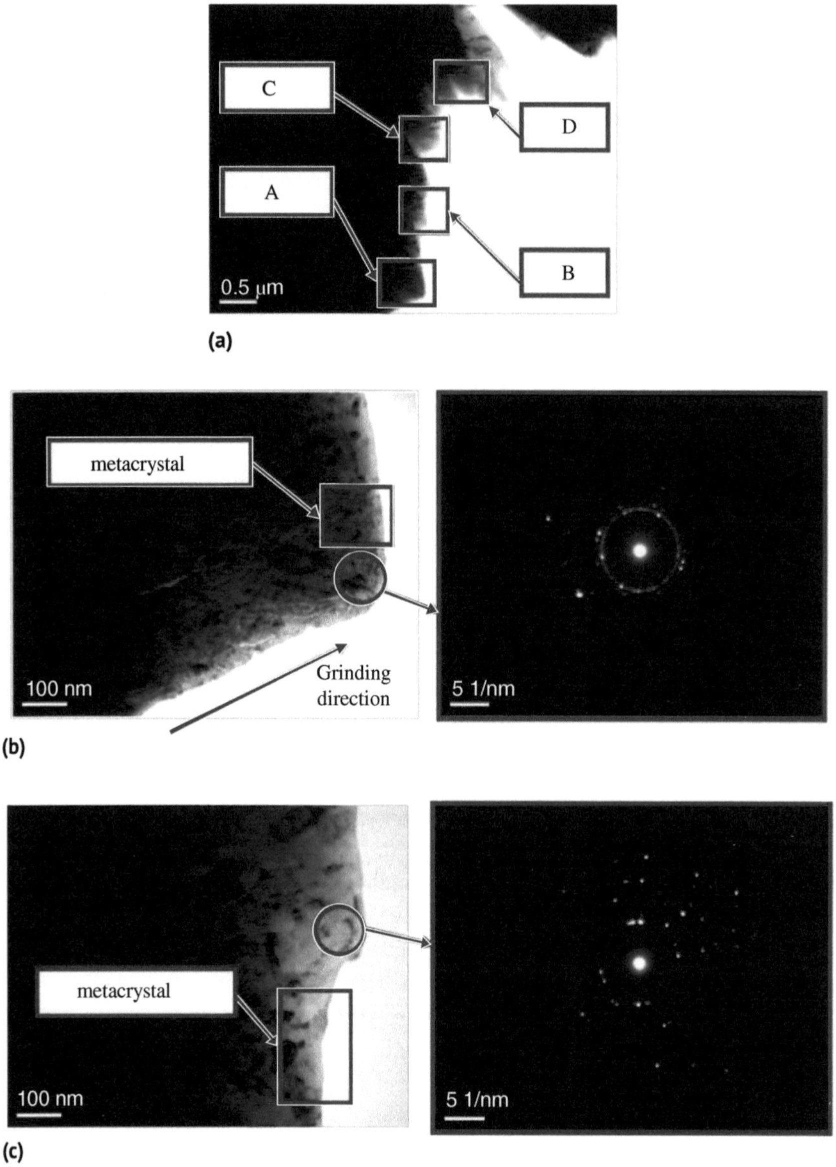

Figure 13.2 TEM bright field with various positions from the grinding surface: (a) subsurface bright-field image of a ground workpiece; (b) bright-field image of region A and electron diffraction pattern of selected area; and (c) bright-field image of region B and electron diffraction map of the selected area.

disparity is pivotal: the low-dislocation-density cell cores maintain relative structural integrity, whereas the high-density walls act as effective barriers to dislocation glide. The entangled dislocations within cell walls generate internal stress fields that impede the motion of incoming dislocations, creating a mechanical interlock where dislocations from adjacent cells interfere with each other's slip trajectories.

This collective resistance to dislocation movement—termed "dislocation–dislocation interaction"—raises the critical resolved shear stress required for plastic deformation, thereby concurrently enhancing material strength and hardness in localized regions. Notably, when the dislocation cell substructure reaches maturity, a synergistic improvement in both strength and plasticity can occur. The well-defined cell boundaries facilitate homogeneous strain distribution across the microstructure, preventing localized necking while maintaining sufficient mobile dislocations in cell interiors to accommodate deformation. This balance between dislocation storage (in cell walls) and dislocation mobility (in cell cores) exemplifies the microstructural basis for optimizing mechanical properties through controlled deformation processing in crystalline materials.

In this manner, the grinding-deformed region emulates a composite structure with alternating soft and hard phases, arising from microstructural domains of varying dislocation densities. The formation of such dislocation configurations within the crystal lattice is analogous to implementing a dispersion-hardening mechanism, where densely packed dislocation structures act as strengthening precipitates—inevitably enhancing the material's overall mechanical performance. TEM observations in Figure 13.2 directly visualize this strengthening architecture on the micro-ground subsurface of nickel-based single-crystal superalloys, revealing abundant dislocation-tangled regions, high-density dislocation walls, and small-angle grain boundaries formed by cumulative plastic deformation. These features constitute a hierarchical substructure: dislocation walls serve as physical barriers to slip, while small-angle boundaries—composed of tightly interwoven dislocations—create localized stress fields that impede dislocation motion, mirroring the strengthening principle of dispersed hard phases in conventional composite materials. The coexistence of low-dislocation-density matrix regions (analogous to "soft" phases) and high-dislocation-density cell walls ("hard" phases) creates a heterogeneous microstructure that resists plastic deformation, thereby synergistically improving the material's strength, hardness, and resistance to fatigue damage under service conditions. This microstructural evidence directly links the observed mechanical property enhancements to the specific dislocation configurations induced by micro-grinding, validating the in situ strengthening mechanism in single-crystal superalloys through controlled deformation processing. At this time, the material structure is in an unstable state, and the storage energy of the plastic-deformation layer on the subsurface of the micro-grinding of single-crystal superalloys is high, which may lead to recrystallization of the parts when they work in a high-temperature environment.

Large plastic deformation and a large number of dislocations occur on the ground subsurface of nickel-based single-crystal superalloys during micro-grinding, but no obvious recrystallization is found. The single-crystal superalloy with plastic deformation is in a metastable high-energy state and stores plastic-deformation energy. Under the action of thermal activation, a series of microstructure changes may occur in single-crystal superalloys. The ground subsurface material transforms from a metastable high-energy configuration to a stable low-energy state, driven by the release of stored mechanical energy within the crystal lattice. This energy—accumulated as dislocations, stacking faults, and other crystallographic defects in the plastic-deformation layer of nickel-based single-crystal superalloys—instigates two sequential microstructural processes: recovery and recrystallization. During recovery, low-energy dislocation rearrangements reduce internal stress without altering grain morphology, while recrystallization involves the nucleation and growth of new strain-free grains, eliminating the deformed substructure entirely. The thermodynamic tendency of the metastable deformed layer to minimize free energy through these processes underscores the critical role of stored defect energy in governing post-machining microstructural evolution, particularly relevant for single-crystal components operating in high-temperature environments where thermal activation accelerates energy dissipation via recovery and recrystallization mechanisms.

The recrystallization temperature of nickel-based single-crystal superalloys is generally higher than 1000°C. For nickel-based single-crystal superalloy micro-grinding parts applied in a temperature environment below 1000°C, the micro-ground single-crystal parts can meet production requirements, and no recrystallization will occur. For single-crystal parts that work at high temperatures above 1000°C for a long time, recrystallization is likely to occur on the ground surface in a high-temperature application environment due to the presence of a plastic-deformation layer on the ground surface.

Therefore, high-temperature working environment treatment was carried out on the workpiece after grinding. The specimens grinding under the same grinding factors (v_s = 30000 r/min, v_w = 100 μm/s, and a_p = 10 μm) were kept at 1050°C and 1100°C for 4 h, respectively, and air cooled. The sectional metallographic structure of the surface of the grinding workpiece after 1050°C high-temperature environment is shown in Figure 13.3. It can be seen from Figure 13.3b that cellular recrystallization occurred on the ground surface after high-temperature treatment.

Figure 13.4 shows the subsurface recrystallization layer thickness (D) of the workpiece grinding along different crystal faces after high-temperature treatment at 1100°C. The grinding factors are v_s = 30000 r/min, v_w = 100 μm/s, and a_p = 10 μm. Figure 13.4 shows the sectional metallographic structure of the workpiece grinding along the (001), (110), and (111) crystal faces after high-temperature treatment.

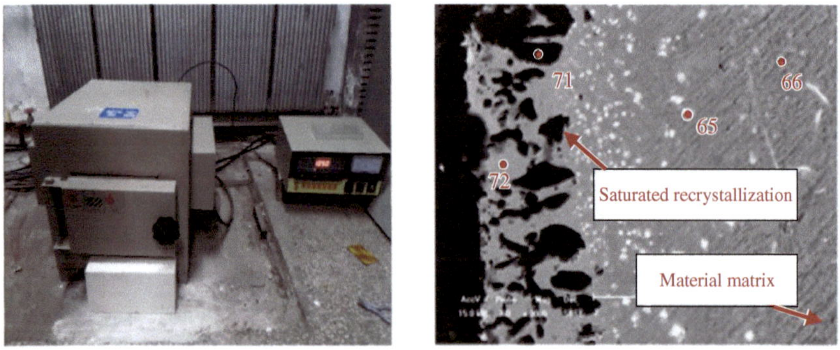

(a) (b)

Figure 13.3 Furnace and subsurface graphical morphology of grinding part experiencing 1050°C: (a) high-temperature electric furnace and (b) subsurface topography after 1050°C.

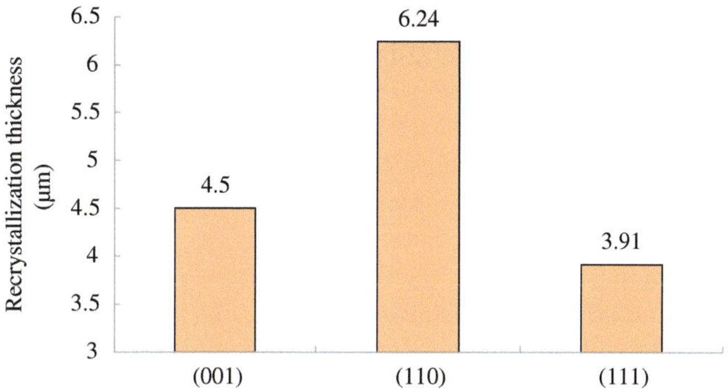

Figure 13.4 Recrystallization thickness of grinding workpiece along different crystal planes after high-temperature treatment.

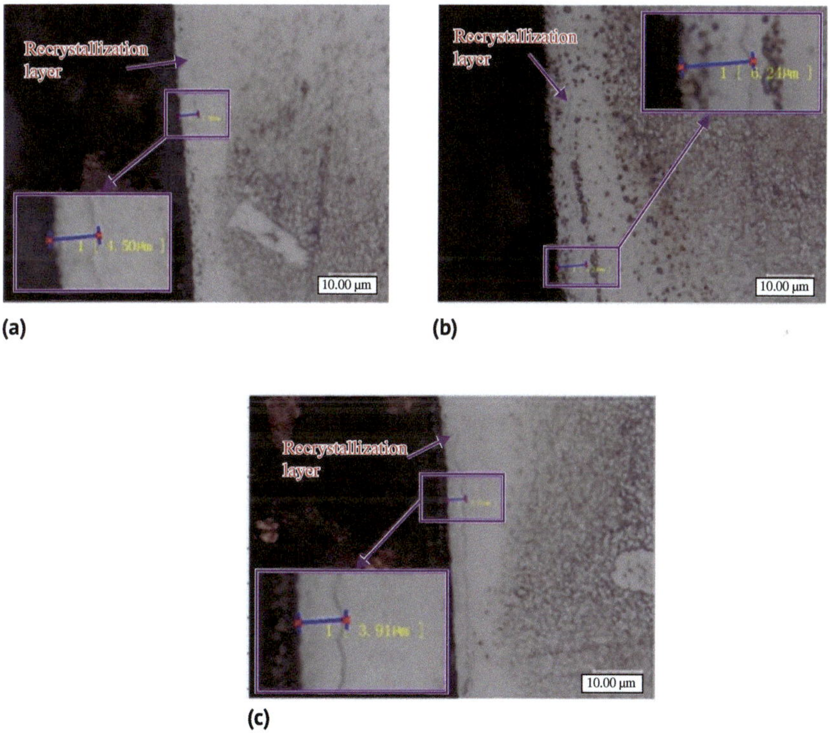

Figure 13.5 Subsurface metallographic structure of nickel-based single-crystal superalloy workpiece grinding along (a) (001), (b) (110), and (c) (111) crystal planes after high-temperature treatment.

Figure 13.5a–c reveals that the subsurface recrystallization thickness (D) is lower when grinding along the (111) crystal face, while it is higher when grinding along the (110) crystal face. This can be attributed to the relationship where the greater the spacing between adjacent crystal faces, the weaker the atomic binding force between them, and the poorer the material's mechanical properties.

Among the three typical growth directions of nickel-based single-crystal superalloys, the surface spacing of the single-crystal material growing along the (111) crystal plane is the largest, followed by that of the (001) crystal plane, and the surface spacing of the single-crystal material growing along the (110) crystal plane is the smallest. The atomic binding force between the crystal faces of the single-crystal material growing along the (111) face is weak, while that of the material growing along the (110) face is strong, resulting in a larger surface deformation storage energy.

Thus, under identical grinding parameters, the thickness of the recrystallization layer varies with crystallographic orientation: it is minimized along the (111) plane and maximized along the (110) plane. A hierarchical resistance to recrystallization exists among these planes: the (111) plane demonstrates the highest resistance, followed by the (001) plane, with the (110) plane exhibiting the lowest resistance.

13.2 Subsurface Recrystallization in Micro-grinding

13.2.1 Effect of Crystal Orientation

Figure 13.6 shows the subsurface recrystallization layer thickness (D) of the workpiece after high-temperature treatment when grinding at different angles θ with [100] crystal direction inside the (001) crystal plane. Grinding factors: $v_s = 30000$ r/min, $v_w = 100$ μm/s, and $a_p = 10$ μm. Figure 13.6 shows the subsurface metallographic structure of the workpiece grinding in the 0°, 10°, 20°, 45°, and 70° directions with the [100] crystal in the (001) crystal plane after treatment in the high-temperature application environment.

As per Figure 13.6, when θ ranges from 0° to 45°, the thickness of the recrystallized layer on the subsurface of the workpiece after high-temperature treatment

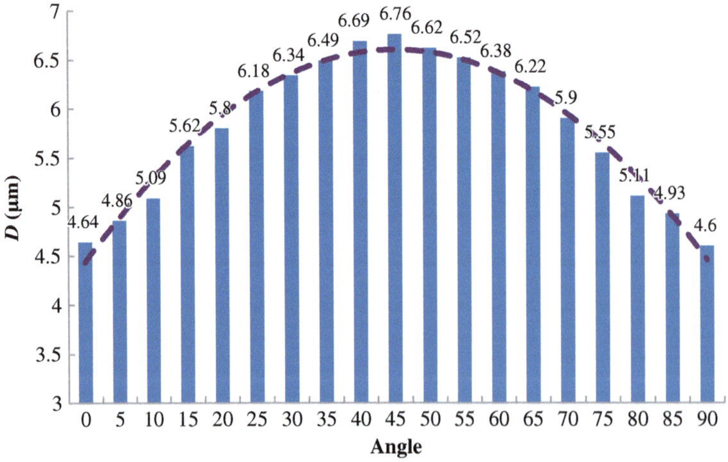

Figure 13.6 Recrystallization thickness of the grinding workpiece along different angles θ in (001) plane after high-temperature treatment.

increases with the increase of θ. Within the range of 45° to 90°, the thickness of the subsurface recrystallization layer decreases as θ increases. When grinding along the 45° direction relative to [100], i.e., along the [110] direction, the thickness of the subsurface recrystallized layer of the nickel-based single-crystal superalloy micro-ground workpiece is at its maximum. The thickness of the subsurface recrystallization layer is minimal when grinding along the [100] and [010] crystal directions.

This indicates that the surface deformation storage energy is lowest when grinding along the [100] and [010] directions and highest when grinding along the [110] direction. The reasons for this are as follows. The elastic modulus of the nickel-based single-crystal superalloy has been derived as shown in Eq. (13.1).

$$E_{(001)} = \frac{1}{\frac{3}{4}S_{11} + \frac{1}{4}S_{12} + \frac{1}{8}S_{44} + \left[(S_{11} - S_{12}) - \frac{1}{2}S_{44}\right]\frac{\cos 4\theta}{2}} \tag{13.1}$$

Based on Eq. (13.1), the elastic modulus along the (001) crystal plane of the nickel-based single-crystal superalloy exhibits periodic variation with the angular parameter θ. As θ increases from 0° to 45°, the modulus rises monotonically, thereby enhancing the material's deformation storage energy during micro-grinding. Conversely, as θ increases beyond 45° to 90°, the modulus decreases with rising θ, leading to a corresponding reduction in deformation-induced energy storage under the same machining conditions.

When grinding along $\theta = 0°$ and $\theta = 90°$, the normal grinding force is small, resulting in little plastic deformation on the grinding subsurface. Thus, the storage energy is low, and the thickness of the recrystallized layer formed after high-temperature treatment is also low. When grinding along $\theta = 45°$, the material's elastic modulus is larger, the normal grinding force is greater, the plastic deformation of the grinding subsurface is more significant, the material's deformation storage energy is higher, and the thickness of the recrystallized layer formed after high-temperature treatment is also higher. The change trend of the thickness of the subsurface recrystallized layer is basically consistent with that of the material's elastic modulus.

13.2.2 Recrystallization Controlling

Recrystallization may occur during the service of nickel-based single-crystal superalloy components, a phenomenon influenced by the deformation level introduced during their manufacturing processes [1, 2]. For single-crystal parts operating in environments exceeding 1000°C, the micro-grinded workpiece will remain free of recrystallization at a given temperature under two conditions: first, the deformation induced by micro-grinding is below the critical threshold for recrystallization initiation; second, the activation energy required to drive recrystallization in the single-crystal superalloy is not supplied [3].

In the production of nickel-based single-crystal superalloy micro-parts, micro-grinding parameters directly influence the thickness of the damaged layer on the surface of the ground parts [4, 5]. To reduce surface recrystallization of micro-single-crystal parts during micro-grinding, measures should be taken to decrease the thickness of the plastic-deformation layer, thereby avoiding or reducing the likelihood of recrystallization of the grinding surface in a high-temperature working environment. The following methods can be adopted:

- First, appropriately increasing the spindle speed, reducing the grinding depth, and decreasing the feed speed are beneficial. This can reduce the plastic deformation on the micro-grinding subsurface and decrease the deformation storage energy of the micro-ground parts.
- Second, process optimization involves refraining from machining components prior to solution treatment. Nickel-based single-crystal superalloy parts should undergo a solid-solution and aging heat-treatment regimen before service to optimize the morphology and distribution of the γ phase, including its size, quantity, and spatial arrangement, thereby enhancing the alloy's mechanical performance. The solution treatment is typically conducted at around 1300°C, a temperature exceeding the material's recrystallization threshold to ensure that any deformation-induced defects are mitigated. This thermal processing step homogenizes the microstructure, promoting a stable γ matrix that resists recrystallization during subsequent high-temperature service by eliminating pre-existing nucleation sites formed from machining-induced stresses. Thus, the material should be subjected to solid-solution treatment first, followed by mechanical processing.
- Third, where feasible, reduce the severity of surface mechanical processing. Implementing a stress-relief annealing process aims to eliminate residual stresses induced by grinding, thereby decreasing the plastic-deformation energy accumulated on the micro-ground part surface. This thermal treatment effectively mitigates the driving force for recrystallization, leading to a thinner recrystallization layer on the ground surface by reducing both stress concentrations and stored deformation energy.

The deformation storage energy is the driving force for recrystallization in nickel-based single-crystal superalloys. Under the action of the grinding force, the crystal lattice of the nickel-based single-crystal superalloy distorts and slips along the grinding direction, resulting in plastic deformation and the accumulation of a certain amount of deformation storage energy [6, 7]. Therefore, the deformation-storage energy can be released through pre-recovery treatment of the micro-ground workpiece at a relatively low temperature, reducing the driving force for recrystallization and achieving the goal of inhibiting or reducing recrystallization.

Based on relevant materials, the stress-removal temperature of the nickel-based single-crystal superalloy is between 700°C and 800°C. All stress-relieved workpieces were kept in a 1100°C high-temperature environment for 3 h, and then the cross-sectional metallographic structure of the processing area was observed under a microscope. Figure 13.7 shows the subsurface microstructure of the grinding area and the thickness of the recrystallization layer after different heat-treatment processes.

As can be seen from Figure 13.7, the thickness of the subsurface recrystallization layer of the DD98 micro-ground workpiece decreases after stress-relief treatment at approximately 760°C, followed by exposure to a high-temperature application

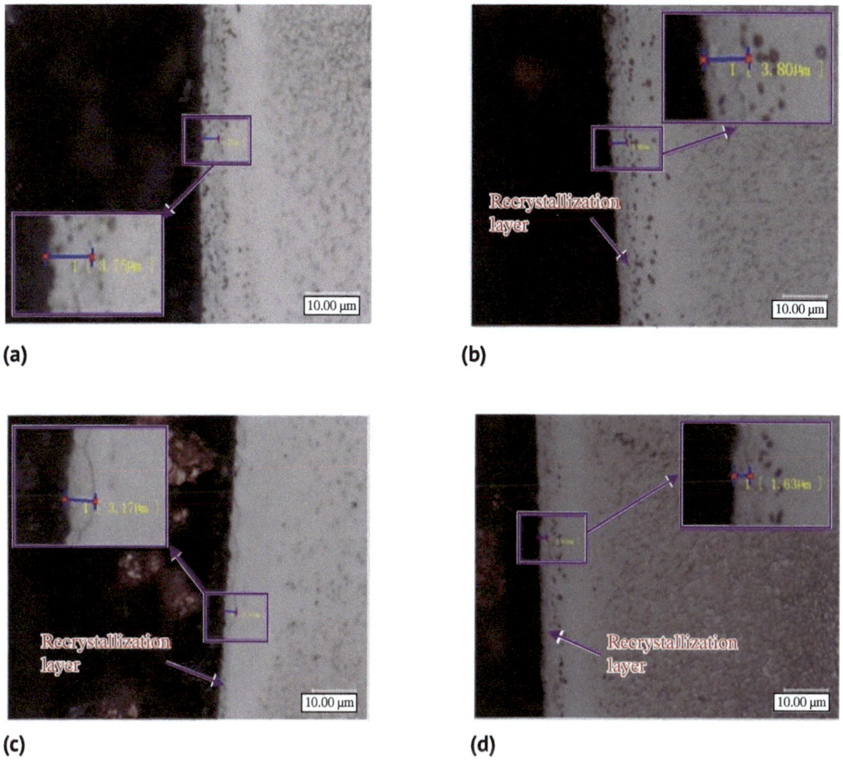

Figure 13.7 Subsurface microstructure of grinding DD98 after different heat treatment method: (a) 70°C, 3 h, air cooling; (b) 76°C, 1.5 h, air cooling; (c) 76 °C, 4 h, air cooling; and (d) 76°C, 4 h, furnace cooling.

environment. The furnace-cooling method is more effective than air cooling in reducing the thickness of the subsurface recrystallization layer. This may be because when a high-heat workpiece is air cooled, it suddenly encounters a lower temperature, resulting in a fast surface-cooling speed. The material shrinks, storing deformation energy. In contrast, furnace-cooling is a slow-cooling process, preventing sudden shrinkage deformation of the material and slowly releasing the stress on the grinding surface. Thus, an appropriate heat-treatment process can reduce the subsurface recrystallization thickness of micro-ground parts in a high-temperature application environment. The thickness of the ground subsurface recrystallization layer of the nickel-based single-crystal superalloy workpiece is 1.63 μm after being kept at 760°C for 4 h and cooled by furnace. The micro-ground workpiece of the nickel-based single-crystal superalloy can be heat-treated by holding it at 760°C for 4 h and then cooling it in the furnace to reduce the thickness of the recrystallization layer in a high-temperature environment.

Removing the surface plastic-deformation layer can also inhibit recrystallization. Salkeld et al. [8] suggested employing electrochemical corrosion to eliminate the surface deformation layer when plastic deformation is moderate, thus preventing recrystallization of superalloy components in high-temperature service environments. The procedure involves two sequential steps. First, the sample is subjected to anodic corrosion in a 70 vol% phosphoric acid solution for 3 minutes under a current density of 620 A/m^2. Subsequently, it is immersed in one of two warm solutions at 65 ± 2 °C: either a mixture containing 2 parts concentrated nitric acid, 80 parts 32% hydrochloric acid, and 11 parts ferric chloride solution for 3–5 minutes, or a solution composed of 90% hydrochloric acid and 10% hydrogen peroxide for the same duration. This method effectively removes a 0.013 – 0.05 mm surface layer from the single-crystal alloy, eliminating deformation-induced defects that could act as recrystallization nucleation sites.

In this chapter, the method of removing the deformation layer of the grinding surface by mechanical machining is adopted. In the finishing part, a micro-grinding rod with a diameter of 0.9 mm and a surface coated with 1000#CBN abrasive particles was selected. The DD98 workpiece was ground with a surface 500# particle micro-grinding rod with $v_s = 50000$ r/min, $v_w = 20$ μm/s, and $a_p = 2$ μm to remove the plastic-deformation layer caused by the 500# grinding rod. Then the polished workpiece is kept at 760°C for 4 h, and the stress is removed by cooling with the furnace. Finally, the workpiece is kept at 1100°C for 4 h to experience high-temperature working environment of superalloy parts. By measuring the thickness of the subsurface recrystallization layer of the workpiece, it can be seen that recrystallization layer thickness of the micro-grinding workpiece after fine grinding and stress relief treatment is significantly reduced to 0.70 μm.

According to relevant research on directionally solidified superalloys at home and abroad, the following points are worth noting:

- The technical document of the French Turbomeca company stipulates that directional solidification alloy parts with a recrystallization layer thicker than 13 μm cannot be used at any time.
- According to relevant Russian literature, directionally solidified superalloy parts are prohibited from further use when the recrystallization thickness exceeds 50 μm.
- Relevant data in China show that DZ4 superalloy cannot be used if the recrystallization thickness exceeds 45 μm.

Relative to directionally solidified superalloys, recrystallization impacts the properties of single-crystal superalloys more profoundly, thus necessitating stricter control criteria for their manufacturing processes. Under current laboratory conditions, this study establishes a minimum allowable recrystallization layer thickness of 0.70 μm to ensure structural integrity and performance consistency in single-crystal components.

13.2.3 Subsurface Microhardness in Micro-grinding

The microhardness of the ground surface is the result of the combined effects of cold work hardening caused by grinding-induced plastic deformation and material structure changes due to grinding heat. Cold work hardening is manifested as shear slip between crystals in the material. Near the grinding surface, the lattice distorts or even refines, strengthening the material and altering its hardness, ultimately influencing the component's performance. Among all the surface quality indicators for the micro-grinding of nickel-based single-crystal superalloys, the subsurface microhardness is a crucial one for evaluating the quality of the ground workpieces. Thus, this study experimentally investigated the microhardness of nickel-based single-crystal superalloy micro-grinding.

During the microhardness test, indentations at varying depths from the ground surface were examined via metallographic microscopy. The diamond-shaped impressions, produced as Vickers indents under a 10 gf load, had their microhardness values determined by measuring the diagonal lengths of these indentations.

In this study, to eliminate the impact of crystallographic anisotropy on microhardness measurement of single-crystal materials, a grinding rod with a diameter of 0.9 mm was used with grinding factors of $v_s = 30$ kr/min, $a_p = 10$ μm, and $v_w = 100$ μm/s to grind along the $< 100 >$ crystal orientation within the (001) crystal plane. When measuring the subsurface hardness of the ground surface, the direction of measurement was along the $< 010 >$ crystal orientation, which is perpendicular to the grinding surface ($< 100 >$ crystal orientation). Since the hardness values along the $< 100 >$ crystal orientation within the unprocessed single-crystal material (001) crystal plane are equal, grinding along the $< 100 >$ crystal orientation eliminates the impact of crystallographic anisotropy on microhardness.

Figure 13.8 shows the variation of micro-grinding nickel-based single-crystal superalloy workpieces with grinding depth. As can be seen from the figure, a hardened layer forms on the ground surface. The microhardness near the grinding surface of the micro-ground single-crystal superalloy workpiece is higher than that of the single-crystal alloy matrix. As the microhardness measurement position moves deeper perpendicular to the grinding surface, the hardness change caused by

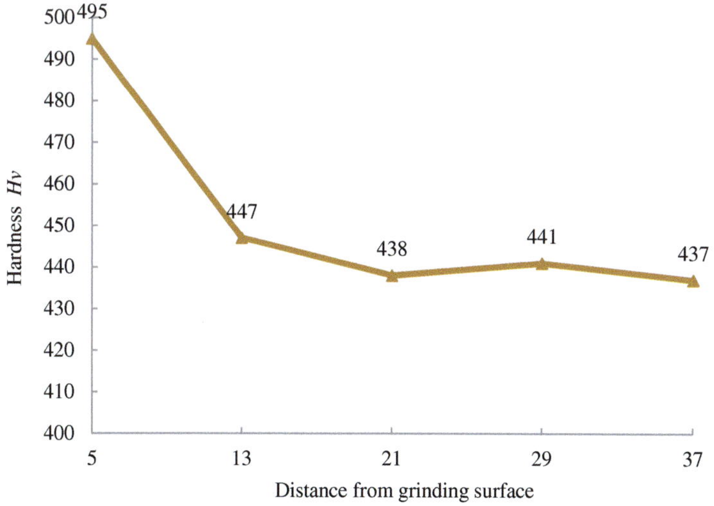

Figure 13.8 Microhardness of different positions in the grinding subsurface.

grinding gradually diminishes, and the microhardness value gradually approaches that of the alloy matrix.

The reason for this is that after plastic deformation of the metal crystal, the resistance to dislocation slip within the crystal increases, making the material more resistant to deformation. From the perspective of dislocation theory, as the material undergoes plastic deformation, the movement of dislocations in the crystal becomes more difficult. The internal stress generated during plastic deformation hinders the movement of dislocations. Explanations of work hardening are based on the difficulty of dislocation movement. The movement of dislocations on the grinding surface alters the crystal material's structure, and the internal stress in the crystal inhibits the continuous movement of dislocations. Evidently, work hardening is caused by the accumulation of elastic interactions between dislocations and deformation induced by dislocations.

The hardness Hv of the deformed region material is proportional to the average stress during deformation and is proportional to the square root of the dislocation density ρ. The hardness Hv can be expressed as Eq. (13.2).

$$Hv = \frac{Gb}{K}\sqrt{\rho} \tag{13.2}$$

where K is a numerical factor related to dislocation structure and distribution.

During the micro-grinding of nickel-based single-crystal superalloys, numerous slip bands appear at the contact position between the grinding rod and the workpiece. Under the grinding-induced stress, dislocations occur within the material and move along the slip planes. At this point, the shear stress exceeds the material's yield limit, and dislocations form at the abrasive grain's cutting edge and then move into the deeper deformed material.

The dislocation density is higher at the grinding surface, and it decreases as the distance from the grinding surface increases. As the dislocation density in the grinding subsurface rises, dislocations entangle at the subsurface of the single-crystal alloy grinding area, forming tangles and cellular substructures. These restrict the movement of dislocations to a smaller range, making it difficult for them to penetrate these obstacles.

Furthermore, as the degree of plastic deformation in the single-crystal superalloy subsurface increases, the dislocation segments that form the three-dimensional dislocation network become shorter, the cell walls develop more completely, and the cell size decreases. This leads to a reduction in the average degree of freedom of dislocations and an increase in the material's resistance to deformation, resulting in an increase in microhardness values.

References

1 Li, Q., Gong, Y.D., Sun, Y. et al. (2018). Milling performance optimization of DD5 Ni-based single-crystal superalloy. *International Journal of Advanced Manufacturing Technology* 94 (5–8): 2875–2894.

2 Xiao, Q.F., Xu, Y.M., Liu, X.L. et al. (2023). Oxidation-induced recrystallization and damage mechanism of a Ni-based single-crystal superalloy during creep. *Materials Characterization* 195: 112465.
3 Sun, Y., Su, Z., Gong, Y. et al. (2012). Analytical and experimental study on micro-grinding surface-generated mechanism of DD5 single-crystal superalloy using micro-diamond pencil grinding tool. *Archives of Civil and Mechanical Engineering* 21, 1–21.
4 Chen, M., Cai, M., Gong, Y. et al. (2023). Study on the removal mechanism in multi-abasive micro-grinding of nickel-based superalloy. *International Journal of Advanced Manufacturing Technology* 128: 2199–2220.
5 Zhu, T., Cai, M., Gong, Y. et al. (2023). Research progress of eco-friendly grinding technology for aviation nickel-based superalloys. *International Journal of Advanced Manufacturing Technology* 126: 2863–2886.
6 Liang, C.Y., Gong, Y.D., Qu, S.S. et al. (2022). Performance of grinding nickel-based single crystal superalloy: effect of crystallographic orientations and cooling-lubrication modes. *Wear* 508–509: 204453.
7 Liang, C., Gong, Y., Li, P. et al. (2022). Subsurface deformation and burr formation in nickel-based single-crystal superalloy under grinding. *Archives of Civil and Mechanical Engineering* 23(2): 126.
8 Salkeld, R.W., Field, T.T., and Ault, E.A. (1995). Preparation of single crystal superalloy for post-casting heat treatment, US patent: 5413648.

Part V

Simulation, Optimization, and Control in Grinding of Single-crystal Turbine Blade Root

The components of single-crystal nickel alloys (e.g., turbine blades) usually possess complex structure shape and raise high requirements in surface quality and shape accuracy. These throw great challenges for mechanical grinding. The profile grinding process removes the materials by the strong thermo-mechanical interactions between abrasive grains and workpiece, leading to the high temperature and stress fields usually. This influences significantly the machined surface quality. Particularly, the complicated structure shape of workpiece makes it difficult for the experimental measurements of temperatures and stresses. Under such conditions, the simulation techniques on grinding temperature and stresses are necessary for optimizing and controlling of profile grinding. Therefore, this part aims to develop the finite element (FE) methods for analyzing the three-dimensional (3D) characteristics of grinding temperature and residual stress fields of single-crystal nickel alloy turbine blade roots. Firstly, the FE model of single-crystal blade root workpiece was built and then the simulated temperature and stresses were verified by experiments. Subsequently, the evolutions of both temperature field and residual stress field were studied. Furthermore, the gradient distributions and the effects of blade root workpiece structure shape were discussed. Finally, for collaborative manufacturing of single-crystal turbine blade roots, the possibility on controlling the grinding temperature and stress fields through reliable predictive models was evaluated.

14

Temperature Field in Grinding of Single-crystal Turbine Blade Root

14.1 FE Model for Grinding Temperature Simulation

In general, both the grinding temperature field and residual stress field are influenced by the interactive coupling effects of grinding heat and grinding force. For cost-effectiveness and easy modification, the existing FE model adopts the indirectly coupled thermomechanical approach. The overall process consists of three steps (Figure 14.1), which are specifically described as follows:

1. Model construction: The temperature-dependent constitutive parameters were calibrated through thermophysical property interpolation from Tables 14.1–14.2. Following this, a parametric CAD model of the turbine root's aerodynamic profile was reconstructed using SolidWorks' surface modelling module. Subsequent domain discretization employed an adaptive hex-dominant meshing strategy, implementing boundary layer refinement (0.1 mm element size) at grinding interfaces for thermal gradient resolution, while applying geometric progression coarsening (max 3.2 mm) in non-critical regions (Figure 14.2). The converged mesh topology comprised 360,148 structured elements with Jacobian determinants exceeding 0.85.
2. Thermal analysis: The computational framework employed SOLID70 hexahedral elements (ANSYS thermoelastic element type) with anisotropic thermal conductivity capabilities. Initial conditions were set to isothermal equilibrium at 20 °C, while boundary conditions incorporated convective heat transfer coefficients per MIL-STD-810G aerospace standards. Transient thermal analysis was executed through adaptive time-stepping algorithms, achieving Courant number stability below 0.8 throughout the temporal domain.

Grinding of Single-Crystal Superalloys: Fundamentals and Technologies, First Edition.
Wenfeng Ding et al.
© 2026 Wiley-VCH GmbH. Published 2026 by Wiley-VCH GmbH.

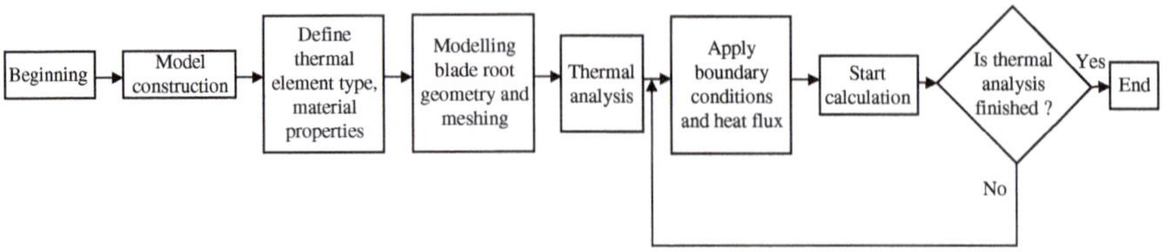

Figure 14.1 Simulation flow chart of grinding temperature of blade root workpiece.

Table 14.1 Physical properties of single-crystal nickel alloy along the principal crystal axes.

Temperature T (°C)	Specific heat C (J/[g · °C])	Thermal expansion coefficient α (10^{-6}/°C)	Thermal conductivity λ (W/[m · °C])	Density ρ (g/cm^3)
100	0.358	11.92	8	8.78
200	0.392	11.92	9.45	8.78

Table 14.2 Mechanical properties of single-crystal nickel alloy along the principal crystal axes.

Temperature T (°C)	Young's Modulus E (GPa)	Poisson ratio v	Tangent modulus G (GPa)	Yield stress σ_s (MPa)
20	134.1	0.419	106	930
650	116.4	0.379	91	940

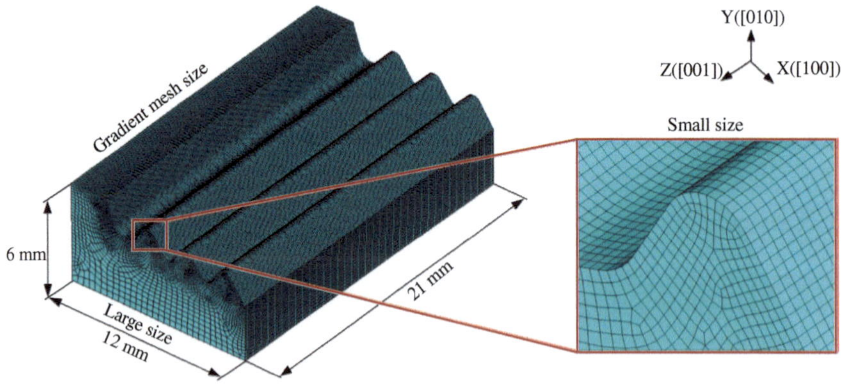

Figure 14.2 Model mesh of the turbine blade root workpiece.

14.2 Thermal Analysis for Grinding Temperature Simulation

14.2.1 Thermal Energy and Energy Partition

While grinding is underway, the vast majority of grinding energy gets transformed into thermal energy (namely, the total heat flux q_t). This thermal energy is distributed among several aspects: the heat flux entering the grinding wheel q_s, the grinding chip q_{ch}, the workpiece q_w, and the heat carried away by the grinding fluid q_f:

$$q_t = q_s + q_{ch} + q_w + q_f \tag{14.1}$$

Furthermore, as per Lavine's report [1], the grinding wheel together with the grinding fluid can be regarded as a composite entity, and the heat entering this composite object is readily obtainable. Rowe [2] demonstrated that the energy distribution ratio ε during grinding operations can be quantified through the following relationship between the abrasive tool and the processed material:

$$\varepsilon = \left(1 + \sqrt{\frac{V_s(k\rho c)_c}{V_w(k\rho c)_w}}\right)^{-1} \tag{14.2}$$

Within this formulation, the thermal conductivity coefficient is denoted by k and the workpiece density by ρ. The composite term $(k\rho c)_c$ characterizes the thermophysical property product of the coolant–abrasive system, whereas $(k\rho c)_w$ quantifies the equivalent thermal effusivity parameter for the workpiece material. Therefore, q_w can be calculated as:

$$q_w = \varepsilon \cdot (q_t - q_{ch}) \tag{14.3}$$

The total heat flux q_t can generally be determined by referencing the relationship between the tangential grinding force F_t, the grinding wheel speed v_s, and the workpiece–wheel contact area.

Within thermal analysis frameworks, the total heat flux generation q_t is derived through the thermomechanical energy conversion relationship expressed as F_t, where F_t denotes the tangential grinding force component, and v_s represents the rotational velocity of the grinding wheel.

$$q_t = \frac{F_t \cdot v_s}{S} = \frac{F_t \cdot v_s}{l_c \cdot b} = \frac{F_t \cdot v_s}{\sqrt{a_p \cdot d_e} \cdot b} \tag{14.4}$$

In Eq. (14.4), a_p represents the grinding depth, d_e stands for the grinding wheel diameter, b indicates the contact area width, S signifies the contact area, and l_c denotes the contact length, which is determined by the grinding depth a_p and the grinding wheel diameter d_e. Malkin [3] has suggested that the heat flux flowing into the grinding chip can be expressed as:

$$q_{ch} = e_{ch} \cdot \frac{a_p \cdot V_w}{l_c} = \rho_w \cdot c_w \cdot T_{mp} \cdot \frac{a_p \cdot V_w}{\sqrt{a_p \cdot d_e}} \tag{14.5}$$

In this context, ρ_w stands for the workpiece density, c_w is the specific heat capacity of the workpiece, T_{mp} is the melting temperature of the workpiece, and e_{ch} is the energy needed to raise the temperature of unit weight chips from room temperature to the workpiece melting temperature.

Therefore, combined with Eqs. (14.1–14.5), the heat flux into the workpiece is given by:

$$q_w = \varepsilon \cdot (q_t - q_{ch}) = \frac{\left(\frac{F_t \cdot V_s}{\sqrt{a_p \cdot d_e} \cdot b} - \rho_w \cdot c_w \cdot T_{mp} \cdot \frac{a_p \cdot V_w}{\sqrt{a_p \cdot d_e}}\right)}{1 + \sqrt{\frac{V_s (k\rho c)_s}{V_w (k\rho c)_w}}} \tag{14.6}$$

14.2.2 Heat Source and Boundary Conditions

In the grinding process, the largest uncut chip typically forms at the leading edge, and the uncut chip thickness for a specific abrasive grain varies as it moves through the grinding contact zone (Figure 14.3). In view of this, the simulation utilized a triangular heat flux distribution model. Figure 14.3 also depicts the movement of the triangular heat flux across the workpiece surface. The contact length l_c defines the interface region between the abrasive wheel and the machined surface, which also represents the effective heating zone. The triangular thermal flux profile induced in the workpiece can be mathematically modelled using the following relationship:

$$q(x) = \frac{2q_w \cdot x}{l_c} = \frac{2q_w \cdot x}{\sqrt{a_p \cdot d_e}} \tag{14.7}$$

In this equation, x denotes the length of the grinding zone.

The fundamental relationship is established based on three-dimensional heat transfer principles in solids, expressed mathematically as:

$$\frac{\partial}{\partial x}\left(k_x \frac{\partial T}{\partial x}\right) + \frac{\partial}{\partial y}\left(k_y \frac{\partial T}{\partial y}\right) + \frac{\partial}{\partial z}\left(k_z \frac{\partial T}{\partial z}\right) = \rho c \frac{\partial T}{\partial t} \tag{14.8}$$

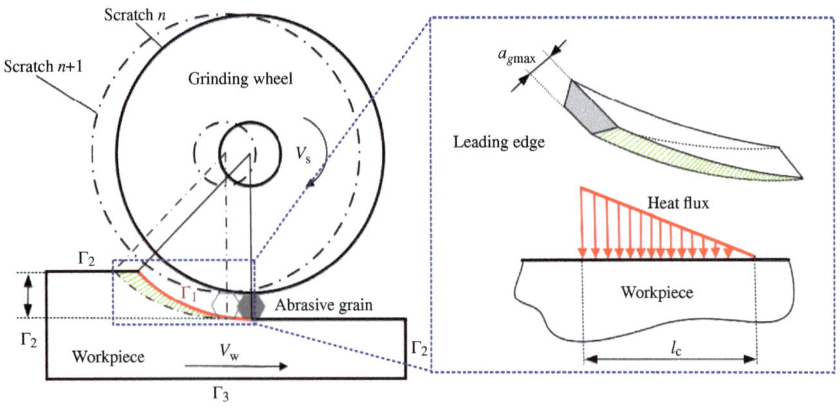

Figure 14.3 Schematic illustration of heat loads and thermal boundary conditions.

Table 14.3 The parameters used for grinding temperature simulation.

Grinding steps	v_s (m/s)	v_w (mm/min)	a_p (mm)	H (mm)	F_t (N)	q_w (J/(m$^2\cdot$s))
i	25	240	1.50	1.50	111.6	1.6×10^5
ii	35	210	1.00	2.50	159.0	1.9×10^5
iii	35	300	0.13	2.63	25.5	0.9×10^5
iv	25	600	1.50	4.13	275.2	2.1×10^6

In this equation, T denotes the temperature, while x, y, and z represent the Cartesian coordinates along their respective axes. The symbol c stands for specific heat capacity, and t represents time. The system's initial temperature is set to $T = T_0 = 20\,°\mathrm{C}$. The thermal modelling framework adopts three well-defined thermomechanical boundary conditions, predicated on the semi-infinite solid assumption for the workpiece while embedding a moving heat source. These governing equations, as rigorously derived in [4], are formulated as:

$$k_x \frac{\partial T}{\partial x} n_x + k_y \frac{\partial T}{\partial y} n_y + k_z \frac{\partial T}{\partial z} n_z = q \text{ (first boundary } \Gamma_1) \tag{14.9}$$

$$k_x \frac{\partial T}{\partial x} n_x + k_y \frac{\partial T}{\partial y} n_y + k_z \frac{\partial T}{\partial z} n_z = \alpha\,(T_a - T) \text{ (second boundary } \Gamma_2) \tag{14.10}$$

$$k_x \frac{\partial T}{\partial x} n_x + k_y \frac{\partial T}{\partial y} n_y + k_z \frac{\partial T}{\partial z} n_z = 0 \text{ (third boundary } \Gamma_3) \tag{14.11}$$

In this context, α represents the thermal diffusivity, T_a indicates the temperature of the grinding fluid, and n is the unit vector perpendicular to the surface. The thermal boundary conditions were established as depicted in Figure 14.3. Since the bottom surface of the workpiece was treated as an insulator, the first boundary Γ_1 was implemented in the grinding zone, the third boundary Γ_3 was applied to the workpiece bottom surface, and the second boundary Γ_2 was assigned to the remaining surfaces. Based on the aforementioned analysis, the key parameters utilized in the temperature simulation are presented in Table 14.3.

14.3 Experimental Validation of Grinding Temperature

In the present study, the turbine blade root has the total allowance of 2.63 mm, which consists of three grinding steps to finish the grinding operation, as shown in Table 14.3. In order to obtain the workpiece surface with burning, the grinding step iv with aggressive parameters are used. The grinding experimental setup and the temperature measuring method based on the semi-artificial thermocouple technique are described in Part III.

Figure 14.4 presents a thermomechanical coupling analysis validation through comparative temperature profiles between numerical predictions and experimental measurements for turbine blade root grinding. Notably, when interfacial temperatures exceed critical thresholds (≥ 800 °C), oxidation-induced discoloration phenomena manifest in workpiece apex zones, as evidenced by localized carbon deposition patterns. This thermal degradation mechanism serves as a diagnostic indicator for grinding burn inception. Scanning electron microscopy (SEM) image in Figure 14.4a further indicated the generation of rough ground surface with severe material tearing. When grinding without burning, the temperature reached a low level of less than 200 °C under which the smooth ground surface was created in

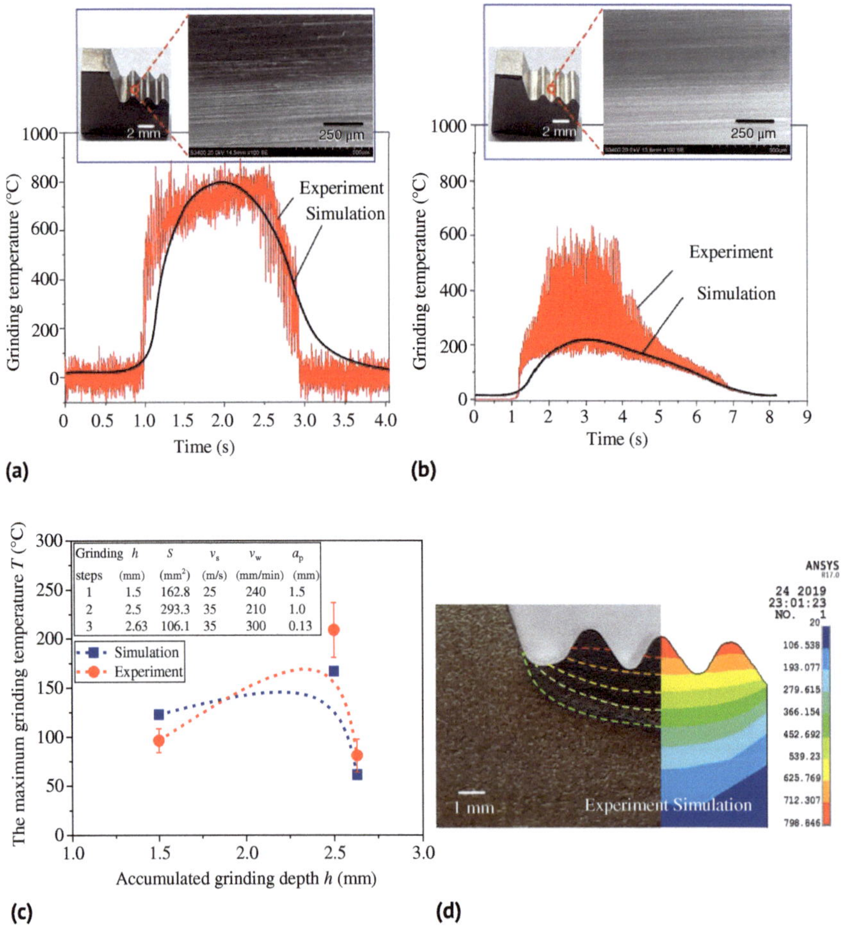

Figure 14.4 Comparison between simulated and experimental grinding temperatures of blade root: (a) temperature signals when $v_s = 25$ m/s, $v_w = 600$ mm/min, $a_p = 1.50$ mm, $h = 4.13$ mm; (b) temperature signals when $v_s = 35$ m/s, $v_w = 210$ mm/min, $a_p = 1.00$ mm, $h = 2.5$ mm; (c) the maximum temperature with accumulated grinding depth; and (d) heat-affected zone.

the workpiece valley region (Figure 14.4b). For both cases, it can be found that the simulated temperature agreed well with the experimental value. What is more, the grinding temperatures obtained by simulation or experiment at different grinding steps showed the identical trend. That is, when the accumulated grinding depth *h* increased from 0 to 2.63 mm, the grinding temperature increased firstly, then reached the maximum value ($h = 2.5$ mm), and finally decreased, as indicated in Figure 14.4c.

Further analysis indicates that the average error of the grinding temperature between the simulation and experimental results is approximately 16%, which is within the acceptable range for predicting grinding process variables [5, 6]. Additionally, the isotherms of the grinding temperature field in the simulation are in good agreement with the observation of the surface color of the grinding heat-affected zone in the experiment, as shown in Figure 14.4d.

14.4 Temperature Simulation Results and Analysis

14.4.1 Effect of Blade Root Forming Process on Temperature Evolution

By employing the aforementioned FE method, the influence of the blade root forming process on the trajectory of grinding temperature evolution and the temperature distribution within the workpiece can be investigated, with the results illustrated in Figure 14.5.

Upon setting *h* at 1.5 mm (Figure 14.5a), the grinding process of the three blade root valleys got underway. The peak temperature values appeared in the blade root peak regions, with the workpiece edge (view A-A) reaching the highest temperature of 119.7 °C. The thermal load distribution exhibited marked disparities compared to conventional surface grinding operations. The observed thermal contrast principally stems from the kinematic configuration featuring discrete tribological interaction zones, where four quasi-distributed thermal excitation sources undergo spatiotemporal superposition across the workpiece surface.

When *h* was increased further to 2.5 mm (Figure 14.5b), the initial formation of the three blade root peaks commenced, each exhibiting temperature peaks of approximately 167.2 °C with a comparable temperature distribution observed in cross-sectional view B-B. The location of maximum temperature exhibited a gradual inward migration from the workpiece periphery toward its core region. This thermal shift correlated strongly with increasing accumulated material removal depth, resulting in progressively nonuniform heat distribution throughout the workpiece.

When the cumulative grinding depth hit 2.7 mm (Figure 14.5c), the fully developed blade root profile was produced. The T_{max} position moved to the blade root peaks, yet the peak temperature declined from 167.2 to 61.1 °C, which might be due to the decreased grinding depth. Moreover, the blade root peak regions showed higher grinding temperatures than the valley regions.

When the aggregate material removal depth reached $h = 4.13$ mm (Figure 14.5d), complete blade root formation was achieved. However, this stage was accompanied

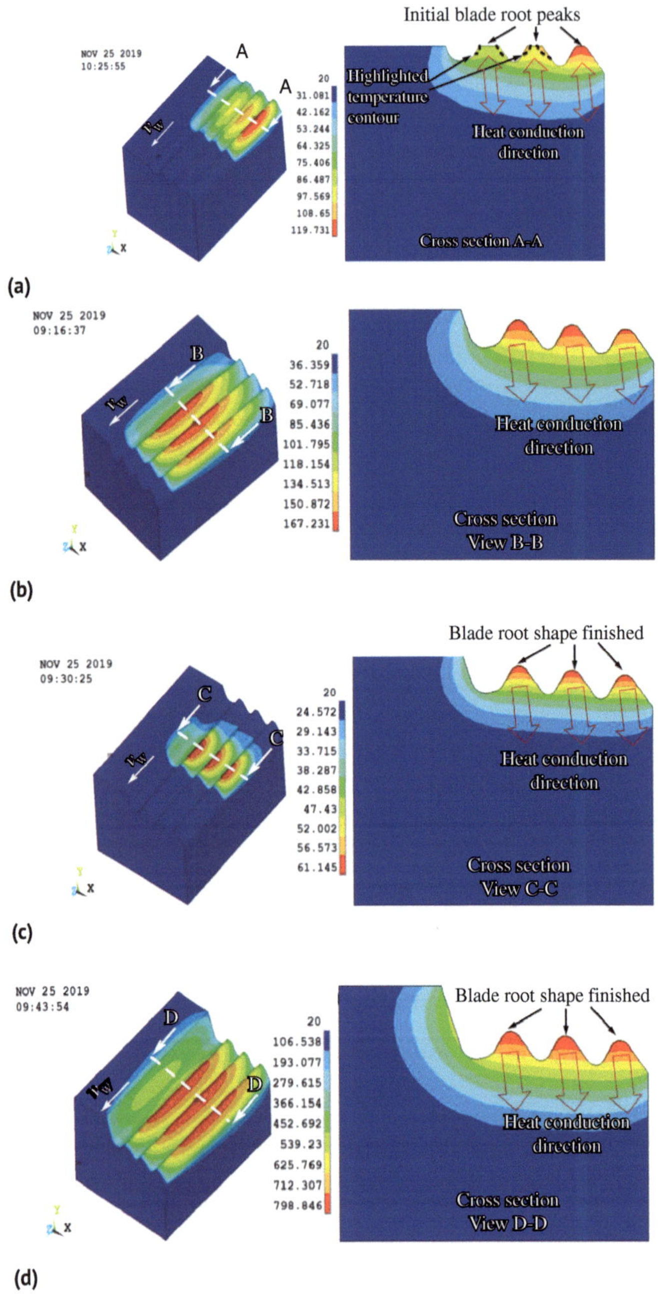

Figure 14.5 Influence of the blade root forming process on grinding temperature evolution: (a) $h = 1.5$ mm; (b) $h = 2.5$ mm; (c) $h = 2.63$ mm; and (d) $h = 4.13$ mm.

by a substantial thermal escalation, with peak temperatures reaching 798.8 °C. At the same time, the root peaks exhibited significantly higher temperatures.

14.4.2 Grinding Temperature Gradient in Cross Section of Blade Root Workpiece

For comprehensive thermal analysis, the cross-sectional temperature gradients are presented in Figures 14.6 (view C-C in Figure 14.5c) and 14.7 (view D-D in Figure 14.5d). In Figure 14.6a, the thermal mapping reveals significant spatial variations, with Peak II and Peak III regions exhibiting approximately 50°C during normal grinding conditions, compared to 32°C in Valley I and Valley II, as indicated in Figure 14.6b and c. Under grinding burn conditions (Figure 14.7a), these thermal characteristics undergo dramatic intensification, reaching 400°C in valley regions (Figure 14.7b) and 650°C in peak regions (Figure 14.7c), representing a 13-fold and 12.5-fold increase respectively over normal operating temperatures. And the temperature gradient can reach 300°C/mm if grinding burn happened. Such a large temperature gradient can easily induce big residual tensile stresses in surface layer of workpiece. Experimental observations in Figure 14.6 demonstrate a consistent thermal disparity between geometric features, with peak regions maintaining 20–35% higher

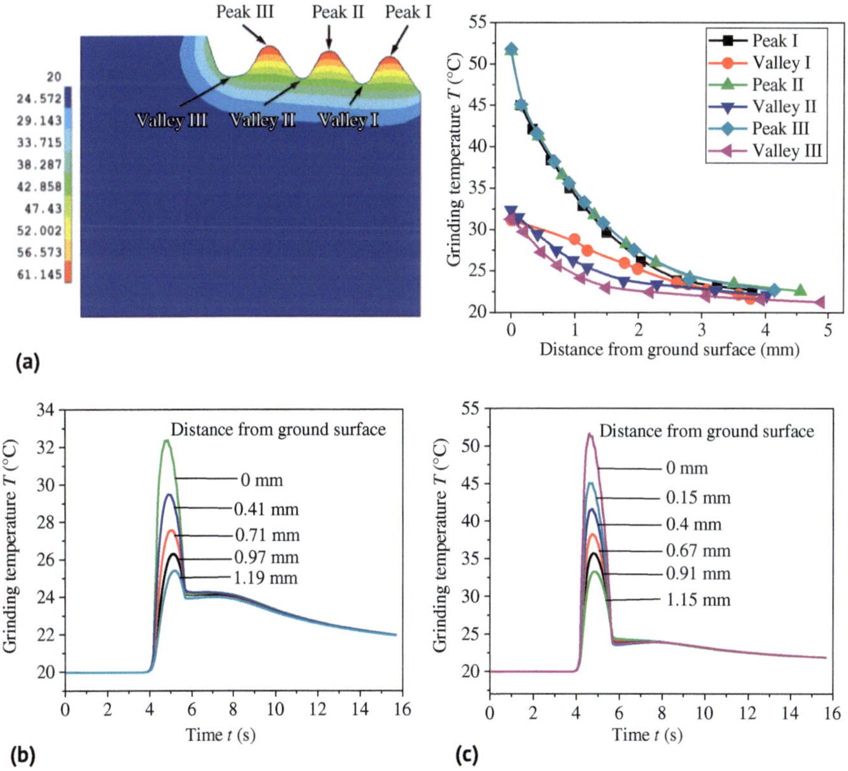

Figure 14.6 Grinding temperature gradient in different regions (grinding step iii): (a) cross section; (b) valley II region; and (c) peak III region.

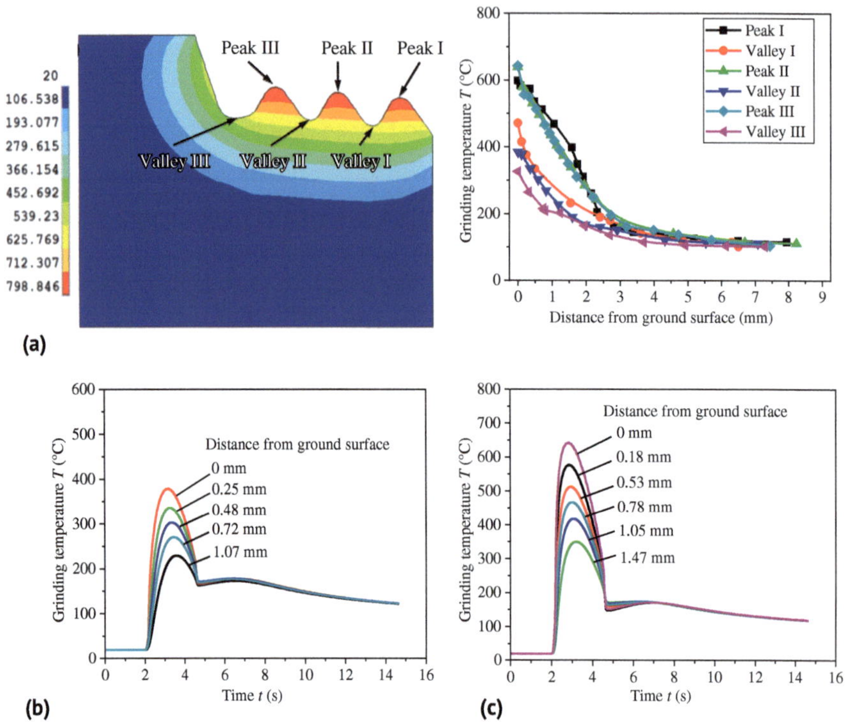

Figure 14.7 Grinding temperature gradient in different regions (grinding step iv): (a) cross section; (b) valley II region; and (c) peak III region.

temperatures than adjacent valleys across all grinding conditions (normal vs. burn scenarios). This thermal differentiation follows a characteristic decay profile, where the temperature gradient attenuates progressively with increasing depth from the ground surface, suggesting a depth-dependent thermal dissipation mechanism.

14.4.3 Effect of Blade Root Shape on Grinding Heat Conduction

The morphology of the blade root can greatly influence the internal heat conduction and distribution in the workpiece [7, 8]. To better understand this phenomenon, this section delves into the process of heat conduction variations and highlights the key characteristics of the workpiece's temperature field. This will be accomplished by examining the temperature distribution across multiple grinding steps depicted in Figure 14.5. The analysis is as follows:

1. During intermediate grinding stages where blade root topography remains incomplete (Figure 14.5a), localized temperature maxima consistently develop at regions experiencing thermal superposition from multiple transient heat sources. These hot spots correlate strongly with areas of maximum resultant temperature generation.

2. The highlighted temperature contours appear to be uniformly distributed around the uncompleted blade root peaks (Figure 14.5b), attributed to the discrete heat sources. The thermal distribution patterns surrounding partially formed blade root protrusions (Figure 14.5b) exhibit symmetrical contour dispersion, resulting from the combined effect of discrete thermal inputs. As the grinding operation proceeds vertically, more blade root profiles emerge. Subsequently, the initial symmetric temperature contours evolve, and areas with temperature maxima start to form.
3. Once the blade root profile reaches its final form (Figure 14.5c and d), the previously distinct heat sources merge, creating a continuous curved surface heat source within the workpiece–wheel contact zone. However, due to the limited material available for heat dissipation, a significant amount of heat tends to accumulate in the blade root peak regions. This accumulation results in the highest temperature values occurring in these peak areas [9].

The described thermal phenomena are quantitatively illustrated in Figure 14.8. During initial grinding phases where blade root geometry is still developing (Figure 14.8a), thermal energy generation predominantly occurs at interfacial contact zones between the abrasive wheel's active grains and the workpiece surface. This localized heat concentration corresponds to regions of maximum material engagement. However, due to heat conduction in two directions and effective fluid convection, heat does not significantly accumulate in either the blade root valleys or peaks. As the blade root shape nears completion (Figure 14.8b), the absence of matrix material around the blade root peaks hinders heat dissipation, and the complex blade root geometry may also impede fluid convection. These factors result in higher temperatures in the blade root peak regions compared to the valleys. It is intriguing to observe that heat conduction evolves as the blade root shape forms. This indicates that the influence of the blade root shape on the temperature field becomes more pronounced as grinding approaches the final shape. If the cutting allowance prior to grinding can be decreased and presents a reasonable distribution along the

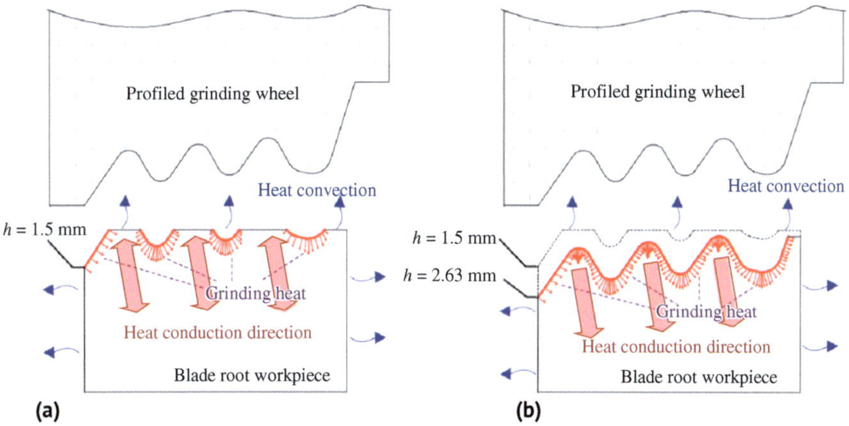

Figure 14.8 Schematic diagram of the heat flux distribution along the blade root profile: (a) $h = 1.5$ mm and (b) $h = 2.63$ mm.

workpiece shape, the improved grinding conditions can reduce effectively the concentration of grinding heat and avoid the occurrence of high grinding temperature [10]. Under such circumstances, the risk of grinding burning can be restrained and the ability to control grinding surface quality can be improved.

References

1 Lavine, A.A. (1988). Simple model for convective cooling during the grinding process. *Journal of Manufacturing Science and Engineering* 110 (1): 1–6.
2 Rowe, W.B. (2001). Temperature case studies in grinding including an inclined heat source model. *Proceedings of the Institution of Mechanical Engineers, Part B: Journal of Engineering Manufacture* 215 (4): 473–491.
3 Malkin, S. and Guo, C. (2007). Thermal analysis of grinding. *CIRP Annals-Manufacturing Technology* 56 (2): 760–782.
4 Nie, Z.G., Wang, G., Liu, D.H. et al. (2018). A statistical model of equivalent grinding heat source based on random distributed grains. *ASME Journal of Manufacturing Science and Engineering* 140 (5): 051016.
5 Ghosh, S., Chattopadhyay, A.B., Paul, S. et al. (2008). Modelling of specific energy requirement during high-efficiency deep grinding. *International Journal of Machine Tools and Manufacture* 48 (11): 1242–1253.
6 Li, H.N. and Axinte, D. (2018). On the inverse design of discontinuous abrasive surface to lower friction-induced temperature in grinding: An example of engineered abrasive tools. *International Journal of Machine Tools and Manufacture* 132: 50–63.
7 Yang, S.Y., Chen, W.F., and Zheng, L. (2024). On the grinding temperature field of helical gears: modeling and experiments. *Journal of Manufacturing Processes* 124: 1010–1024.
8 Yin, Y.K., Xu, J.Y., Shi, L.Y. et al. (2024). Study on thermal damage formation mechanisms and surface integrity in creep-feed profile grinding of superalloys. *Journal of Manufacturing Processes* 120: 28–50.
9 Li, B.K., Ding, W.F., Zhu, Y.J. et al. (2024). Thermal analysis on profile grinding of turbine disc slots of powder metallurgy superalloy FGH96. *International Communications in Heat and Mass Transfer* 159: 108207.
10 Mao, C., Zhou, Z.X., Guo, K.X. et al. (2010). Numerical simulation and experimental validation of the temperature field in surface grinding. *Machining Science and Technology* 14 (3): 344–364.

15

Residual Stress Field in Grinding of Single-crystal Turbine Blade Root

15.1 Mechanical Analysis for Residual Stress Simulation

15.1.1 Thermal Analysis Prior to Stress Simulation

Once the thermal analysis was complete, its results served as loads for the mechanical analysis aimed at calculating the grinding-induced residual stresses in the turbine blade root workpiece [1]. The procedure is outlined in Figure 15.1: (i) Solid185 elements were utilized as the structural elements. (ii) The thermal simulation outcomes were initially transformed into partial mechanical loads. (iii) The constraints and boundary conditions were specified. Notably, an innovative method was employed, which incorporates the micro-interaction effect between the workpiece material and each separate grain into the formation of part of the mechanical loads (as detailed in Section 15.1.3). (iv) Finally, the transient and residual stresses were computed, enabling the observation of their distribution variations over time.

Note that, in the creep-feed grinding of turbine blade root, due to the complex shape structure of the workpiece and the large grinding depth, the residual stresses of the final grinding step are the main concern in engineering applications. Combined with the above analysis and mentioned method, the parameters

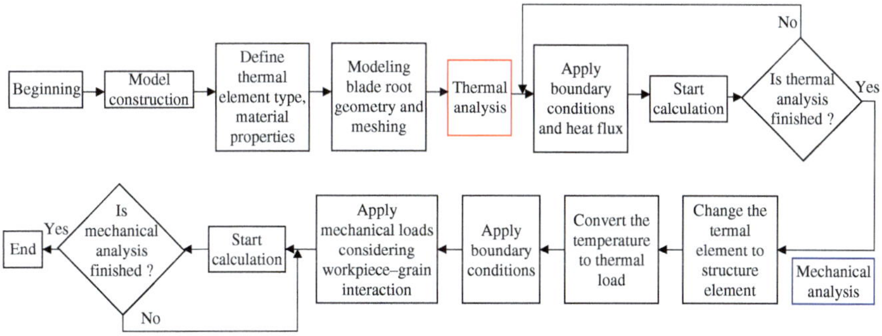

Figure 15.1 Simulation flow chart of grinding-induced residual stresses.

Grinding of Single-Crystal Superalloys: Fundamentals and Technologies, First Edition.
Wenfeng Ding et al.
© 2026 Wiley-VCH GmbH. Published 2026 by Wiley-VCH GmbH.

used for thermal analysis prior to simulating the final residual stress field are listed in Table 15.1. The temperature simulation results of the final grinding step of the blade root workpiece are shown in Figure 15.2.

The highest grinding temperature, around 85 °C, was typically found in the blade root's peak region (Figure 15.2). The temperature gradually decreased from the peak to the valley region. This suggests that the complex root geometry specifically affects grinding temperature distribution, such as by causing significant heat accumulation in the peak region, which has less matrix material than other areas.

15.1.2 Yield Function and Elastoplastic Constitutive

In this study, considering the workpiece material is orthotropic single-crystal nickel alloy, Hill's yield criterion [2] was chosen as the yield function to analyze mechanical plastic strain.

$$\sigma_S^2 = F(\sigma_{22} - \sigma_{33})^2 + G(\sigma_{33} - \sigma_{11})^2 + H(\sigma_{11} - \sigma_{22})^2 + 2L\sigma_{23}^2 + 2M\sigma_{31}^2 + 2N\sigma_{12}^2 \quad (15.1)$$

Table 15.1 Parameters used in the simulation of grinding temperature prior to simulating stress.

Parameters	Values
Total heat flux q_t	9.0×10^6 J/(m$^2 \cdot$ s)
Heat flux flow into the chip q_{ch}	4.6×10^5 J/(m$^2 \cdot$ s)
Energy partition ε	1.3%
Heat flux flow into the workpiece q_w	1.1×10^5 J/(m$^2 \cdot$ s)
Abrasive grain size d_{mean}	180 μm

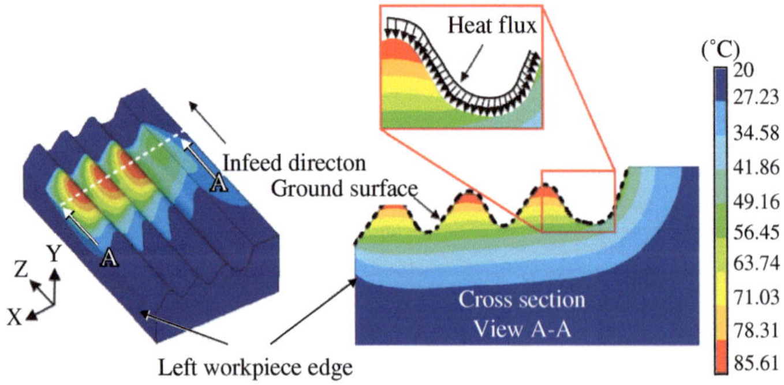

Figure 15.2 Simulated temperature results prior to stress simulation.

In the equation, σ_S denotes the reference yield stress, and F, G, H, L, M, and N are independent parameters used to describe the material's yielding behavior under various stress conditions.

Notably, the single-crystal nickel alloy has the same mechanical properties along the three principal crystal axes, the [001], [010], and [100] directions. This means that the parameters F, G, and H are equal, as are the parameters L, M, and N. As a result, Eq. (15.1) can be simplified to the form presented in Eq. (15.2).

$$\bar{\sigma} = \frac{1}{\sqrt{2}}\left[(\sigma_{22} - \sigma_{33})^2 + (\sigma_{33} - \sigma_{11})^2 + (\sigma_{11} - \sigma_{22})^2 + 2K\left(\sigma_{23}^2 + \sigma_{31}^2 + \sigma_{12}^2\right)\right]^{1/2} \tag{15.2}$$

The generalized Hooke's law, which incorporates a temperature-dependent Young's modulus and Poisson ratio, was employed to model the elastic strain, as follows:

$$\begin{aligned} \{\varepsilon\} &= [S]\{\sigma\} \\ \{\sigma\} &= [S]^{-1}\{\varepsilon\} = [C]\{\varepsilon\} \end{aligned} \tag{15.3}$$

$$\begin{aligned} \{\varepsilon\} &= \left[\varepsilon_{11}\varepsilon_{22}\varepsilon_{33}\gamma_{12}\gamma_{23}\gamma_{31}\right]^T \\ \{\sigma\} &= \left[\sigma_{11}\sigma_{22}\sigma_{33}\tau_{12}\tau_{23}\tau_{31}\right]^T \\ [S] &= [C]^{-1} \end{aligned} \tag{15.4}$$

$$[S] = \begin{bmatrix} 1/E & -\mu/E & -\mu/E & 0 & 0 & 0 \\ -\mu/E & 1/E & -\mu/E & 0 & 0 & 0 \\ -\mu/E & -\mu/E & 1/E & 0 & 0 & 0 \\ 0 & 0 & 0 & 1/G & 0 & 0 \\ 0 & 0 & 0 & 0 & 1/G & 1/G \\ 0 & 0 & 0 & 0 & 0 & 0 \end{bmatrix} \tag{15.5}$$

In this context, $\{\sigma\}$ and $\{\varepsilon\}$ represent the stress and strain vectors, respectively. The symbols S_{ij} and C_{ij} stand for the elastic compliance and stiffness constants. Additionally, E denotes Young's modulus, μ represents the Poisson ratio, and G is the shear modulus.

By incorporating the associated flow rule and yield function into a plastic potential function, the elastoplastic constitutive relationship is formulated as shown in Eq. (15.6) [3].

$$d\{\sigma\} = [C]_{\text{ep}} d\{\varepsilon\} \tag{15.6}$$

$$[C]_{\text{ep}} = [S]^{-1} - \frac{[S]^{-1}\left\{\frac{\partial\bar{\sigma}}{\partial\sigma}\right\}\left\{\frac{\partial\bar{\sigma}}{\partial\sigma}\right\}^T[S]^{-1}}{H' + \left\{\frac{\partial\bar{\sigma}}{\partial\sigma}\right\}^T[S]^{-1}\left\{\frac{\partial\bar{\sigma}}{\partial\sigma}\right\}} \tag{15.7}$$

where, $[C]_{\text{ep}}$ signifies the elastic–plastic matrix, $[S]^{-1}$ denotes the elastic matrix, and H' corresponds to the isotropic hardening modulus.

As reported in Ref. [4], the total strain increment is split into five components: i.e., elastic ($\dot{\varepsilon}^E$), plastic ($\dot{\varepsilon}^P$), thermal loading ($\dot{\varepsilon}^{Th}$), volumetric change, and transformation-induced plasticity. However, in the stable creep-feed profile grinding process, the grinding temperature is low (e.g., below 130 °C [5, 6]), so the thermal strain of the machined workpiece material is negligible. Consequently, the turbine blade root shows no volume change or transformation-induced plasticity during and after creep-feed profile grinding. In this case, the total strain increment of each element can be written as follows:

$$\dot{\varepsilon} = \dot{\varepsilon}^E + \dot{\varepsilon}^P + \dot{\varepsilon}^{Th} \tag{15.8}$$

15.1.3 Mechanical Loads Based on Discrete Distribution of Abrasive Grain

In the conventional approach, loads are typically distributed uniformly across the grinding zone. The equation for calculating pressure is as follows:

$$\begin{cases} P_\mathrm{n} = F_\mathrm{n}/A \\ P_\mathrm{t} = F_\mathrm{t}/A \end{cases} \tag{15.9}$$

In the equation, P_n and P_t signify the normal and tangential pressure, respectively. A denotes the entire area of the grinding zone, and F_n represents the normal grinding force. As per Eq. (15.9), the force-induced pressure is typically rather minimal. This occurs because the conventional method presumes continuous contact in the grinding zone and overlooks the true intermittent interaction between the workpiece and each grain. In such cases, if the grinding temperature stays low (e.g., below 130 °C during creep-feed grinding), the simulated residual stress values tend to be relatively low, like below 10 MPa or nearly zero. This discrepancy indicates that the conventional method does not align well with actual grinding operations.

Unlike the conventional model, mechanical loads are applied directly to a few active grains actively engaged in the grinding process, instead of across the entire grinding zone, as illustrated in Figure 15.3. In other words, the interaction between the abrasive wheel and the workpiece is sporadic. As a result, the micro-interaction between the workpiece and individual grains can be reflected by linking the grinding force to the active grain's penetration depth, the actual contact zone, and the count of active grains [7, 8]. The relevant calculations (Eqs. (15.10)–(15.12)) can be expressed as follows:

$$\begin{cases} P'_\mathrm{n} = F'_\mathrm{n}/A_\mathrm{cs} \\ P'_\mathrm{t} = F'_\mathrm{t}/A_\mathrm{cs} \end{cases} \tag{15.10}$$

$$\begin{cases} F'_\mathrm{n} = F_\mathrm{n}/N \\ F'_\mathrm{t} = F_\mathrm{t}/N \end{cases} \tag{15.11}$$

$$A_\mathrm{cs} = \pi d_\mathrm{mean} \cdot h_\mathrm{cu} \tag{15.12}$$

In the new simulation approach, P_n' and F_n' represent the normal contact pressure and grinding force per active grain, while P_t' and F_t' denote their tangential counterparts. N indicates the number of active grains, d_mean is the average abrasive grain size, H_cu represents the grain penetration depth, and A_cs is the real contact

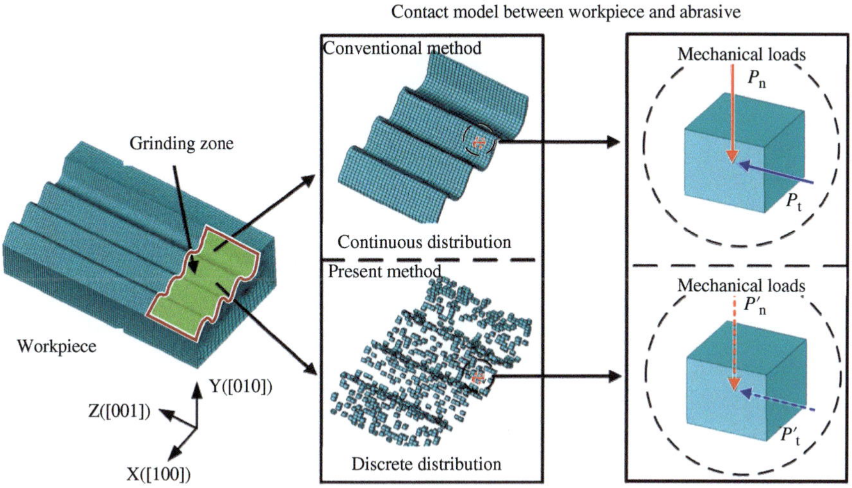

Figure 15.3 Illustration of mechanical loads by using present methods.

Table 15.2 Parameters used in the simulation of grinding-induced stresses.

Parameters	Values
Abrasive grain size d_{mean}	180 μm
Number of active grains N	5
Penetration depth of grain h_{cu}	20 μm
Normal grinding force F_n	28 N
Tangential grinding force F_t	61 N
Normal contact pressure P_n'	2.5×10^9 Pa
Tangential contact pressure P_t'	5.4×10^9 Pa

area. The grains are assumed to be spherical, with size details calculable via Ref. [9]. By utilizing Eqs. (15.10)–(15.12), the real contact pressure per active grain can be evaluated. Given the small workpiece–wheel contact area, the mechanical loads inducing stresses can become very large. From this analysis, the critical simulation parameters are listed in Table 15.2.

In the conventional simulation model, the nodes on the bottom surface of the blade root workpiece are restricted, since the workpiece remains stationary during grinding. In the current simulation model, the nodes on the workpiece top surface were also constrained. In response to the grinding wheel's pressure on the workpiece surface, Figure 15.4 presents the constrained nodes located on the workpiece's top surface. Additionally, in the assumption of the present method, the top surface element contacts the abrasive grain, and the workpiece material is also removed through grinding. Once the simulation calculation is complete, the top surface elements must be hidden to reveal the residual stresses in the blade root (refer to Figure 15.4).

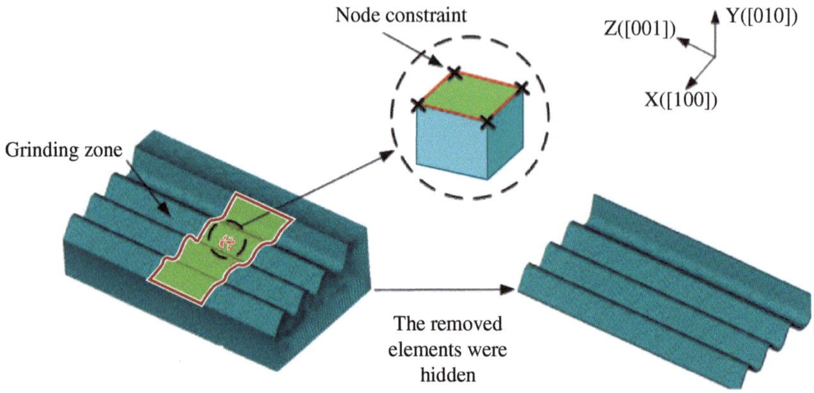

Figure 15.4 Illustration of the hidden and constrained elements on the top workpiece surface.

15.2 Experimental Verification of Residual Stresses

In the present study, measuring residual stresses in the valley regions of the blade root workpiece was particularly challenging due to its small size, complex shape, and measurement method limitations. The measurement of residual stresses was confined to the root peak regions. As depicted in Figure 15.5a, the measured stresses were −95 MPa perpendicular to the grinding direction and −167 MPa parallel to the grinding direction. The stress values obtained through the present method showed good agreement with experimental results, with an average error below 20%. This was deemed acceptable for the feasibility and accuracy of residual stress simulations [10]. This suggests that the present method is better suited than the conventional approach for predicting residual stresses in creep-feed profile grinding of single-crystal turbine blade roots.

These results demonstrate that the proposed methodology achieves sufficient accuracy for residual stress prediction while maintaining computational feasibility [10]. Comparative analysis confirms its superior performance over conventional approaches for creep-feed profile grinding applications, particularly in addressing critical challenges specific to single-crystal turbine blade roots.

Figure 15.5 also illustrates the changes in residual stress along the cross-sectional depth in both the peak and valley regions of the workpiece surface. At the blade root peak region (Figure 15.5a), the compressive stresses in both directions reach maxima of −157 MPa (parallel) and −119 MPa (vertical) at the start point and then diminish with increasing depth. Conversely, at the blade root valley region (Figure 15.5b), the maximum tensile stresses are 87 MPa (parallel) and 105 MPa (vertical) at the start point and then decrease with depth. Additionally, the layer affected by thermal and mechanical factors extends to less than 2 mm in depth parallel to the grinding direction and about 1.25 mm vertically to it (Figure 15.5), a trend consistent with findings in Ref. [11]. The significant mechanical loads during grinding lead to intense workpiece–grain interaction. This interaction induces serious plastic deformation in the ground surface layer and leads to relatively large residual stresses near the workpiece surface.

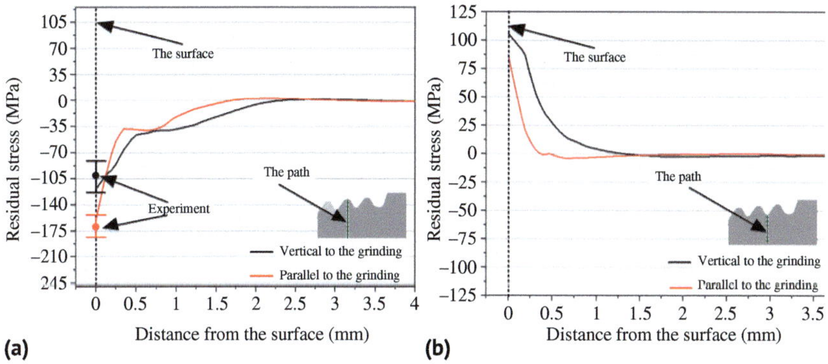

Figure 15.5 Experimental verification of residual stresses in grinding of blade root workpiece: (a) peak region and (b) valley region.

15.3 Residual Stress Simulation Results and Analysis

15.3.1 Residual Stress Maps

As shown in Figure 15.6, during the stable grinding stage, the instantaneous stress mapping (perpendicular to the grinding direction, i.e., the [010] orientation, and not grinding-induced residual stresses) was obtained using the present simulation method. As the abrasive grains traversed the workpiece surface, compressive stresses were caused to form in the grinding zone. When the simulation was conducted using the conventional method, the stress values ranged from −97.2 to 46.6 MPa (Figure 15.6a). The simulated stresses were so minimal that they had little impact on the workpiece.

This phenomenon occurred as a result of the comparatively modest grinding temperature (for instance, 85 °C during creep-feed profile grinding operations). Conversely, when employing the novel simulation approach devised in this research, notable tensile stresses (reaching up to 1590 MPa) were identified and localized in the valley areas, as depicted in Figure 15.6b. Under these circumstances, the workpiece material can be extracted from the matrix since the shearing stress induced by grinding surpasses the material's strength. This suggests a vigorous interaction between the workpiece and grains, with a random distribution of grains within the grinding zone, which accounts for the formation of a uniform stress field during the grinding process. These sporadic interactions underscore the intricate nature of the wheel–workpiece contact, as documented by Li and Axinte [12], who theoretically constructed the workpiece–grain contact model.

After grinding, the residual stress distribution maps were generated, as demonstrated in Figure 15.7. When employing the conventional model (Figure 15.7a), the residual stresses induced in the workpiece surface layer were relatively low. Specifically, from the peak area to the valley area, the tensile residual stresses generally escalated until the maximum tensile stress (e.g., 12.2 MPa) was attained in the flat area. Conversely, when utilizing the current model, the tensile residual stresses predominantly emerged in the valley regions, whereas compressive residual stresses were observed in the peak and flat regions. The residual stress values on the blade root surface encompassed a range from −350 to 179 MPa, as evidenced in Figure 15.7b.

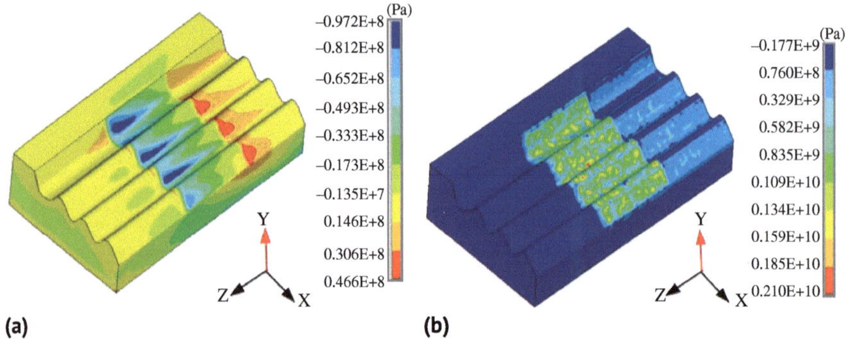

Figure 15.6 Instantaneous stress maps generated in the grinding process of turbine blade root: (a) by conventional simulation model and (b) by present simulation model.

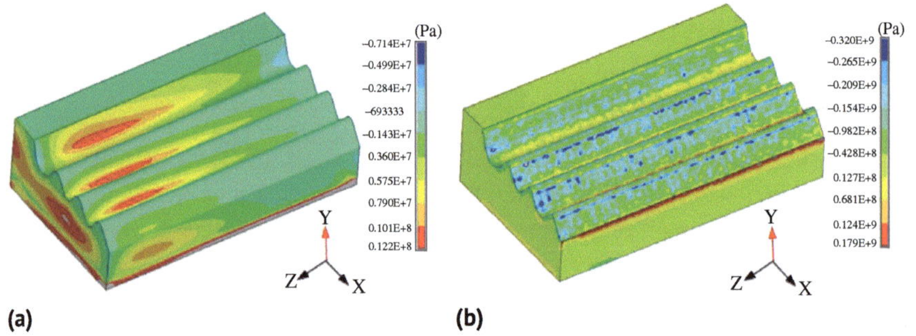

Figure 15.7 Residual stress maps of blade root workpiece (vertical to the grinding direction): (a) by conventional simulation model and (b) by present simulation model.

15.3.2 Residual Stress Contours

Figure 15.8 presents the residual stress values in the peak regions of the blade root workpiece. Along the parallel grinding direction ([010] orientation), the smallest compressive stress was observed at the grinding start point, after which the compressive stress gradually increased from −150 to −100 MPa with increasing distance, as shown in Figure 15.8a. Along with the vertical grinding direction ([001] orientation) (Figure 15.8b), as the distance increased, the compressive stress experienced a slight increase before stabilizing. Approaching the grinding end point, the residual stresses rebounded from −200 to −150 MPa.

In the blade root valley region, residual stresses measured perpendicular to the grinding direction typically fall within the range of −30 to 125 MPa (Figure 15.9a), whereas those parallel to the grinding direction exhibit values between −50 and 100 MPa (Figure 15.9b). Tensile stresses mainly develop during the stable grinding phase, whereas compressive residual stresses appear in Valley II and Valley III near the endpoint. Additionally, Valley III records the lowest residual stress values during stable grinding 20 MPa in the parallel direction and 25 MPa in the vertical direction. A notable trend is the gradual decline in residual stresses as the distance increases from 16 to 20 mm, which coincides with the grinding wheel's contact

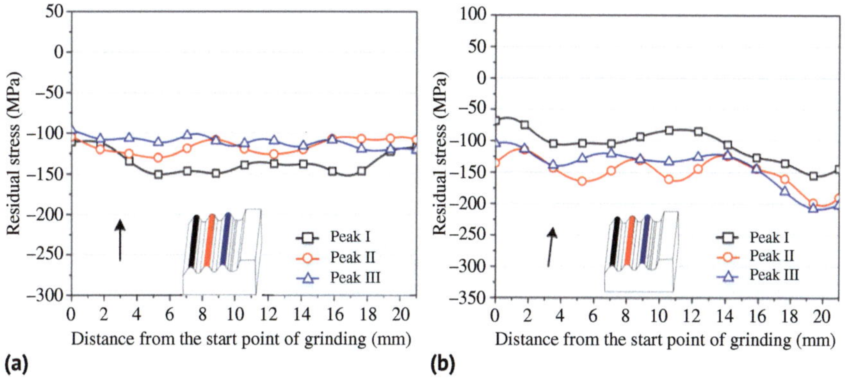

Figure 15.8 Residual stresses in different peak regions of blade root workpiece: (a) vertical to grinding direction and (b) parallel to grinding direction.

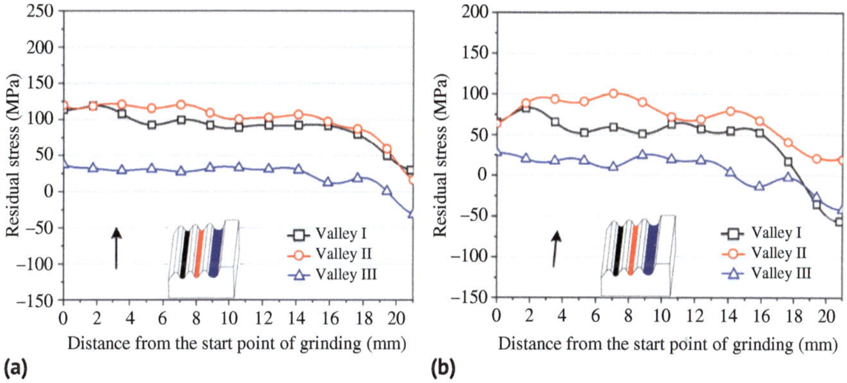

Figure 15.9 Residual stresses in different valley regions of blade root workpiece: (a) vertical to grinding direction and (b) parallel to grinding direction.

with the workpiece. For a comprehensive visualization of residual stress distribution across various blade root regions, see Figure 15.10a and b. The residual stresses transitioned from compressive in the peak regions to tensile in the valley regions, with the maximum tensile residual stresses generated in the valley regions.

Figure 15.11 presents the residual stresses across different flat regions of the turbine blade root. In the parallel direction to grinding (Figure 15.11b), the least compressive stress is at the grinding start, slightly rising with distance before stabilizing between −175 and −50 MPa. In the vertical direction (Figure 15.11a), the maximum compressive stress is at the grinding end. As distance increases, the compressive stresses first stabilize (from −130 to −25 MPa), then rise between 16 and 20 mm.

Because of the complex structure of the turbine blade root workpiece, the grinding temperature and stress fields show uneven distributions inside the workpiece during grinding. These two factors together influence the final residual stress distribution. For instance, along the vertical grinding direction, the residual stresses are between −100 and −150 MPa in the peak regions, between −30 and 125 MPa in the flat regions, and between −175 and −50 MPa in the valley regions. These

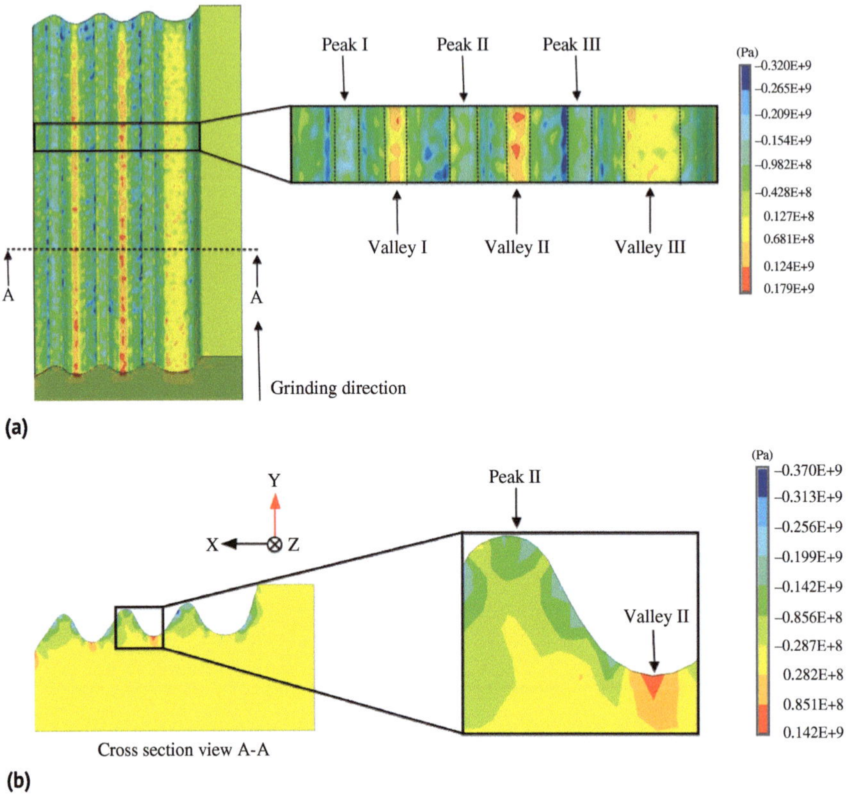

Figure 15.10 Detailed distribution of residual stresses of blade root workpiece (vertical to the grinding direction): (a) on the workpiece surface and (b) on the workpiece cross section.

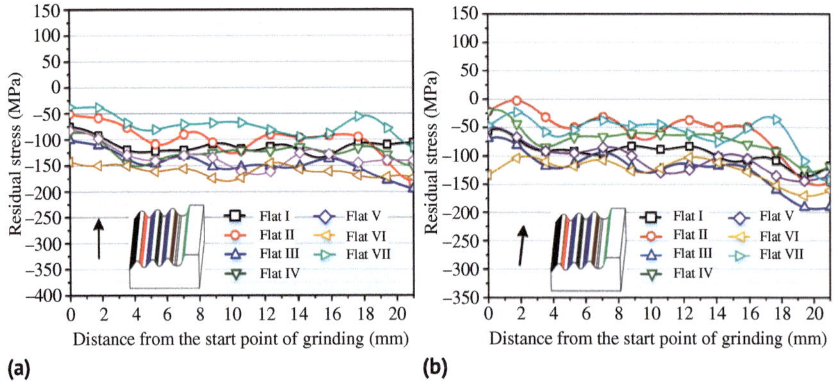

Figure 15.11 Residual stresses in different flat regions of the blade root workpiece: (a) vertical to grinding direction and (b) parallel to grinding direction.

results show that the change in residual stresses from the root peak to the root valley can cause different levels of surface integrity in the ground workpiece. Such differences in the distribution of grinding-induced residual stresses may reduce the fatigue strength, wear resistance, and service reliability of the turbine blade root. So, controlling the consistency of residual stresses is crucial for enhancing the service performance of turbine blade components.

15.4 Collaborative Manufacturing of Structure Shape and Surface Integrity

To achieve high-performance manufacturing of single-crystal nickel alloy turbine blade fir-tree roots, it is crucial to reconcile shape and geometric accuracy with surface integrity, as depicted in Figure 15.12. This calls for a comprehensive approach to grinding inputs, such as parameters, tool specifications, and cooling conditions, to guarantee favorable grinding outputs like tool wear, surface roughness, subsurface microstructure, and mechanical properties. Effective management of process variables such as grinding force and temperature [13, 14] is pivotal for regulating these outputs. Shape and geometric accuracy are mainly dictated by tool wear behavior, whereas surface integrity is considerably influenced by surface roughness, subsurface microstructure, and mechanical properties. These variables are of great significance in the vast majority of grinding outputs. In summary, the profile grinding of single-crystal nickel alloy turbine blade fir-tree roots involves substantial thermo-mechanical-structural coupling effects [15, 16].

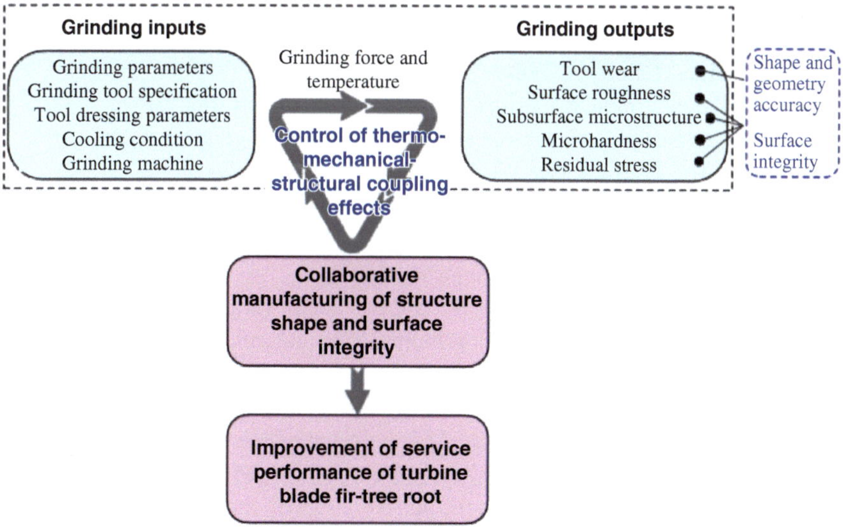

Figure 15.12 Correlation between the grinding inputs and outputs during the grinding of turbine blade root.

Consequently, achieving simultaneous control of structural geometry and surface integrity in manufacturing processes continues to pose significant challenges without a comprehensive understanding and precise regulation of thermo-mechanical-structural interactions. Current research gaps persist in developing reliable predictive models for these multiphysics coupling phenomena.

References

1 Yu, X.X. and Lau, W.S. (1999). A finite-element analysis of residual stress in stretch grinding. *Journal of Materials Processing Technology* 94 (1): 13–22.
2 Hill, R. (1948). A theory of the yielding and plastic flow of anisotropic metals. *Proceedings of The Royal Society A* 193: 281–297.
3 Han, S.L., Li, S.X., and Smith, D.J. (2001). Comparison of phenomenological and crystallographic models for single crystal nickel base superalloys. I. Analytical identification. *Mechanics of Materials* 33 (5): 251–266.
4 Sun, G.F., Wang, Z.D., Lu, Y. et al. (2018). Numerical and experimental investigation of thermal field and residual stress in laser-MIG hybrid welded NV E690 steel plates. *Journal of Manufacturing Processes* 34: 106–120.
5 Kim, N.K., Guo, C., and Malkon, S. (1997). Heat flux distribution and energy partition in creep-Feed grinding. *CIRP Annals—Manufacturing Technology* 46 (1): 227–232.
6 Grimmert, A., Pachnek, F., and Wiederkehr, P. (2023). Temperature modeling of creep-feed grinding processes for nickel-based superalloys with variable heat flux distribution. *CIRP Journal of Manufacturing Science and Technology* 41: 477–489.
7 Liu, M.Z., Li, C.H., Zhang, Y.B. et al. (2023). Analysis of grinding mechanics and improved grinding force model based on randomized grain geometric characteristics. *Chinese Journal of Aeronautics* 36 (7): 160–193.
8 Chu, Y., Yan, S.J., Yang, Z.Y. et al. (2024). Grain shape-protrusion-based modeling and analysis of material removal in robotic belt grinding. *Journal of Manufacturing Processes* 110: 211–223.
9 Jiang, J.L., Ge, P.Q., Bi, W.B. et al. (2013). 2D/3D ground surface topography modeling considering dressing and wear effects in grinding process. *International Journal of Machine Tools and Manufacture* 74: 29–40.
10 Huang, Y., Liu, S., Xiao, G.J. et al. (2020). Prediction of surface residual stress on titanium alloy generated by belt grinding using molecular system dynamics. *Procedia CIRP* 87: 480–484.
11 Ge, J.M., Deng, Z.H., Cao, D.H. et al. (2023). Study on modelling simulation and distribution characteristics of residual stress during robotic weld end grinding of high strength modulated steel. *Journal of Manufacturing Processes* 108: 98–113.
12 Li, H.N. and Axinte, D. (2018). On the inverse design of discontinuous abrasive surface to lower friction-induced temperature in grinding: an example of engineered abrasive tools. *International Journal of Machine Tools and Manufacture* 132: 50–63.

13 Zhao, B., Ding, W.F., Shan, Z.D. et al. (2023). Collaborative manufacturing technologies of structure shape and surface integrity for complex thin-walled components of aero-engine: status, challenge and tendency. *Chinese Journal of Aeronautics* 36 (7): 1–24.
14 Gong, P., Zhang, Y.B., Wang, C.J. et al. (2024). Residual stress generation in grinding: mechanism and modeling. *Journal of Materials Processing Technology* 324: 118262.
15 Liao, Z.R., Schoop, J.M., Saelzer, J. et al. (2024). Review of current best-practices in machinability evaluation and understanding for improving machining performance. *CIRP Journal of Manufacturing Science and Technology* 50: 151–184.
16 Aspinwall, D.K., Soo, S.L., Curtis, D.T. et al. (2007). Profiled superabrasive grinding wheels for the machining of a nickel based superalloy. *CIRP Annals—Manufacturing Technology* 56 (1): 335–338.

Index

Note: Page numbers in *italics* and **bold** refers to figures and tables respectively.

Grinding of Single-Crystal Superalloys: Fundamentals and Technologies, First Edition.
Wenfeng Ding et al.

© 2026 Wiley-VCH GmbH. Published 2026 by Wiley-VCH GmbH.